T5-DGG-439

Biological Membranes

Biological Membranes

By 9 Authors

Edited by
Robert M. Dowben
Department of Biology
Massachusetts Institute of Technology
Cambridge

Little, Brown and Company
Boston

Library of Congress catalog card No. 68-55585

First Edition

Published in Great Britain
by J. & A. Churchill Ltd., London
British Standard Book No. 7000 0137 9

Printed in the United States of America

Contributing Authors

ROBERT P. DAVIS
Division of Biological and Medical Sciences, Brown University;
Miriam Hospital, Providence, Rhode Island 02906

ROBERT M. DOWBEN
Department of Biology, Massachusetts Institute of Technology,
Cambridge, Massachusetts 02139

I. ROBERT FENICHEL
Laboratory of Cellular Biophysics, Albert Einstein Medical Center,
Philadelphia, Pennsylvania 19141

SAMUEL B. HOROWITZ
Laboratory of Cellular Biophysics, Albert Einstein Medical Center,
Philadelphia, Pennsylvania 19141

KARL KORNACKER
Faculty of Biophysics, Ohio State University,
Columbus, Ohio 43216

JOHN P. REEVES
Department of Biology, Massachusetts Institute of Technology,
Cambridge, Massachusetts 02139

LEON E. ROSENBERG
Department of Medicine, Yale University School of Medicine,
New Haven, Connecticut 06510

DAVID SCHACHTER
Department of Physiology,
Columbia University College of Physicians and Surgeons,
New York, New York 10032

STANLEY G. SCHULTZ
Department of Physiology,
The University of Pittsburgh School of Medicine,
Pittsburgh, Pennsylvania 15213

Preface

A NUMBER OF VITAL PROCESSES have been thought, at one time or another, to be functions of the cell membrane. In complex organisms the surface membrane acts as a physical boundary for each cell, permitting the cell to maintain internal conditions most appropriate for carrying out its life processes as an individual unit. In addition, cell membranes appear to facilitate the integration of function among groups of cells, enabling the whole organism to operate effectively. This topic of biology has attracted continued interest, and a vast and almost unmanageable literature has accumulated. Yet our knowledge of the cell membrane is still fragmentary; many uncertainties exist concerning its origin, its structure, its exact functional roles, and the mechanisms by which it fulfills them. During the past decade or two, research on biological membranes has intensified, and there is a widely held conviction that important new discoveries concerning membranes will soon be forthcoming—discoveries that will have important implications for biology generally. These considerations led us to write this volume.

In its broadest form, the cell membrane concept is more than a century old. In 1855 Karl Wilhelm Nägeli described the formation of a firm film when the protein-rich cell sap of an injured cell came in contact with water [7]. He called this film the *Plasmamembran* and recognized that it protected the remaining cytoplasm from the deleterious effects of the water. Nägeli found the membrane impermeable to naturally occurring plant pigments, which escaped, however, after a severe, penetrating injury to the cell. He was able to show that the membrane was semipermeable and probably responsible for the osmotic phenomena shown by living cells. It is difficult to believe that Nägeli actually observed the cell membrane through the microscope;

rather, he probably detected a boundary of cytoplasm and inferred the presence of a membrane from the properties of semipermeability.

It is a curious fact that after Nägeli the cell membrane was not regarded as a real structure with any degree of certainty until the 1920's, when Robert Chambers reported his microdissection studies, in which the cell membrane was manipulated directly under the microscope by means of fine needles [1]. In the meantime Moritz Traube, an amateur chemist, discovered the semipermeable properties of copper ferrocyanide films. These films were used by Wilhelm Pfeffer in exhaustive studies of osmotic phenomena [9] that provided data for van't Hoff's theory of solutions. Pfeffer proposed that an invisible membrane surrounding plant cells accounted for the similarity of their osmotic behavior with that of his artificial osmometers. However, he withdrew this hypothesis in the face of insistent objections by colloid chemists of the time that plant cells failed to display true osmotic phenomena. After Pfeffer, Hamburger [4] studied them extensively in plant cells, and Overton [8] measured the permeability of living cells to over 500 substances. Although the idea of a cell membrane is implicit in the work of these investigators, it is remarkable that none of them felt the necessity for invoking the presence of a cell membrane as a substantive structure to account for their findings.

The striking correlation between the solubility of various low-molecular-weight substances in lipid solvents and the rate of their permeation into cells led Overton to postulate that cells were covered with lipid material. In 1910 Rudolph Höber [5] showed that intact erythrocytes have a high electrical resistance, while the material filling the interior of the cell is a good conductor. He concluded that the high electrical resistance resided in a thin film, probably lipid, at the surface of the cells. It was disquieting to a number of early observers that cells were permeable to several hydrophilic compounds—glycerol, for example—that were rather insoluble in lipid solvents. Similarly, inorganic electrolytes, which are not lipid soluble, penetrate rather readily into living cells. Such findings led Ruhland [10] to conclude that the cell membrane had pores through which many substances passed. To reconcile the observations that small hydrophilic molecules penetrate cells more quickly than might be expected from the lipid-solubility theory and that large fat-soluble molecules penetrate more easily than might be expected from the pore theory, Collander [2] combined both concepts into a lipoid-filter theory.

Contemporary ideas of membrane structure have been profoundly influenced by the ideas of Langmuir [6] on the arrangement of amphiphilic lipids at an air-water interface. A monolayer of lipids at such an interface assumes a configuration in which the lipid molecules are parallel to one another with their polar ends submerged in the aqueous phase and their hydrophobic ends projecting upward into the air.

We have come a long way from these beginnings. In view of the great interest in biological membranes, particularly in the last few years, it seemed worthwhile to present the current thinking in this domain of physiology in a series of thoughtful and critical essays. The contributors to this volume have aimed to approach it from several different vantage points, and in addition, to present a point of view. We hope that the book will fill the need for an overview of this important subject and that it will be useful to a wide range of scientists.

R. M. D.

Cambridge, Mass.

REFERENCES

1. Chambers, R. The nature of the living cell as revealed by microdissection. *Harvey Lect.* 22:41, 1926.
2. Collander, R., and Bärlund, H. Permeabilitätsstudien an *Chara ceratophylla:* II. Die Permeabilität für Nichtelektrolyte. *Acta Botan. Fennica* 11:1, 1933.
3. de Vries, H. Plasmolytische Studien über die Wand der Vascuolen. *Jahrb. Wiss. Botan.* 16:465, 1885.
4. Hamburger, H. J. *Osmotischer Drück und Ionenlehre.* Wiesbaden: Bergmann, 1902.
5. Höber, R. Eine Methode, die elektrische Leitfähigheit im Innern von Zellen zu messen. *Pflueger Arch. Ges. Physiol.* 133:237, 1910.
6. Langmuir, I. The constitution and fundamental properties of solids and liquids: II. Liquids. *J. Amer. Chem. Soc.* 39:1848, 1917.
7. Nägeli, K., and Cramer, K. *Pflanzenphysiologische Uentersuchungen.* Zürich: Schultess, 1855.
8. Overton, E. Ueber die allgemeinen osmotischen Eigenschaften der Zelle, ihre vermutlichen Ursachen und ihre Bedeutung für die Physiologie. *Vischr. Naturforsch. Ges. Zürich* 44:88, 1899.
9. Pfeffer, W. *Osmotische Uentersuchungen.* Leipzig: Engelmann, 1877.
10. Ruhland, W., and Hoffman, C. Die Permeabilität von *Beggiatoa mirabilis. Planta Med.* 1:1, 1925.

Contents

Biological Membranes

1

Composition and Structure of Membranes

Robert M. Dowben

ISOLATION OF CELL MEMBRANES

Some of the recent developments in our concepts of biological membranes have been prompted by studies of membrane constituents. To study the composition of cell membranes, it is necessary to isolate reasonably pure membranes from disrupted cells in quantities sufficiently large for chemical analyses. In scrutinizing the experimental validity of isolation procedures, two crucial points should be borne in mind which at present are unsettled. First, it is not known whether the limiting surface membrane or plasma membrane of the cell resembles the intracellular membranes in composition and structure or whether important differences exist. Second, it is not clear whether membranes are composed of a single type of structural subunit or are a mosaic of a number of different types of structural subunits.

All preparative procedures begin with the disruption of cells, during which there is at least a temporary discontinuity in the structural integrity of the cell membranes. In general, cellular disruption is brought about by the application of mechanical shearing forces. Curtis [30] frequently found a remarkable drop in the surface viscosity of cells immediately after they were subjected to shearing

This work was supported in part by Grant AM-09175 from the National Institute of Arthritis and Metabolic Diseases and Contract F44620-67-C-0031 from the U.S. Air Force Office of Scientific Research.

forces. This thixotropic response reflects a decrease or loosening in the molecular packing upon the application of shearing stress. It is reasonable to assume that the molecular expansion within the membrane during homogenization occurs more readily under conditions in which electrostatic interactions of the membrane constituents with the solvent tend to counteract the internal cohesive forces. The most important factors decreasing electrostatic interactions with the solvent are low ionic strength, low concentrations of divalent cations, and high pH. Therefore, it is not surprising that the size and morphology of the membrane fragments depend not only on the mechanical means used to disrupt the cells but also on the composition, ionic strength, and pH of the suspension medium. While the cell membrane is disrupted during homogenization, electron micrographs show that the junctional specializations between cells, the tight junctions and desmosomes, frequently remain intact [48].

Mature mammalian erythrocytes have been the cell type most widely used as a source for membranes because they are obtained easily in suspension by venipuncture and because they lack a nucleus, intracellular organelles, and an intracellular membrane system. Membranes of erythrocytes, containing less than 2 percent hemoglobin, can be prepared by lysis in hypotonic media. Control of ionic strength and pH is required for reproducible preparations. Lysis in 0.02 M phosphate buffer at pH 7.4 produces flattened ghosts of uniform appearance with about the same diameter as that of the intact erythrocyte [37], while lysis in media of lower ionic strength or higher pH may result in the formation of small fragments. During hemolysis, the membranes become transiently permeable to large molecules [80] but subsequently recover their relative impermeability to macromolecules and ions.

The fat globules in milk are surrounded by an envelope which appears to be derived from the surface membrane of the mammary cells [11]. Preparations of milk fat globule membranes [38] contain enzymes characteristic of other membrane preparations.

A method for the preparation of plasma membranes from liver and other solid tissues was developed by Neville [134] and later modified by Emmelot and co-workers [42]. Livers from exsanguinated rats were homogenized in 0.001 M sodium bicarbonate, pH 7.2, with disruption of all cells and, in addition, about 90 percent of the nuclei. The membrane fragments, as well as intact cells, intact nuclei, and connective tissue debris, were sedimented. The sediment was washed and partially separated by two successive differential centrifugations. The purified sediment was homogenized in sucrose of density 1.22 and overlayered with sucrose of density 1.16. After centrifugation at 100,000 × *g* for 75 minutes, the membranes were collected from the interface. This technique has also been applied to thyroid [193] and HeLa cells [85].

Plasma membranes prepared in this manner appear as sheets or as spherical vesicles with a trilaminar structure in electron photomicrographs. The sheets are often still connected by desmosomes. Presumably the vesicles are formed from microvilli late in the preparative procedure. Wallach [197] has argued that the yield of membranes by this technique is low, 14 percent or less of the theoretical amount calculated for liver, and that, in addition to membrane fragments derived from the cell surface, membrane fragments of intracellular origin, from nuclei, lysosomes, and endoplasmic reticulum, are also present. Finean [55] has estimated that cell membranes prepared in this manner contain 1–30 percent cytoplasmic contamination. It seems plausible that the surface membrane not recovered in the procedure was fragmented into tiny vesicles which would sediment in the microsomal fraction upon differential centrifugation of a cell homogenate. This may account for the regular appearance in the microsomal fraction of sialic acid, antigens characteristic of the cell surface, viral binding sites, the sodium, potassium-activated adenosine triphosphatase (ATPase), and other enzymes thought to be characteristic of the surface membrane.

Muscle cell membrane preparations have been described by Kono and Colowick [99] and by McCollester [113]. The muscle fibers are homogenized in these procedures and the internal proteins are extracted, leaving tubules presumably composed of the tough outer coat of the myofibrils, to which fragments of the lipoprotein plasma membrane may be attached. The protein of the tubules is rich in glycine, proline, and hydroxyproline (amino acids present in large amounts in collagen), and the tubules disintegrate when treated with collagenase.

Wallach and co-workers [199] pioneered the disruption of Ehrlich ascites tumor cells or tissue-cultured cells by intracellular cavitation which follows sudden reduction in the ambient pressure. Suspensions of cells in a buffered, isotonic medium are equilibrated with an inert gas, usually nitrogen, under 50–75 atmospheres pressure in a Parr hydrogenation bomb. After equilibration, the pressure is suddenly reduced to 1 atm, resulting in rupture of almost all cells without disruption of nuclei or mitochondria. The plasma membrane fragments are purified from the microsomal fraction by centrifugation in sucrose or Ficoll density gradients containing divalent cations. In Ficoll gradients the surface membrane fragments, identified by cell surface antigens or by the presence of sodium, potassium–activated ATPase, collect in the fractions around density 1.07. Sodium, potassium–activated ATPase activity is also found in the pellet which is thought to represent intracellular membranes.

Warren and his collaborators [203] have isolated surface membranes from tissue-cultured cells after treatment of the cells with acetic acid, or by reacting the sulfhydryl groups on the surface with 5,5′-dithiobis(2-nitrobenzoic acid) or fluorescein-dimercuriacetate, or by treating the cells with heavy metal ions such as lead, mercury, or zinc. Any of these treatments "tans" the membrane, rendering it less fragile. Undoubtedly the protein is altered by these reactions, and the preparations must be regarded as a derivative product of membranes.

Preparations of specialized membranes have been useful for certain types of studies. For example, mitochondria may be isolated in pure form by differential centrifugation of cell homogenates, and membranes may be prepared from the washed mitochondria [28]. If the mitochondria are permitted to swell in 0.02 M phosphate buffer, pH 7.2, containing 0.02 percent bovine serum albumin, the inner and outer membranes separate, giving a mixture of inner membrane ghosts and broken outer membrane fragments. The inner membrane and outer membrane fractions can be separated by centrifugation in a sucrose gradient [142]. The density of the inner membrane is about 1.21 while that of the outer membrane is about 1.12.

In electron micrographs the outer membranes appear as collapsed, folded bags with a single membrane layer showing a surface ultrastructure pattern thought to represent patches of small particles about 45 Å in diameter. The outer membranes contain the citric acid cycle enzymes and the fatty acid oxidation enzymes [69]. The inner membranes appear as vesicles in which the cristae have been largely smoothed out, thereby permitting swelling to occur without breakage. The inner membranes contain the oligomycin-sensitive ATPase and at least some of the enzymes of oxidative phosphorylation. The outer membranes seem to be permeable to both water and sucrose, while the inner membranes are permeable to water but relatively impermeable to sucrose. Another type of specialized membrane preparation is that obtained from Mycoplasma (pleuropneumonia-like organisms, PPLO) [148].

MEMBRANE LIPIDS

Lipids account for 25 to 50 percent of the dry weight of membrane preparations. Erythrocyte ghosts, the most thoroughly studied preparation, contain 95 to 100 percent of the lipids in the whole cells. The fact that almost all the red cell lipids can be recovered in the ghosts in spite of repeated washing means that the lipid must be firmly incorporated in the membrane. In the case of most other cells, considerable amounts of phospholipids are found in cell components other than the plasma membrane; it is not clear whether the other lipids are restricted to intracellular membranous structures or whether they also occur free.

Cholesterol, mostly free, accounts for 20 to 30 percent of the lipids extracted from the mammalian erythrocyte ghosts and other plasma membrane preparations, except mitochondrial membranes, which contain very little cholesterol. The phospholipids as a class account for almost all the remaining membrane lipids. Frequently, the molar ratio of cholesterol to phospholipids approaches unity. Cholesterol and phospholipids appear to interact, as judged by the fact that the mean area per molecule of mixed monolayers is less than would be expected from the composition were no interaction present (Fig. 1-1). Zwitterionic phospholipids, such as phosphatidylcholine, make up half, or slightly more, of the total phospholipids present in membranes from most sources [35]. The remaining phospholipids are principally sphingophospholipids, which have a net negative charge at physiological pH. The ratio between phosphatidylcholine and sphingophospholipids differs greatly from species to species and even in membrane preparations from different tissues of a given organism. The composition of lipids from erythrocyte ghosts of various mammals is depicted in Figure 1-2.

It has been shown [73, 149] that erythrocyte cholesterol and phospholipids exchange rapidly in vivo and in vitro with serum cholesterol and serum lipoprotein lipids. The time of half-exchange is approximately one hour. Turnover studies in adult rat brain using ^{32}P incorporation indicate a half-life of about 11 days for phosphoinositides,

17 days for phosphatidylethanolamine, and 40 days for the sphingophosphatides [59, 118]. Cell membranes may very well have the capacity for synthesizing and renewing their complement of phospholipids. Erythrocyte ghosts, for example, are able to incorporate ^{14}C-labeled fatty acids into phospholipids in the presence of ATP, Mg^{++}, and coenzyme A [128, 138].

A variety of fatty acid residues are found in the lipids which range in carbon chain length from 16 to 22 carbon atoms and in the degree of unsaturation from 0 to 4 double bonds. The distribution of fatty acid residues is not fixed for any given tissue and species but depends upon the physiological conditions of the environment, such as ambient temperature and diet. Table 1-1 illustrates this point by listing the variation of the major species of liver phosphatidylcholine and the compositional variation when the rats are fed a fat-free diet, a diet containing coconut oil, and a diet containing corn oil. Of course, the physical characteristics of phospholipids depend upon the nature of the fatty acid residues. For example, the force-area characteristics of monomolecular films of synthetic phosphatidylcholine at an air-water interface are shown in Figure 1-3. As the degree of saturation increases, the fit between neighboring fatty acid side chains is improved, thereby permitting a closer approach and increase in the weak intermolecular attractive forces. Similarly, the London-van der Waals

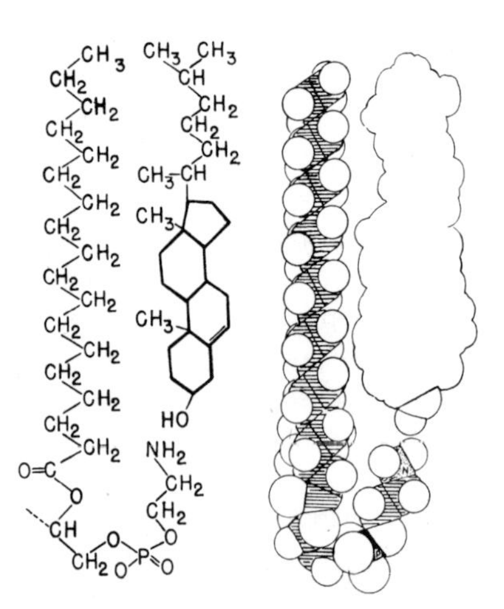

FIG. 1-1. The structure and possible conformation of a molecule of phosphatidylethanolamine and one of cholesterol. (From Finean [53].)

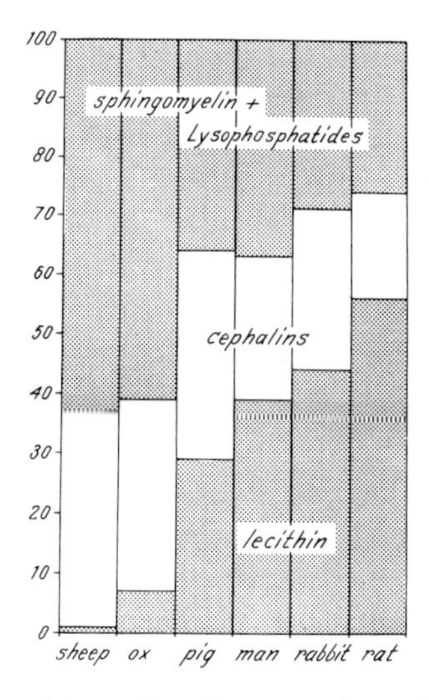

FIG. 1-2. Lipid composition of erythrocyte ghosts from various mammals. (From Elbers [41].)

TABLE 1-1. *Some Major Molecular Species of Lecithin of Liver from Rats Fed on Different Diets*

Diunsaturated			Saturated/unsaturated						Disaturated		
20:4 — [18.1 / P–N]			20:4 — [16:0 / P–N]			20:4 — [18:0 / P–N]			16:0 — [16:0 / P–N]		
—	1%	4%	3.5%	9%	11%	4%	15%	23%	+	+	+
20:3 — [18:1 / P–N]			20:3 — [16:0 / P–N]			20:3 — [18:0 / P–N]					
3.5%	—	—	11%	5%	—	12.5%	8%	—			
18:2 — [18:1 / P–N]			18:2 — [16:0 / P–N]			18:2 — [18:0 / P–N]					
—	2%	4%	2%	7%	12.5%	2%	13%	11%			
18:1 — [18:1 / P–N]			18:1 — [16:0 / P–N]			18:1 — [18:0 / P–N]					
4%	—	—	6%*	11%	5.5%	5%*	5%	2%			

Note: The percentages of each lecithin species correspond to (from left to right): a fat-free diet and diets containing coconut oil and corn oil, respectively.

* These are minimum values only.

From Deenen, L. L. M. van. Some structural and dynamic aspects of lipids in biological membranes. *Ann. N.Y. Acad. Sci.* 137:717, 1966.

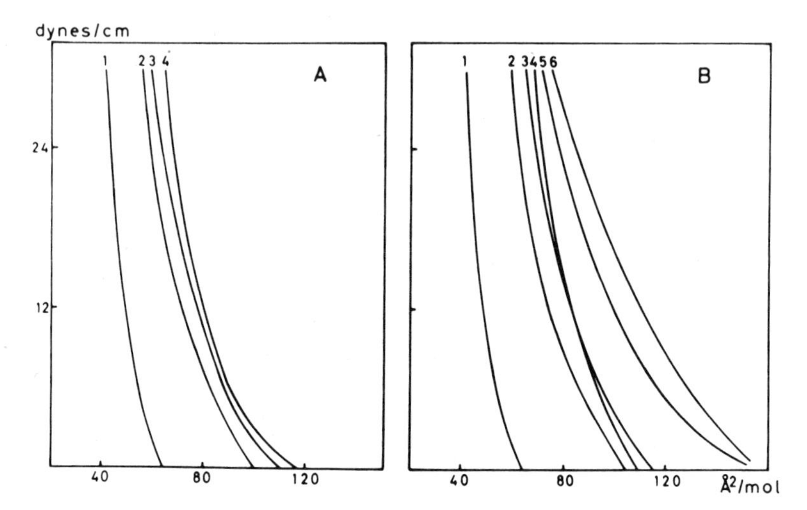

FIG. 1-3. Force-area curves for monomolecular films of synthetic lecithins at the air-water interface. (A) The effect of chain length of saturated fatty acid constituents: (1) 1,2-distearoyl-lecithin, (2) 1,2-dimyristoyl-lecithin, (3) 1-stearoyl,2-lauroyl-lecithin, and (4) 1,2-didecanoyl-lecithin. (B) The effect of the degree of unsaturation of fatty acid constituents: (1) 1,2-distearoyl-lecithin, (2) 1-oleoyl, 2-stearoyl-lecithin, (3) 1-linoleoyl,2-stearoyl-lecithin, (4) 1,2-dioleoyl-lecithin, (5) 1,2-dilinoleoyl-lecithin, and (6) 1-linoleoyl,2-palmitoyl-lecithin. (From Deenen [36].)

attractive forces increase with longer fatty acid chains. Cells appear to utilize the ability to vary the composition of the fatty acid residues as a means of maintaining the physical properties of membrane phospholipids within certain limits (Fig. 1-4). The properties of micelles formed from synthetic phosphatidylcholine also depend upon the species of fatty acid residues [5].

MEMBRANE PROTEINS

Less is known about the proteins associated with biological membranes than about the lipids. One of the early investigators, Jorpes [90], extracted washed erythrocyte ghosts with phosphate buffer at pH 5.5 and then removed the lipid with an ethanol-ether mixture. Jorpes called the protein that remained *stromatin*. Moskowitz and Calvin [126] washed freeze-dried ghosts with buffer at pH 9 until a colorless supernatant fraction was obtained. They called the sediment *stromin* and considered it the main structural framework of the membrane. After removal of the cholesterol and about half of the cephalins by ether extraction, stromin appeared to consist of two components, a lipoprotein, *elinin,* which was soluble at pH 9, and an insoluble protein, *stromatin.* A third protein, "*s*" *protein,* could be precipitated from the elinin solution by lowering the pH to 6.4. The elinin fraction contained the blood group antigens; these were absent from the "s" protein. Elinin was fibrous in character and was visualized in the electron microscope as rodlike particles 5μ–11μ long and 0.3μ–1.3μ wide. Sedimentation studies gave a molecular weight of 4.0×10^7. A 1 percent solution of elinin in water showed marked streaming birefringence, and 5 percent solutions formed a gel.

Some progress has been made toward isolating a structural protein from mitochondrial membranes solubilized by sodium deoxycholate [27, 28, 151]. Protein was precipitated fractionally from the solubilized mitochondrial membrane by adding increments of ammonium sulfate. The lipid moiety was removed from the precipitated lipoprotein fraction by extraction with acetone and water, 9:1. An aggregate of insoluble protein was obtained which appeared to combine stoichiometrically with phospholipid by noncovalent hydrophobic in-

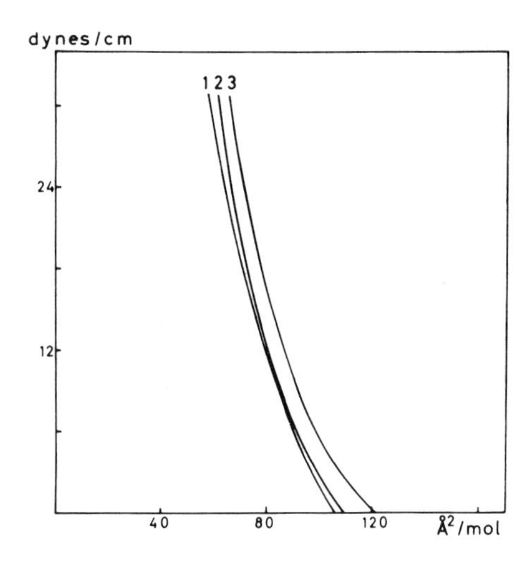

FIG. 1-4. Force-area curves for lecithins from livers of rats fed various diets: (1) fat-free diet, (2) diet with coconut oil as principal fat, and (3) diet with corn oil as principal fat. (From Deenen [36].)

teractions. Examination under the electron microscope (see below) revealed that such reconstituted lipoprotein tended to form vesicles bounded by a trilaminar unit membrane. Structural protein interacted with various cytochromes to form water-soluble adducts. After succinylation [114] the structural protein became water soluble and lost its tendency to aggregate. The monomeric species of the structural protein appeared to have a molecular weight in the neighborhood of 22,000. Similar proteins may have been prepared from liver microsomes, bovine erythrocytes, and spinach leaf chloroplasts [152].

Solubilized lipoproteins have been isolated from erythrocyte ghosts treated with *n*-butanol, sonication, detergents, or a combination of these [3, 8, 115, 116, 121, 125]. Solubilization of membrane protein by extraction of lipid with butanol indicates that nonionic, noncovalent bonds are important in maintaining the structural integrity of the membrane. *In spite of considerable interest and effort, well-characterized proteins have not yet been isolated.* Various fractions have been segregated which differ in their sedimentation characteristics, amino acid composition, and migration in electrophoresis. Starch-gel or acrylamide-gel electrophoresis reveals that these fractions are probably heterogeneous because they give rise to multiple bands. Similarly [6], starch-gel electrophoresis of erythrocyte ghost protein solubilized by 8 M urea and 0.16 M mercaptoethanol after extraction of the lipids gives 10 discrete bands.

Treatment of erythrocyte ghosts with hypotonic buffers extracts aldolase, glyceraldehyde phosphate dehydrogenase, carbonic anhydrase, and adenosine deaminase without damage to the structural integrity of the ghost membrane. Further extraction of erythrocyte ghosts with hypertonic NaCl solutions (salt concentration 0.2–1.4 M) extracts a portion of the stromal lipids, acetylcholinesterase, and other ghost proteins. Solubilization of stromal proteins by concentrated salt solutions suggests that electrostatic protein-protein interactions are also important in maintaining membrane structure.

Proteins have been isolated from erythrocyte ghosts, mitochondrial membranes, chloroplasts, and liver cell membranes by extraction with hypertonic buffers which bear some resemblance to muscle actomyosin [133, 135, 136, 141]. These proteins were isolated from membrane preparations by extraction with slightly alkaline, high-ionic-strength buffers, from which they were precipitated by lowering the ionic strength. The proteins showed Mg-activated ATPase activity, and the addition of ATP to a solution caused a prompt fall in viscosity.

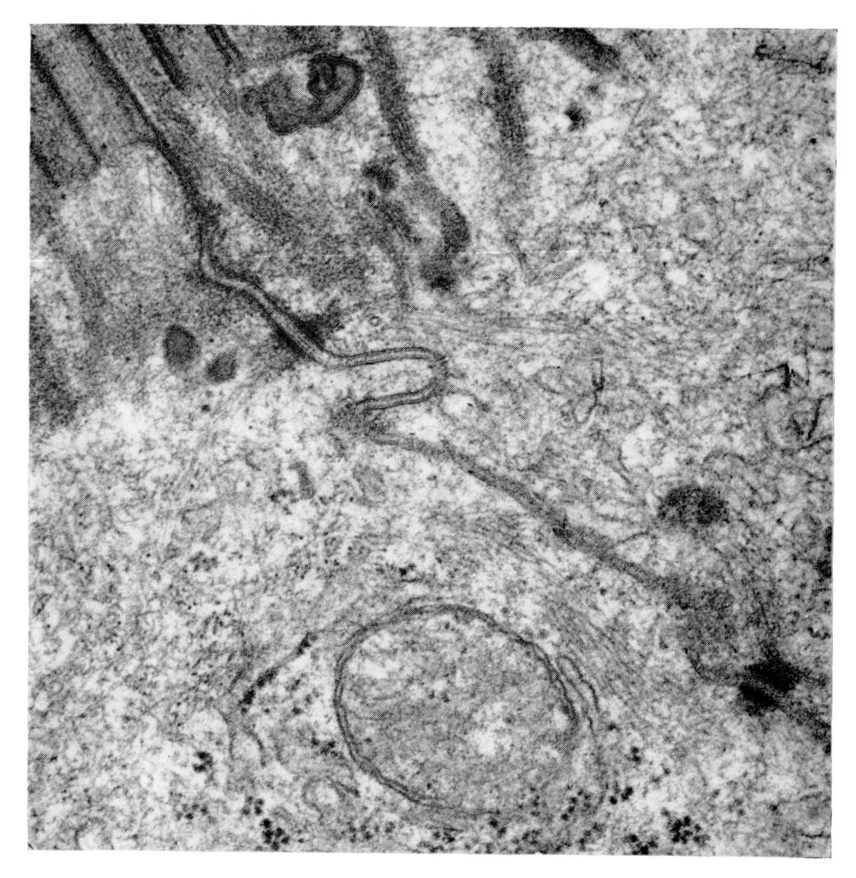

FIG. 1-5. Apposing cell membranes between two adjoining intestinal epithelial cells, including the specialized junctional complex. Note the double trilaminar structure.

The actomyosin-like proteins could be separated into two protein fractions which resembled muscle myosin and actin and formed mixed actomyosins with the protein counterpart prepared from muscle. A similar protein, *dynein,* has been prepared from cilia of *Tetrahymena* [61] which has some myosin-like properties.

MEMBRANE STRUCTURE

Many of the contemporary ideas about the molecular structure of biological membranes can be traced to the classic experiments of Gorter and Grendel [65], who extracted the lipid from washed erythrocytes and spread it as a monomolecular film on a Langmuir trough. At low pressure, Gorter and Grendel found that the film covered an area which was almost exactly twice the total surface area calculated for the intact red cells used in the experiment. The individual molecules were presumably packed together in the monomolecular film in the Langmuir trough with the polar, hydrophilic ends all submerged in the water and the nonpolar hydrophobic chain projecting upward into the air. It was suggested that the cell membrane consisted of a bimolecular layer of lipids, the molecules standing on end and packed with the hydrophobic ends apposed at the center and the hydrophilic ends facing inward and outward. Actually, the erythrocyte surface area is probably 50 percent greater than that calculated by Gorter and Grendel [205, 211]. It also appears that the acetone extraction used by these investigators failed to solubilize about one-third of the lipid present in the ghosts [204]. Their figures, therefore, were a happy accident, the two errors tending to offset each other [10].

The surface tension of marine eggs and other types of cells is as low as 0.1 to 0.2 dyne/cm, a value lower than would be expected from a lipid-water interface (about 9 dynes/cm). Danielli and Harvey [31] found that the interfacial tension between oil and water could be lowered to about 0.6 dyne/cm by adding some of the egg contents to the aqueous phase. Davson and Danielli [34] postulated the presence of a protein layer attached to the polar head groups on either side of Gorter and Grendel's bimolecular lipid leaflet (Fig. 1-5) in order to explain the low surface tension of the interface and satisfy the requirements of mechanical strength. The presence of protein might also be expected to contribute to the stability and strength of the lipid film comprising the membrane.

The universality of the membrane concept depends to a considerable extent upon direct observations made with the electron microscope. The cell membrane appears as a thin band approximately 75–100 Å thick, consisting of two dark lines, each about 25–30 Å thick on either side of a light zone (Fig. 1-6). This ubiquitous trilaminar structure, commonly referred to as the *unit membrane*, has been taken as strong support for the Davson-Danielli model. Unit membranes occur not only at the cell surface but also in the interior of the cells as part of the *endoplasmic reticulum* and at the boundary of intracellular organelles.

Much of the early electron microscopic work was done on the myelin sheath which surrounds peripheral nerve fibers. The studies of Geren [60] have shown that myelin is laid down around a nerve fiber during embryonic development. At first the nerve axon is enveloped by a Schwann cell. The axon is now covered by a double membrane, its own and that of the Schwann cell in which it is embedded. As development proceeds, there is rotation, either of the axon or of the Schwann cell, which results in a spiral winding of Schwann cell membranes around the axon. As the membranes wind around the axon, they come to adhere firmly to one another, obliterating the space between them. Actually, two decades before, Schmidt [160] had shown by polarization optical studies that the lipid molecules in myelin were arranged in a radial array of alternating layers of protein and lipid, flat laminae of protein lying in a plane at right angles to stacked lipid molecules. It was deduced that the lipid-protein layers were thin compared to the wavelength of light and constituted a Wiener mixed body. The small-angle x-ray diffraction pattern showed that myelin consisted of concentric layers with a unit repeat of about 171 Å in myelin from cold-blooded animals and about

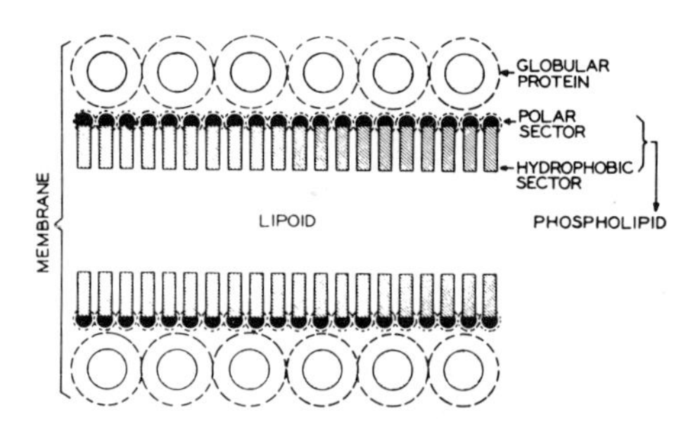

FIG. 1-6. The Davson-Danielli model of the plasma membrane.

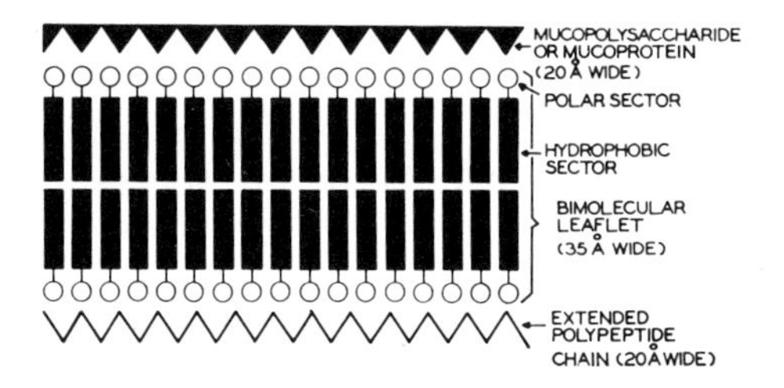

FIG. 1-7. Robertson's concept [154] of the structure of the unit membrane.

186 Å in myelin from warm-blooded animals [159]. Myelin is rich in cholesterol and has a high water content. In further x-ray diffraction studies Finean [54] showed that there was a radial repeat of about 85.5 Å in frog sciatic nerve in addition to the 171 Å, giving further evidence of a double membrane layer. Upon drying, the myelin repeat falls from 171 Å to 146 Å, owing partly to the loss of water and partly to a rearrangement of the lipid molecules.

The electron microscope studies of Fernández-Morán [49] and Sjöstrand [170] provided direct confirmation for the concentric laminar structure of myelin. The structure shows concentric major dense lines approximately 25 Å thick, with a center-to-center distance of 120 Å and intervening thinner and less dense intraperiod lines. The difference between 171 Å and 120 Å was shown to result from manipulations during the processing of the tissue prior to examination in the electron microscope (fixation, dehydration, and embedding). It was in myelin that Robertson [153] first employed potassium permanganate fixation, which enabled him to show that each of the single dense lines seen upon osmium fixation could be resolved into two distinct dense lines, each about 25 Å thick, separated by a light zone of about the same thickness.

Robertson proposed that the membrane had structural polarity, i.e., that the inside and outside surfaces were different. It has been conjectured that the outer surface is covered with mucoprotein while unconjugated protein covers the internal surface of the membrane (Fig. 1-7). As the Schwann cell winds around the axon, the inside surfaces of the membrane come to lie in apposition, giving rise to the major dense lines, and the outside surfaces similarly come to lie in apposition, giving rise to the intraperiod line. Although the extensive work with myelin sheaths has been extrapolated to plasma membranes generally, it should be noted that myelin has many unique features. Myelin is poor in protein, with a protein/lipid ratio of about 0.5, in contrast to other membrane preparations, which have a protein/lipid ratio of 2 or more. In addition, the phospholipid composition of myelin differs from that of other membrane preparations.

While a trilaminar structure often can be demonstrated after fixation and staining with osmium tetroxide alone, it appears with better clarity in tissue fixed with potassium permanganate, in which it is consistently and almost universally seen at cell boundaries [154]. The reactions of potassium permanganate with tissues that result in consistently clear trilaminar structures are unknown [51, 103]. Although osmium tetroxide is known to react with proteins, its principal reaction in membranes is with the olefinic bonds in the fatty acid moieties to form an osmic acid diester of the glycol derivative [29]. Recently, these compounds have been isolated and characterized chemically in model reactions using fatty acid esters [101, 102, 103]. A reduction product is formed which is thought to migrate to the polar ends of the phospholipids [184].

Bimolecular lipid leaflets formed across an aperture in a Teflon diaphragm have been fixed in osmium tetroxide, and these specimens show a trilaminar structure in the electron microscope even though no protein is present [77]. The osmium is presumed to be located at the polar end groups on the outer surface of the leaflet. On the other hand, a trilaminar structure was observed in mitochondria fixed in osmium after all lipid had been removed by extraction with acetone [56]. Hollow spherules formed from proteinoid material can be stained with osmium tetroxide; these preparations too show a trilaminar unit membrane type of structure even though they contain no lipid [57]. In addition to the uncertainties concerning the identity of the chemical groups which are responsible for the observed electron densities, the appearance of the unit membrane may reflect, in part, artifacts produced by the dehydration and extraction of the tissue with solvents during the preparation of sections for electron microscopy. In view of the ease with which artifacts can be produced and the uncertainties of correlating electron densities with chemical structure, care should be exercised in interpreting electron micrographs.

Unit membranes show considerable variation in total thickness and in the relative thicknesses of their members [41, 172, 213]. The cell membrane is thicker at the free surfaces of cells than where it is in contact with other cells in tissue, and it is thicker than the unit membrane of endoplasmic reticulum. To what extent these variations should be ascribed to differences in structure and to what extent they should be ascribed to artifacts of preparation is not clear.

The paucimolecular concept of membrane structure envisions the protein layers as existing in the extended or β-conformation. This arrangement tends to minimize nonpolar or hydrophobic interactions between the protein and lipid. Recent studies using infrared absorption [117] and optical rotatory dispersion [201] indicate that very little membrane protein is in the extended β-configuration but that it is largely randomly coiled with a significant percentage in the α-helical form.

Several electron microscopists have observed repeating subunits in various membrane preparations, particularly in mitochondrial membranes and in chloroplast membranes [50, 52, 155, 171, 183]. The subunits appear as regular arrays of beads or transverse densities in the trilaminar structure with a unit repeat of approximately 40–60 Å [156]. Robertson has argued that many of these subunit structures are probably artifacts produced when sections for examination are cut slightly askew. In the case of retinal rods [13], at least, the globular pattern in electron micrographs corresponds to a periodicity observed by low-angle x-ray diffraction.

Recently, Benedetti and Emmelot [12] examined negatively stained preparations of rat liver membranes in which the staining procedure was carried out at 37° and observed an array of hexagons with occasional pentagons about 80–90 Å apart, in addition to the 50–60 Å globular knobs scattered over the membrane surface. The hexagonal patterns were not observed in membranes prepared at low temperatures.

Electron micrographs of mixed lipid films also show a similar hexagonal pattern of ultrastructure [9, 111]. Evidence for a phase transition from a lamellar structure to inverted cylindrical micelles was obtained with model phospholipid systems examined by x-ray diffraction [112] or electron microscopy [182]. These observations bring to mind the suggestion by Hechter [76] that the membrane is composed of structural units in the form of hexagons with a few pentagons to permit three-dimensional flexibility, the arrangement showing icosahedral symmetry.

The observation of subunits prompted Green [70] to put forward the proposal that biological membranes are always formed by the association of repeating proteolipid structural units (Fig. 1-8). According to this view, each membrane has a specific repeating unit

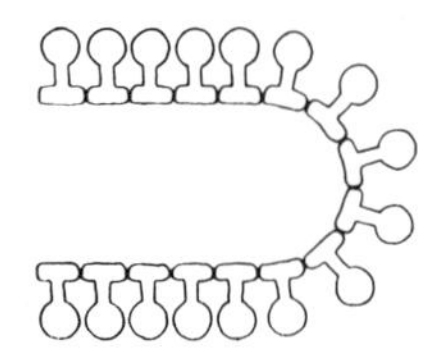

FIG. 1-8. Green's concept of the organization of elementary particles into a membrane. (From Green and Perdue [70].)

which is characteristic in form, size, and composition. The repeating units always contain a membrane-forming portion; at times there may also be a detachable portion. The detachable head and stalk of the elementary particle of the inner membrane of mitochondria can be removed from the membrane-forming portion by sonication [181] or by treatment with bile salts [100] without disrupting the arrangement of the structural portion as a continuous membrane. Thus, the detachable portions are an intrinsic part of the membrane, but they are not essential for the integrity of the membrane structure. Supposedly, the detachable portions contain the membrane-associated enzymes. At least in mitochondria, the membrane-forming portion of the repeating unit is judged to be identical with the structural protein isolated by chemical means (see above). The membrane-forming portions of the structural repeating units have an inherent tendency to form membrane-bounded vesicular structures.

A new technique has been applied recently to the study of membrane ultrastructure. The specimen to be examined is frozen rapidly in liquid freon and fractured with a microtome knife. The exposed surfaces may be "etched" by exposure to a high vacuum for a few minutes while frozen. The surfaces are then shadowed, replicated, and examined in the electron microscope. This technique tends to split membranes, revealing their inner aspects in addition to their surfaces to examination. Small particles, about 85 Å in diameter, are seen frequently in such fractured-membrane preparations, although depressions into which such particles might fit are few in number [18, 19]. The relative lack of "etching" indicates that little water is present at the exposed surfaces. These observations are consistent with the idea that it is the lipid which is exposed. The globular substructures revealed in this type of preparation could be structural subunits or enzymes associated with the membranes.

An emerging view of membranes regards the structural arrange-

+ ions [146, 208]. NH_4^+, Rb^+, Cs^+, and Li^+ can replace
tion. These cations can be ranked in order of activating
$NH_4^+ > K^+ > Rb^+ > Cs^+ > Li^+$. The same order of
on increasing sodium efflux from squid axon is observed
cations are used in the bathing solution [169]. Cardiac
hibit the phosphatase reaction, competing with potas-
Oligomycin inhibits the sodium, potassium–activated
the phosphatase step, but other inhibitors of oxidative
lation do not [72, 120]. Particularly at low temperatures,
ified enzyme preparations exhibit a slow sodium-activated
of ATP that does not appear to proceed through the phos-
d intermediate stage.
mary, the following reaction sequence can be put forward
be this reaction:

$$E + ATP \xrightarrow[Mg^{++}]{Na^+} E{\sim}P + ADP \qquad (1)$$

$$E{\sim}P \xrightarrow{K^+} E + P_i \quad \text{(rapid)} \qquad (2)$$

$$E + ATP \xrightarrow[Mg^{++}]{Na^+} E + ADP + P_i \quad \text{(slow)} \qquad (3)$$

HER ENZYMES ASSOCIATED WITH MEMBRANES

membrane preparations contain phosphatase activity probably
resenting several different enzymes with different properties
4, 33, 78]. Most of the phosphatase activity results from an enzyme
hich has an optimum pH in the acid region and which is activated
magnesium ions and weakly inhibited by fluoride. In addition,
maller amounts of a potassium-activated phosphatase with a slightly
lkaline pH optimum and a requirement for Mg^{++} are found when
p-nitrophenylphosphate and related substances are used as substrates.
The potassium-activated alkaline phosphatase is inhibited by cardiac
glycosides. The latter activity may represent an enzyme which is
identical to the sodium, potassium–activated ATPase. Further, membranes contain phosphodiesterases with an alkaline pH optimum,
particularly an enzyme which splits 2,3-diphosphoglyceric acid
[44, 147].

The Hokins [81, 82, 84] have isolated two enzymes from avian salt
glands, brain microsomes, erythrocyte ghosts, and other membrane
preparations: diglyceride kinase, which catalyzes the formation of

ment of the constituents as dynamic rather than static. The membranes are conceived to be planar aggregates of micellar subunits which are neither constant in their physical dimensions nor collectively arranged in a fixed array [94]. Instead, the subunits are thought to undergo reversible structural changes possibly corresponding to phase transitions. The rapid turnover of membrane constituents which has been observed can be better reconciled with this view of the membrane than with earlier theories of its structure [139]. It may be that the membrane proteins are synthesized by polyribosomes and first form a lipid-free membrane, and that the phospholipids are subsequently synthesized in situ.

SODIUM, POTASSIUM–ACTIVATED ATPase

A characteristic property of many types of cells from a variety of organisms is the ability to transport sodium ions from the cytoplasm to the exterior of the cell against an electrochemical gradient. This phenomenon is commonly called the active transport of sodium or the sodium pump. Experiments on nerve [20] and on erythrocyte membranes [39, 206] show that ATP provides the energy required to transport the sodium ions against the electrochemical gradient. ITP, GTP, and UTP cannot replace ATP as an energy source [79]. There appears to be a stoichiometry to cation transport, with 2 K^+ ions entering the cell for every 3 Na^+ transported outward. Biochemists interested in transport phenomena have focused attention on an ATPase found in membrane-rich fractions of cell homogenates which requires the presence of both sodium and potassium ions for full activity [173, 174]. While the requirement for sodium is specific, other monovalent cations can replace potassium.

The sodium, potassium–activated ATPase is found characteristically in the microsomal fraction [106, 165] of cell homogenates in heavy microsomes [22], in erythrocyte ghosts [145], or in cell membranes or membrane fragments [43, 200]. The enzyme appears to be a lipoprotein [190], and treatment with lipolytic agents destroys the enzyme activity. The sodium, potassium–activated ATPase requires magnesium ions, and the enzyme is inhibited by calcium. Among monovalent and divalent anions, only fluoride is an inhibitor [140]. The enzyme is also inhibited by basic proteins such as protamines, histones, and polylysine [45, 164, 214].

Most preparations contain a magnesium–activated ATPase which is not further activated by the presence of sodium and potassium ions. Frequently there also are present a sodium, potassium-activated adenylate kinase [4] and two enzymes responsible for an ADP-ATP exchange reaction [47, 186], one requiring only Mg^{++} ions and the other activated by Na^{+} and Mg^{++} ions. Aging causes the ratio of sodium, potassium-activated to sodium-independent ATPase activity to increase [106, 175]. The sodium, potassium-activated ATPase can be separated from other proteins and concentrated by extraction with 2 M NaI [131, 180]. After the salt concentration is lowered to 0.6 M, the sodium, potassium-activated ATPase is precipitated while most other enzymes remain in solution. Another purification method utilizes chromatography on glass beads [161].

The sodium, potassium-activated ATPase has been isolated from many different tissues including erythrocyte membranes, brain, nerve, kidney, muscle, liver, intestine, electric organ, parotid gland, frog skin, ciliary body, lens, retina, thyroid, and toad bladder [1, 15, 17, 177]. For a variety of tissues the ratio of sodium flux to the amount of sodium, potassium-activated ATPase is very nearly the same, although, for example, the net sodium efflux is about 25,000 times as great in the electric organ as in erythrocytes [14].

From experiments using erythrocyte ghosts filled with various media by "reversible" hemolysis [20, 207], it was found that ATP was hydrolyzed only when it was on the inside of the membrane and the ADP liberated also remained on the interior. Maximum ATPase activity was found with high intracellular sodium and high extracellular potassium concentrations. These experiments lead to the conclusion that the ATPase is oriented across the cell membrane with the sodium site facing the interior and the potassium site facing the exterior. It has been shown that ^{22}Na is bound by microsomes, erythrocyte ghosts, and perhaps the partially purified enzyme in the presence of ATP [23, 89, 202]. Ouabain inhibited the ATP-dependent binding of sodium ions.

Low concentrations of cardiac glycosides are specific inhibitors of sodium transport [63, 98, 158]. The cardiac glycosides inhibit the activity of the sodium, potassium-activated ATPase but not the activity of the ATPase depending upon magnesium ions alone. There is a close correlation between the inhibitory potency of different glycosides on cation transport and that on the sodium, potassium-activated ATPase [40]. In the squid axon, cardiac glycosides are effective only on the outside of the membrane [21]. In part, the inhibition by cardiac

glycosides can be ofi
ions [40, 144]. The
very different from tl
in pharmacological ac
activated ATPase [16].
The sodium, potassiun
of sulfhydryl reagents in
maleimide, and 2,4-dinitr
preparations, sodium-indep
idly than Na,K-activated AT
potassium-activated enzyme
ence of ATP + Na^{+} + K^{+} + M
ments the presence of a sulfhy
inferred.
In the presence of γ-^{32}P-ATP,
sium-activated ATPase incorpora
can be precipitated by perchloric o
maximal phosphorylation of beef ki
ATPase is achieved at a 1.6 mM N
dependent incorporation of ^{32}P into
a few seconds at 5° C. The bound ^{32}P i
If a small amount of K^{+} is present in a
lyzed, but the phosphorylated interme
posure of the ^{32}P-labeled intermediate to
in the formation of phosphate esters of
interest that no other cation can substitute
of the phosphorylated intermediate, and pa
mimics Na^{+} during excitation, is not able to s
of the phosphorylated intermediate is inhib
calcium ions but not by cardiac glycosides [45,
The phosphorylated intermediate is hydro
than neutral and in strong acid. Hydrolysis is g
hydroxylamine or molybdate ions. Upon digest
enzymes, a phosphorylated peptide fragment is c
erties like those of the phosphorylated intermedia
the phosphorylated intermediate is an acylphosp
130, 146], and Kahlenberg and Hokin [91] have b
some direct evidence that it is, in fact, a phosphorylg
Under physiological conditions the hydrolysis of th
ated intermediate of the enzyme is markedly enh

phosphatidic acid from α,β-diglycerides and ATP, and phosphatidic acid phosphatase, which catalyzes the hydrolysis of phosphatidic acid. The former enzyme appears to be activated by acetylcholine while the latter enzyme appears to be inhibited by cardiac glycosides. A complete cycle through both reactions, during which phosphatidic acid is synthesized and broken down, adds up to the net hydrolysis of ATP. In the avian salt gland the intracellular level of these enzymes is stimulated by acetylcholine and is accompanied by the excretion of NaCl. The relationship of the enzymes to the sodium, potassium–activated ATPase is unknown.

Erythrocyte ghosts and other membrane preparations contain acetylcholinesterase, a different enzyme from the pseudocholinesterase found in plasma. Acetylcholinesterase activity has been localized in the cell membrane by electron microscopic examination after histochemical staining [168] and appears to be an integral structural component of the membrane [122, 195].

Erythrocyte ghosts and other membranes contain catalase, a heme-containing enzyme which decomposes hydrogen peroxide and also catalyzes the oxidation of alcohol and aldehydes [58, 95].

Cell membrane preparations contain enzyme systems which couple the reduction cytochromes *c* or dyes such as methylene blue to the oxidation of NADH (reduced nicotinamide dinucleotide) [62]. The NAD-diaphorase system has been found in the same deoxycholate-treated submicroscopic particles isolated from brain, kidney, erythrocyte ghosts, and cardiac muscle [176] that contain the sodium, potassium–activated ATPase. Both the NAD-diaphorase and the cation-activated ATPase are inhibited by N-ethylmaleimide, and ATP affords protection to both enzymes against the action of N-ethylmaleimide. Kamat and Wallach [92], on the other hand, have been able to separate the microsomal fraction into two subfractions, one rich in Na, K–activated ATPase and one rich in NAD-diaphorase. Because the cation-activated ATPase fraction also is rich in surface antigens, these investigators thought that it might represent cell membrane fragments, while the NAD-diaphorase–rich subfraction contained little surface antigen and in their opinion represented endoplasmic reticulum fragments.

Sodium, potassium–activated ATPase obtained from beef heart microsomes is unaffected by aurovertin, atractyloside, dinitrophenyl, or azide, inhibitors of mitochondrial ATPase and electron transport; however, it is inhibited by oligomycin, octylguanidine, and tributyltin [120]. Whether a relation exists between the electron transport NAD-diaphorase system and the cation-activated ATPase is not clear at the present time. Enzymes associated with plasma membranes are listed in Table 1-2.

Erythrocyte ghosts appear to be able to carry on the complete sequence of glycolytic reactions leading from glucose to lactic acid [110]. Schrier [162, 163] has shown that triose phosphate dehydrogenase and phosphoglycerate kinase are particularly tightly bound to erythrocyte ghosts. Schrier has proposed that the enzymes may be organized into a system which utilizes triose phosphate and NAD as substrates, leading to 3-phosphoglyceric acid and NADH as products in a reaction coupled to the formation of ATP from ADP. The ATP so generated then can be used directly by the sodium, potassium-activated ATPase. In carefully prepared erythrocyte ghosts and yeast membranes the entire complex of glycolytic enzymes is found largely associated with the membranes [68]. The loosely held enzymes of the sequence could be eluted from the membrane system and recombined by a cycle of change in pH. The organization of enzymes

TABLE 1-2. *Enzymes Found in Cell Membrane Preparations*

Enzyme	Rat Liver*	Erythrocyte†	Milk Fat‡
	(μmoles product/mg protein/hr)		
Na, K-activated ATPase	11.7	1.6	0.91
Alkaline phosphatase (PNPP)	1.5	1.5	0.63
Alkaline phosphatase K-activated	1.5		0.63
Acid phosphatase (PNPP)	5.8	2.5	0.09
Alkaline phosphodiesterase	3.6		0.134
Acid phosphodiesterase	0.7		0.008
Glucose-6-phosphatase	1.4	0.7	0.064
NADH-cytochrome *c* reductase	7.68	1.47	0.001
Triose phosphate dehydrogenase	2.04	6.5	
Acetylcholinesterase		19.06	0.17

* Summarized in Emmelot, P., and Benedetti, E. L. On the Possible Involvement of the Plasma Membrane in the Carcinogenic Process. In *Carcinogenesis: A Broad Critique*. Baltimore: Williams & Wilkins, 1967. P. 512.

† Summarized in Pennell, R. B. Composition of Normal Human Red Cells. In Bishop, C., and Surgenor, D. M. (Eds.), *The Red Blood Cell*. New York: Academic, 1964. P. 35.

‡ From Dowben et al. [38].

in a membrane may facilitate interactions arising from Onsager coupling between the fluxes of various components of a metabolic system.

PROPERTIES OF THE CELL SURFACE

The surface of erythrocytes and suspensions of other cells shows the properties of a polyanion. Cell suspensions migrate toward the anode in a direct current field, indicating that they possess a negative net surface charge. The electrophoretic mobility appears to be characteristic of all cells of a given species and is unaffected by the shape of the cells, indicating that the density of net surface charge is constant. Because erythrocyte ghosts migrate with the same velocity as intact cells, the surface charge is evidently not lost upon rupture [143]. The electrophoretic mobility is constant for a large range of salt concentrations, and the isoelectric point is quite low, in the vicinity of pH 2.

It appears that the negative surface charge of cells is largely due to the carboxyl group of sialic acids (N-acetylneuraminic, N-glycolylneuraminic acid, etc.) [25, 26, 46, 67], which has a pK_a of about 2.5. Exposure of cells to the enzymes neuraminidase or trypsin results in cleavage of a sialomucopeptide which contains the N-acetylneuraminic acid residues. After such treatment more than 95 percent of the sialic acid residues are removed from the cell surface and there is a marked drop in the net negative surface charge. With erythrocytes the sialic acid appears to be restricted to the plasma membrane, but in the case of other cell types sialic acid may occur in the interior. Marcus and Salb [119] offered evidence for the presence of sialic acid in the nuclear membranes of HeLa cells and Ehrlich ascites tumor cells.

Because of the polyanionic property of cells it is not surprising that they interact with polycations such as polylysine. Erythrocytes can be agglutinated by the addition of polylysine and the process reversed by the subsequent addition of polyglutamate. The amount of polylysine required to cause agglutination is 6 percent or less of the total that can be adsorbed on the surface.

Before it was established that the polyanionic character of the cell surface was due largely to sialic acid residues, it was found that the electrophoretic mobility of erythrocytes could be reduced by treatment with influenza virus [74]. Subsequently, the viral principle responsible was shown to be a neuraminidase [66]. Animal viruses enter susceptible cells by attaching themselves to specific receptor sites on the cell surface. Purified receptor substance for the influenza virus was isolated and shown to be a complex macromolecule containing sialic acid [93]. Complexes can be formed between polylysine and the virus receptor glycoprotein which frequently are insoluble [32]. Such insoluble complexes can be redissolved by the addition of polyaspartate.

Erythrocytes have a variety of distinct antigenic groups on their surface. The individual members of a species may be divided into blood groups according to their complement of cell surface antigens. The first blood group system to be discovered [107], the ABO system, is the most significant one in connection with the compatibility of transfused blood. Individuals are divided into four blood groups, A, B, AB, and O, according to the presence or absence of two antigens, A and B, on the erythrocyte surface. In the *absence* of either or both antigens, the corresponding anti-A and anti-B antibodies are found in the serum. A similar system, the MN system [108, 109], depends on the presence or absence of two antigens, M and N. Although the MN system is less important in connection with blood transfusion than the ABO system, it has been used extensively as a marker in genetic and anthropological studies. Another important blood group system is the rhesus factor (rh factor), which is genetically determined by the CDE substances.

Because the A and B and the M and N antigens were found in erythrocytes and because they are important in blood transfusions, they have been called blood group substances. However, it appears that they are present on the cell surfaces of most epithelial cells [105, 187, 188, 209]. Szulman [189] examined the distribution of blood group antigens in the tissues of fetuses. In an elegant study he showed that the distribution of A and B antigens is very widespread early in embryonic life. After the twelfth week, however, the blood group antigens become restricted to simple or stratified confining epithelium, while epithelium which proceeds to differentiate further tends to lose these antigens. The A and B antigens, but not the M and N antigens, also are found in a water-soluble form in tissue fluids and secretions [212].

A rather unique blood group system is the Lewis antigen system [127, 179]. The antigens of the Lewis system are present in the plasma and appear to be adsorbed by erythrocytes, rather than being more integrated into the membrane structure.

The purified blood group substances are glycoproteins with molecular weights ranging from 2×10^5 to 1×10^6. Actually, the protein moiety of the molecules represents a quantitatively minor portion, accounting for only about 15 percent of the molecular weight. The detailed molecular structures have not yet been elucidated, but they appear to consist of short oligosaccharide chains joined at intervals to a peptide backbone. The carbohydrate moiety contains five principal sugars, L-fucose, D-galactose, N-acetyl-D-glucosamine, N-acetyl-D-galactosamine, and N-acetylneuraminic acid. One difference observed consistently between A and B antigens is that the ratio of galactosamine to glucosamine is higher in A than in B [157]. The antigenic specificity appears to reside in the carbohydrate moiety and depends upon the nature of the sugars at the nonreducing ends of the carbohydrate chain.

In addition to the blood group antigens, the cell membrane appears to possess a diverse constellation of antigenic determinants. One type which has received considerable attention is the histocompatibility antigens, those resulting in host rejection responses to transplants. The H-2 isoantigen system produces the most intense reactions against homotransplantation in mice. The localization of the H-2 system in the surface membrane is supported by the fact that the entire H-2 activity of isoantisera can be adsorbed by suspensions of intact cells [75]. Furthermore, if cells in suspension are exposed to H-2 isoantiserum and then reacted with fluorescein-conjugated rabbit-antimouse globulin, the fluorescent marker clearly is localized on the cell surface [123]. A comparison of the reaction of whole and disrupted cells with isoantisera indicates that 80 percent or more of the H-2 antigens are located on the cell surface [210].

The cell surface also appears to possess specific antigenic sites capable of stimulating the production of specific heterologous antibodies. Such heterologous antibodies result in agglutination of cell suspensions or in cytolysis [64, 71]. Heterologous antibodies to Ehrlich ascites tumor cells can be absorbed almost completely by suspensions of tumor cells. Reaction of disrupted cells with heterologous antibodies indicates that the antigen is found in the microsomal fraction [92, 196, 198].

A number of tumor-specific transplantation antigens have been studied which may be regarded as weak histocompatibility antigens [96, 137]. Evidence that the tumor-specific transplantation antigens are located on the cell surface includes the absorption of humoral antibodies by Moloney lymphoma cells, by immunofluorescence localization on the cell surface, and by the interference with the adherence of host lymphoid cells to target cells by isoantibodies [97, 124].

The reactive sulfhydryl groups on cell surfaces seem to be essential for functional integrity. The human erythrocyte contains about 4×10^{-17} moles of sulfhydryl groups per cell. Reaction of approximately 5 percent of these groups with an organic mercurial results in a marked diminution in the facilitated diffusion of glucose [194]. Cation permeability is markedly increased by reaction of the surface sulfhydryl groups with organomercurials, an effect which can be largely reversed by incubation with reduced glutathione [166]. Reaction of the surface sulfhydryl groups of erythrocytes with organomercurials results in increased fragility and decreased survival in vivo [86, 87, 104, 167, 185].

Lastly, mention should be made of the effects of treating the cell surface with proteases or phospholipases upon permeability and excitation. Treatment of lobster giant axon with trypsin, chymotrypsin, or collagenase failed to alter significantly the transmembrane resting potential or excitability, even though the externally applied proteases appeared to penetrate through the membrane and liquefy the intracellular protein. The preparations treated with proteases were short lived. On the other hand, treatment with phospholipase A, which catalyzes the hydrolysis of a fatty acid from a phospholipid leaving a lysolecithin, or treatment with phospholipase C, which catalyzes the hydrolysis of a phospholipid to an α,β-diglyceride, results in a fall in membrane resistance and resting transmembrane potential as well as a decrease in the magnitude of the action potential [132, 191, 192]. These results support the view that the phospholipids are essential for the functional integrity of cell membrane.

REFERENCES

1. Albers, R. W. Biochemical aspects of active transport. *Ann. Rev. Biochem.* 36:727, 1967.
2. Albers, R. W., Fahn, S., and Covell, G. J. The role of sodium ions in the activation of *Electrophorus* electric organ adenosine triphosphatase. *Proc. Nat. Acad. Sci. U.S.A.* 50:474, 1963.
3. Andersen, V. Extraction of erythrocyte proteins with *n*-butanol. *Clin. Chim. Acta* 8:454, 1963.
4. Askari, A., and Fratantoni, J. C. Effect of Na^+ and K^+ and adenylate kinase on human erythrocytes. *Proc. Soc. Exp. Biol. Med.* 116:751, 1964.
5. Attwood, D., Saunders, L., Gammack, D. B., deHaas, G. H., and van

Deenen, L. L. M. The size and shape of micelles of some synthetic phospholipids in aqueous dispersions. *Biochim. Biophys. Acta* 102:301, 1965.
6. Azen, E. A., Orr, S., and Smithies, O. Starch-gel electrophoresis of erythrocyte stroma. *J. Lab. Clin. Med.* 65:440, 1965.
7. Bader, H., Sen, A. K., and Post, R. L. Isolation and characterization of a phosphorylated intermediate in the ($Na^+ + K^+$) system-dependent ATPase. *Biochim. Biophys. Acta* 118:106, 1966.
8. Bakerman, S., and Wasemiller, G. Studies on structural units of human erythrocyte membrane: I. Separation, isolation, and partial characterization. *Biochemistry* (Washington) 6:1100, 1967.
9. Bangham, A. D., and Horne, R. W. Action of saponin on biological cell membranes. *Nature* (London) 196:952, 1962.
10. Bar, R. S., Deamer, D. W., and Cornwell, D. G. Surface area of human erythrocyte lipids: Reinvestigation of experiments on plasma membrane. *Science* 153:1010, 1966.
11. Bargmann, W., and Knoop, A. Über die Morphologie der Milchsekretion. Licht- und elektronenmikroskopische Studien an der Milchdrüse der Ratte. *Z. Zellforsch.* 49:344, 1959.
12. Benedetti, E. L., and Emmelot, P. Electron microscopic observations on negatively stained plasma membranes isolated from rat liver. *J. Cell Biol.* 26:299, 1965.
13. Blasie, J. K., Dewey, M. M., Blaurock, A. E., and Worthington, C. R. Electron microscope and low-angle x-ray diffraction studies on outer segment membranes from the retina of the frog. *J. Molec. Biol.* 14:143, 1965.
14. Bonting, S. L., and Caravaggio, L. L. Studies on sodium-potassium-activated adenosinetriphosphatase: V. Correlation of enzyme activity with cation flux in six tissues. *Arch. Biochem.* 101:37, 1963.
15. Bonting, S. L., Caravaggio, L. L., and Hawkins, N. M. Studies on sodium-potassium-activated adenosinetriphosphatase: IV. Correlation with cation transport sensitive to cardiac glycosides. *Arch. Biochem.* 98:413, 1962.
16. Bonting, S. L., Hawkins, N. M., and Canady, M. R. Studies of sodium-potassium activated adenosine triphosphatase: VII. Inhibition by erythrophleum alkaloids. *Biochem. Pharmacol.* 13:13, 1964.
17. Bonting, S. L., Simon, K. A., and Hawkins, N. M. Studies on sodium-potassium-activated adenosine triphosphatase: I. Quantitative distribution in several tissues of the cat. *Arch. Biochem.* 95:416, 1961.
18. Branton, D. Fracture faces of frozen membranes. *Proc. Nat. Acad. Sci. U.S.A.* 55:1048, 1966.
19. Branton, D. Fracture faces of frozen myelin. *Exp. Cell Res.* 45:703, 1967.
20. Caldwell, P. C., Hodgkin, A. L., Keynes, R. D., and Shaw, T. J. The effects of injecting 'energy-rich' phosphate compounds on the active transport of ions in the giant axons of *Loligo*. *J. Physiol.* (London) 152:561, 1960.
21. Caldwell, P. C., and Keynes, R. D. The effect of ouabain on the efflux

of sodium from a squid giant axon. *J. Physiol.* (London) 148:8P, 1959.

22. Charnock, J. S., and Post, R. L. Studies of the mechanism of cation transport: I. The preparation and properties of a cation stimulated adenosine-triphosphatase from guinea pig kidney cortex. *Aust. J. Exp. Biol. Med. Sci.* 41:547, 1963.
23. Charnock, J. S., and Post, R. L. Evidence of the mechanism of ouabain inhibition of cation activated adenosine triphosphatase. *Nature* (London) 199:910, 1963.
24. Clarkson, E. M., and Maizels, M. Distribution of phosphatases in human erythrocytes. *J. Physiol.* (London) 116:112, 1952.
25. Cook, G. M. W. Linkage of sialic acid in the human erythrocyte ultrastructure. *Nature* (London) 195:159, 1962.
26. Cook, G. M. W., Heard, D. H., and Seaman, G. V. F. Sialic acids and the electrokinetic charge of the human erythrocyte. *Nature* (London) 191:44, 1961.
27. Criddle, R. S., Bock, R. M., Green, D. E., and Tisdale, H. Specific interaction of mitochondrial structural protein with cytochromes and lipid. *Biochem. Biophys. Res. Commun.* 5:75, 81, 109, 1961.
28. Criddle, R. S., Bock, R. M., Green, D. E., and Tisdale, H. Physical characteristics of proteins of the electron transfer system and interpretation of the structure of the mitochondrion. *Biochemistry* (Washington) 1:827, 1962.
29. Criegee, R. Osmic esters as intermediate products in oxidation. *Ann. Chem.* 522:75, 1936.
30. Curtis, A. S. G. Cell contact and adhesion. *Biol. Rev.* 37:82, 1962.
31. Danielli, J. F., and Harvey, E. N. The tension at the surface of mackerel egg oil, with remarks on the nature of the cell surface. *J. Cell. Comp. Physiol.* 5:483, 1935.
32. Danon, D., Howe, C., and Lee, L. T. Interaction of polylysine with soluble components of human erythrocyte membranes. *Biochim. Biophys. Acta* 101:201, 1965.
33. Davies, D. R. The phosphatase activity of spleen extracts. *Biochem. J.* 28:529, 1934.
34. Davson, H., and Danielli, J. F. *The Permeability of Natural Membranes* (2nd ed.). London: Cambridge University Press, 1952.
35. Deenen, L. L. M. van. Phospholipids and Biomembranes. In Holman, R. (Ed.), *Progress in the Chemistry of Fats and Related Lipids.* Oxford: Pergamon, 1965. Vol. 8, p. 1.
36. Deenen, L. L. M. van. Some structural and dynamic aspects of lipids in biological membranes. *Ann. N.Y. Acad. Sci.* 137:717, 1966.
37. Dodge, J. T., Mitchell, C., and Hanahan, D. The preparation and chemical characteristics of hemoglobin-free ghosts of human erythrocytes. *Arch. Biochem.* 100:119, 1962.
38. Dowben, R. M., Brunner, J. R., and Philpott, D. E. Studies on milk fat globule membranes. *Biochim. Biophys. Acta* 135:1, 1967.
39. Dunham, E. T. Linkage of active cation transport to ATP utilization. *Physiologist* 1:23, 1957.

40. Dunham, E. T., and Glynn, I. M. Adenosinetriphosphatase activity and the active movements of alkali metal ions. *J. Physiol.* (London) 156:274, 1961.
41. Elbers, P. F. The cell membrane: Image and interpretation. *Recent Progr. Surface Sci.* 2:443, 1964.
42. Emmelot, P., and Bos, C. J. Adenosine triphosphatase in the cell-membrane fraction from rat liver. *Biochim. Biophys. Acta* 58:374, 1962.
43. Emmelot, P., and Bos, C. J. Studies on plasma membranes: III. Mg^{2+}-ATPase, (Na^{+}-K^{+}-Mg^{2+})-ATPase and 5′-nucleotidase activity of plasma membranes isolated from rat liver. *Biochim. Biophys. Acta* 120: 369, 1966.
44. Emmelot, P., Bos, C. J., Benedetti, E. L., and Rümke, P. Studies on plasma membranes: I. Chemical composition and enzyme content of plasma membranes isolated from rat liver. *Biochim. Biophys. Acta* 90:126, 1964.
45. Epstein, F. H., and Whittam, R. The mode of inhibition by calcium of cell-membrane adenosine-triphosphatase activity. *Biochem. J.* 99:232, 1966.
46. Eylar, E. H., Madoff, M. A., Brody, O. V., and Oncley, J. L. The contribution of sialic acid to the surface charge of the erythrocyte. *J. Biol. Chem.* 237:1992, 1962.
47. Fahn, S., Koval, G. J., and Albers, R. W. Sodium-potassium-activated adenosine triphosphatase of *Electrophorus* electric organ: I. An associated sodium-activated transphorylation. *J. Biol. Chem.* 241: 1882, 1966.
48. Farquhar, M. G., and Palade, G. E. Junctional complexes in various epithelia. *J. Cell Biol.* 17:375, 1963.
49. Fernández-Morán, H. Sheath and axon structures in the internode portion of vertebrate myelinated nerve fibres. *Exp. Cell Res.* 1:309, 1950.
50. Fernández-Morán, H. Cell membrane ultrastructure. *Circulation* 26: 1039, 1962.
51. Fernández-Morán, H., and Finean, J. B. Electron microscope and low-angle x-ray diffraction studies of the nerve myelin sheath. *J. Biophys. Biochem. Cytol.* 3:725, 1957.
52. Fernández-Morán, H., Oda, T., Blair, P. V., and Green, D. E. A macromolecular repeating unit of mitochondrial structure and function. *J. Cell Biol.* 22:63, 1964.
53. Finean, J. B. Phospholipid-cholesterol complex in the structure of myelin. *Experientia* 9:17, 1953.
54. Finean, J. B. The nature and stability of nerve myelin. *Int. Rev. Cytol.* 12:303, 1961.
55. Finean, J. B., Coleman, R., and Green, W. A. Studies of isolated plasma membrane preparations. *Ann. N.Y. Acad. Sci.* 137:414, 1966.
56. Fleischer, S., Fleischer, B., and Stoeckenius, W. Fine structure of whole and fragmented mitochondria after lipid depletion. *Fed. Proc.* 24:296, 1965.
57. Fox, S. W., Harada, K., and Kendrick, J. Production of spherules from synthetic proteinoid and hot water. *Science* 129:1221, 1959.

58. Francoeur, N., and Denstedt, O. F. Metabolism of mammalian erythrocytes: V. Role of catalase in the oxidation of ribose-5-phosphate by the erythrocyte. *Canad. J. Biochem. Physiol.* 32:644, 1954.
59. Freysz, L., Bierth, R. I., and Mandel, P. Étude cinétique du renouvellement des divers phospholipides du cerveau de rat adulte. *Bull. Soc. Chim. Biol.* (Paris) 47:1441, 1965.
60. Geren, B. B. The formation from the Schwann cell surface of myelin in the peripheral nerves of chick embryos. *Exp. Cell Res.* 7:558, 1954.
61. Gibbons, I. R., and Rowe, A. J. Dynein: A protein with adenosine triphosphatase activity from cilia. *Science* 149:424, 1965.
62. Gibson, Q. H. The reduction of methaemoglobin in red blood cells and studies on the cause of idiopathic methaemoglobinaemia. *Biochem. J.* 42:13, 1948.
63. Glynn, I. M. The action of cardiac glycosides on sodium and potassium movements in human red cells. *J. Physiol.* (London) 136:148, 1957.
64. Goldberg, B., and Green, H. The cytotoxic action of immune gamma globulin and complement on Krebs ascites tumor cells: I. Ultrastructure studies. *J. Exp. Med.* 109:505, 1959.
65. Gorter, E., and Grendel, F. On bimolecular layers of lipoids on the chromocytes of the blood. *J. Exp. Med.* 41:439, 1925.
66. Gottschalk, A. Neuraminidase: The specific enzyme of influenza virus and *Vibrio cholerae. Biochim. Biophys. Acta* 23:645, 1957.
67. Gottschalk, A. Correlation between composition, structure, shape and function of a salivary mucoprotein. *Nature* (London) 186:949, 1960.
68. Green, D. E., Murer, E., Hultin, H. O., Richardson, S. H., Salmon, B., Brierley, G. P., and Baum, H. Association of integrated metabolic pathways with membranes: I. Glycolytic enzymes of the red blood corpuscle and yeast. *Arch. Biochem.* 112:635, 1965.
69. Green, D. E., and Perdue, J. F. Correlation of mitochondrial structure and function. *Ann. N.Y. Acad. Sci.* 137:667, 1966.
70. Green, D. E., and Perdue, J. F. Membranes as expressions of repeating units. *Proc. Nat. Acad. Sci. U.S.A.* 55:1295, 1966.
71. Green, H., Barrow, P., and Goldberg, B. Effect of antibody and complement on permeability control in ascites tumor cells and erythrocytes. *J. Exp. Med.* 110:699, 1959.
72. Groningen, H. E. M. van, and Slater, E. C. The effect of oligomycin on the ($Na^+ + K^+$)-activated magnesium-ATPase of brain microsomes and erythrocyte membrane. *Biochim. Biophys. Acta* 73:527, 1963.
73. Hagerman, J. S., and Gould, R. G. The *in vitro* interchange of cholesterol between plasma and red cells. *Proc. Soc. Exp. Biol. Med.* 78:329, 1951.
74. Hanig, M. Electrokinetic change in human erythrocytes during adsorption and elution of PR 8 influenza virus. *Proc. Soc. Exp. Biol. Med.* 68:385, 1948.
75. Haughton, G. Transplantation antigen of mice: Cellular localization of antigen determined by the H-2 locus. *Transplantation* 4:238, 1966.
76. Hechter, O. Role of water structure in the molecular organization of cell membranes. *Fed. Proc.* 24 (Suppl. 15):S-91, 1965.

77. Henn, F. A., Decker, G. L., Greenwalt, J. W., and Thompson, T. E. Properties of lipid bilayer membranes separating two aqueous phases: Electron microscope studies. *J. Molec. Biol.* 24:51, 1962.
78. Herbert, E. A study of the liberation of orthophosphate from adenosine triphosphate by the stromata of human erythrocytes. *J. Cell. Comp. Physiol.* 47:11, 1956.
79. Hoffman, J. F. The link between metabolism and the active transport of Na in human red cell ghosts. *Fed. Proc.* 19:127, 1960.
80. Hoffman, J. F. The active transport of sodium by ghosts of human red blood cells. *J. Gen. Physiol.* 45:837, 1962.
81. Hokin, L. E., and Hokin, M. R. The mechanism of phosphate exchange in phosphatidic acid response to acetylcholine. *J. Biol. Chem.* 234:1387, 1959.
82. Hokin, L. E., and Hokin, M. R. Diglyceride kinase and phosphatidic acid phosphatase in erythrocyte membranes. *Nature* (London) 189:836, 1961.
83. Hokin, L. E., Sastry, P. S., Galsworthy, P. R., and Yoda, A. Evidence that a phosphorylated intermediate in a brain transport adenosine triphosphatase is an acyl phosphate. *Proc. Nat. Acad. Sci. U.S.A.* 54:177, 1965.
84. Hokin, M. R., and Hokin, L. E. The synthesis of phosphatidic acid from diglyceride and adenosine triphosphate in extracts of brain microsomes. *J. Biol. Chem.* 234:1381, 1959.
85. Holland, J. J. Irreversible eclipse of Poliovirus by HeLa cells. *Virology* 16:163, 1962.
86. Jacob, H. S., and Jandl, J. H. Effects of sulfhydryl inhibition on red blood cells: I. Mechanism of hemolysis. *J. Clin. Invest.* 41:779, 1962.
87. Jacob, H. S., and Jandl, J. H. Effects of sulfhydryl inhibition on red blood cells: II. Studies *in vivo*. *J. Clin. Invest.* 41:1514, 1962.
88. Järnefelt, J. Properties and possible mechanism of the Na^+ and K^+-stimulated microsomal adenosinetriphosphatase. *Biochim. Biophys. Acta* 59:643, 1962.
89. Järnefelt, J., and Stendingk, L. V. von. Some properties of the ATP dependent Na^+ binding system of rat brain microsomes. *Acta Physiol. Scand.* 57:328, 1963.
90. Jorpes, E. The protein component of the erythrocyte membrane or stroma. *Biochem. J.* 26:1488, 1932.
91. Kahlenberg, A., Galsworthy, P. R., and Hokin, L. E. Sodium-potassium adenosine triphosphatase: Acyl phosphate "intermediate" shown to be L-glutamyl-γ-phosphate. *Science* 157:434, 1967.
92. Kamat, V. B., and Wallach, D. F. H. Separation and partial purification of plasma-membrane fragments from Ehrlich ascites carcinoma microsomes. *Science* 148:1343, 1965.
93. Kathan, R. H., Winzler, R. J., and Johnson, C. A. Preparation of an inhibitor of viral hemagglutination from human erythrocytes. *J. Exp. Med.* 113:37, 1961.
94. Kavanau, J. L. *Structure and Function in Biological Membranes.* San Francisco: Holden-Day, 1965.

95. Keilin, D., and Hartree, F. Purification of horse-radish peroxidase and comparison of its properties with those of catalase and methaemoglobin. *Biochem. J.* 49:88, 1951.
96. Klein, E. Le Problème de l'Auto-antigénecité. In Grabar, P. (Ed.), *Immunopathology.* Basel: Schwabe, 1966. P. 20.
97. Klein, G., Klein, E., and Haughton, G. Variation of antigenic characteristics between different mouse lymphomas induced by the Moloney virus. *J. Nat. Cancer Inst.* 36:607, 1966.
98. Koefoed-Johnsen, V. The effect of g-strophanthin (ouabain) on the active transport of sodium through the isolated frog skin. *Acta Physiol. Scand.* 42 (Suppl. 145):87, 1958.
99. Kono, T., and Colowick, S. P. Isolation of skeletal muscle cell membrane and some of its properties. *Arch. Biochem.* 93:520, 1961.
100. Kopaczyk, K., Perdue, J., and Green, D. E. The relation of structural and catalytic protein in the mitochondrial electron transfer chain. *Arch. Biochem.* 115:215, 1966.
101. Korn, E. D. II. Synthesis of bis(methyl 9,10-dihydroxy-stearate)osmate from methyl oleate and osmium tetroxide under conditions used for fixation of biological material. *Biochim. Biophys. Acta* 116:317, 1966.
102. Korn, E. D. III. Modification of oleic acid during fixation of amoebae by osmium tetroxide. *Biochim. Biophys. Acta* 116:325, 1966.
103. Korn, E. D., and Weisman, R. A. I. Loss of lipids during preparation of amoebae for electron microscopy. *Biochim. Biophys. Acta* 116:309, 1966.
104. Korst, D. R., Nixon, J. C., Boblitt, D. E., and Quirk, J. Studies of selective splenic sequestration of erythrocytes labeled with radioactive mercurihydroxypropane. *J. Lab. Clin. Med.* 66:788, 1965.
105. Kosjakov, P. N., and Tribulev, G. P. Group-specific differentiation of the organs of man. *J. Immun.* 37:283, 1939.
106. Landon, E. J., and Norris, J. L. Sodium- and potassium-dependent adenosine triphosphatase activity in rat-kidney endoplasmic reticulum fraction. *Biochim. Biophys. Acta* 71:266, 1963.
107. Landsteiner, K. Zur Kenntnis der antifermentativen, lytischen und agglutinierenden Wirkungen des Blutserums und der Lymphe. *Cent. Bakteriol. Parasiten Infektionskrankheiten* 27:357, 1900.
108. Landsteiner, K., and Levine, P. A new agglutinable factor differentiating individual human bloods. *Proc. Soc. Exp. Biol. Med.* 24:600, 1927.
109. Landsteiner, K., and Levine, P. Further observations on individual differences of human blood. *Proc. Soc. Exp. Biol. Med.* 24:941, 1927.
110. Lionetti, F., Rees, S. B., Healy, W. B., Walker, B. S., and Gibson, J. G. The effect of adenosine upon esterification of phosphate by erythrocyte ghosts. *J. Biol. Chem.* 220:467, 1956.
111. Lucy, J. A., and Glauert, A. M. Structure and assembly of macromolecular lipid complexes composed of globular micelles. *J. Molec. Biol.* 8:727, 1964.
112. Luzatti, V., and Husson, F. The structure of the lipid-crystalline phases of lipid-water systems. *J. Cell Biol.* 12:207, 1962.
113. McCollester, D. L. A method for isolating skeletal-muscle cell-membrane components. *Biochim. Biophys. Acta* 57:427, 1962.

114. MacLennan, D. H., Tzagoloff, A., and Rieske, J. S. Studies on the electron transfer system: LXIII. Solubilization and fractionation of mitochondrial protein by succinylation. *Arch. Biochem.* 109:383, 1965.
115. Maddy, A. H. The solubilization of the protein of the ox-erythrocyte ghost. *Biochim. Biophys. Acta* 88:448, 1964.
116. Maddy, A. H. The properties of the protein of the plasma membrane of ox erythrocytes. *Biochim. Biophys. Acta* 117:193, 1966.
117. Maddy, A. H., and Malcolm, B. R. Protein conformations in biological membranes. *Science* 153:212, 1966.
118. Mandel, P., and Nussbaum, J. L. Incorporation of ^{32}P into the phosphatides of myelin sheaths and of intracellular membranes. *J. Neurochem.* 13:629, 1966.
119. Marcus, P. I., and Salb, J. N. Molecular basis of interferon action: Inhibition of viral RNA translation. *Virology* 30:502, 1966.
120. Matsui, H., and Schwartz, A. Purification and properties of a highly active ouabain-sensitive Na^+, K^+-dependent adenosinetriphosphatase from cardiac tissue. *Biochim. Biophys. Acta* 128:380, 1966.
121. Mitchell, C. D., and Hanahan, D. J. Solubilization of certain proteins from the human erythrocyte stroma. *Biochemistry* (Washington) 5:51, 1966.
122. Mitchell, C. D., Mitchell, W. B., and Hanahan, D. J. Enzyme and hemoglobin retention in human erythrocyte stroma. *Biochim. Biophys. Acta* 104:348, 1965.
123. Möller, E. Isoantigenic properties of tumors transgressing histocompatibility barriers of the H-2 system. *J. Nat. Cancer Inst.* 33:979, 1964.
124. Möller, E. Antagonistic effects of humoral isoantibodies on the *in vitro* cytotoxicity of immune lymphoid cells. *J. Exp. Med.* 122:11, 1965.
125. Morgan, T. E., and Hanahan, D. J. Solubilization and characterization of a lipoprotein from erythrocyte stroma. *Biochemistry* (Washington) 5:1050, 1966.
126. Moskowitz, M., and Calvin, M. On the components and structure of the human red cell membrane. *Exp. Cell Res.* 3:33, 1952.
127. Mourant, A. E. A "new" human blood group antigen of frequent occurrence. *Nature* (London) 158:237, 1946.
128. Mulder, E., and van Deenen, L. L. M. Metabolism of red-cell lipids: I. Incorporation *in vitro* of fatty acids into phospholipids from mature erythrocytes. *Biochim. Biophys. Acta* 106:106, 1965.
129. Nagano, K., Kanazawa, T., Mizuno, N., Tashima, Y., Nakao, T., and Nakao, M. Some acyl phosphate-like properties of P^{32}-labeled sodium-potassium-activated adenosine triphosphatase. *Biochem. Biophys. Res. Commun.* 19:759, 1965.
130. Nagano, K., Mizuno, N., Fugita, M., Tashima, Y., Nakao, T., and Nakao, M. On the possible role of the phosphorylated intermediate in the reaction mechanism of (Na^+-K^+)-ATPase. *Biochim. Biophys. Acta* 143:239, 1967.
131. Nakao, T., Tashima, Y., Nagano, K., and Nakao, M. Highly specific sodium-potassium-activated adenosine triphosphatase from various tissues of rabbit. *Biochem. Biophys. Res. Commun.* 19:755, 1965.

132. Narahashi, T., and Tobias, J. M. Properties of axon membrane as affected by cobra venom, digitonin, and proteases. *Amer. J. Physiol.* 207:1441, 1964.
133. Neifakh, S. A., and Vasilets, I. M. Actomyosin-like protein in outer membrane of liver cells. *Vop. Med. Khim.* 10:326, 1964.
134. Neville, D. M., Jr. The isolation of a cell membrane fraction from rat liver. *J. Biophys. Biochem. Cytol.* 8:413, 1960.
135. Ohnishi, T. Extraction of actin- and myosin-like proteins from erythrocyte membrane. *J. Biochem.* (Tokyo) 52:307, 1962.
136. Ohnishi, T., and Ohnishi, T. Extraction of contractile protein from liver mitochondria. *J. Biochem.* (Tokyo) 51:380, 1962.
137. Old, L. J., and Boyse, E. A. Immunology of experimental tumors. *Ann. Rev. Med.* 15:167, 1964.
138. Oliviera, M. M., and Vaughan, M. Incorporation of fatty acids into phospholipids of erythrocyte membranes. *J. Lipid Res.* 5:156, 1964.
139. Omura, T., Siekevitz, P., and Palade, G. E. Turnover of constituents of the endoplasmic reticulum membranes of rat hepatocytes. *J. Biol. Chem.* 242:2389, 1967.
140. Opit, L. J., Potter, H., and Charnock, J. S. The effect of anions on (Na^+ + K^+)-activated ATPase. *Biochim. Biophys. Acta* 120:159, 1966.
141. Packer, L., Marchant, R. H., and Mukohata, Y. Observations on the control of chloroplast structure by adenosine triphosphate. *Biochem. Biophys. Res. Commun.* 11:429, 1963.
142. Parsons, D. F., Williams, G. R., and Chance, B. Characteristics of isolated and purified preparations of the outer and inner membranes of mitochondria. *Ann. N.Y. Acad. Sci.* 137:643, 1966.
143. Ponder, E., and Ponder, R. V. The electrophoretic velocity of human red cells, of their ghosts and mechanically produced fragments, and of certain lipid complexes. *J. Gen. Physiol.* 43:503, 1960.
144. Portius, H. J., and Repke, K. Die Wirkung W. von Herzglykosiden auf verschiedene ATPasen des Herzmuskels in Abhängigkeit vom Ionenmilieu. *Arch. Exp. Path. Pharmakol.* 243:335, 1962.
145. Post, R. L., Merritt, C. R., Kinsolving, C. R., and Albright, C. D. Membrane adenosine triphosphatase as a participant of active transport of sodium and potassium in the human erythrocyte. *J. Biol. Chem.* 235:1796, 1960.
146. Post, R. L., Sen, A. K., and Rosenthal, A. S. A phosphorylated intermediate in adenosine triphosphate-dependent sodium and potassium transport across kidney membranes. *J. Biol. Chem.* 240:1437, 1965.
147. Rapoport, S., and Leubering, J. Glycerate-2,3-diphosphatase. *J. Biol. Chem.* 189:683, 1951.
148. Razin, S. Factors influencing osmotic fragility of Mycoplasma. *J. Gen. Microbiol.* 36:451, 1964.
149. Reed, C. F. Studies of *in vivo* and *in vitro* exchange of erythrocyte and plasma phospholipids. *J. Clin. Invest.* 38:1032, 1959.
150. Rendi, R. Sodium, potassium-requiring adenosine triphosphatase activity: II. Mechanism of inhibition by sulfhydryl reagents. *Biochim. Biophys. Acta* 99:564, 1965.
151. Richardson, S. H., Hulton, H. O., and Fleischer, S. Interactions of

mitochondrial structural protein with phospholipids. *Arch. Biochem.* 105:254, 1964.

152. Richardson, S. H., Hulton, H. O., and Green, D. E. Structural proteins of membrane systems. *Proc. Nat. Acad. Sci. U.S.A.* 50:821, 1963.
153. Robertson, J. D. New observations on the ultrastructure of the membranes of frog peripheral nerve fibers. *J. Biophys. Biochem. Cytol.* 3:1043, 1957.
154. Robertson, J. D. The ultrastructure of cell membranes and their derivatives. *Sympos. Biochem. Soc.* 16:3, 1959.
155. Robertson, J. D. The occurrence of a subunit pattern in the unit membranes of club endings in Mauthner cell synapses in goldfish brains. *J. Cell Biol.* 19:201, 1963.
156. Robertson, J. D. Granulo-fibrillar and globular substructure in unit membranes. *Ann. N.Y. Acad. Sci.* 137:421, 1966.
157. Rondle, C. J. M., and Morgan, W. T. J. A recording dielectrometric method for column chromatography of the neutral amino acids. *Biochem. J.* 59:xiii, 1955.
158. Schatzmann, J. H. Herzglykoside als Hemmstoffe für den aktiven Kalium- und Natriumtransport durch die Erythrocytenmembran. *Helv. Physiol. Pharmacol. Acta* 11:346, 1953.
159. Schmidt, F. O., Bear, R. S., and Palmer, K. J. X-ray diffraction studies on the structure of the nerve myelin sheath. *J. Cell. Comp. Physiol.* 18:31, 1941.
160. Schmidt, W. J. *Die Doppelbrechung von Karyoplasma, Zytoplasma und Metaplasma.* Berlin: Borntraeger, 1937.
161. Schoner, W., von Ilberg, C., Kramer, R., and Seubert, W. On the mechanism of Na^+-K^+-stimulated hydrolysis of adenosine triphosphate: I. Purification and properties of a Na^+- and K^+-activated ATPase from ox brain. *Europ. J. Biochem.* 1:334, 1967.
162. Schrier, S. L. Studies of the metabolism of human erythrocyte membranes. *J. Clin. Invest.* 42:756, 1963.
163. Schrier, S. L. Organization of enzymes in human erythrocyte membranes. *Amer. J. Physiol.* 210:139, 1966.
164. Schwartz, A. A sodium and potassium-stimulated adenosine triphosphatase from cardiac tissues: IV. Localization and further studies of a basic protein inhibitory factor. *Biochim. Biophys. Acta* 100:202, 1965.
165. Schwartz, A., Bachelard, H. S., and McIlwain, H. The sodium-stimulated adenosine-triphosphatase activity and other properties of cerebral microsomal fractions and subfractions. *Biochem. J.* 84:626, 1962.
166. Shapiro, B., Kollman, G., and Asnen, J. Mechanism of the effect of ionizing radiation on sodium uptake by human erythrocytes. *Radiat. Res.* 27:139, 1966.
167. Sheets, R. F., Hamilton, H. E., and DeGowin, E. L. Hemolysis of human erythrocytes by a sulfhydryl inhibitor, *p*-chloromercuribenzoic acid. *Proc. Soc. Exp. Biol. Med.* 91:423, 1956.
168. Shinagawa, Y., and Ogura, M. Cholinesterase in erythrocyte membrane. *Kagaku* 31:554, 1961.
169. Sjodin, R. A., and Beaugé, L. A. The ion selectivity and concentration

dependence of cation coupled active sodium transport in squid giant axons. *Currents Mod. Biol.* 1:105, 1967.

170. Sjöstrand, F. S. The lamellated structure of the nerve myelin sheath as revealed by high resolution electron microscopy. *Experientia* 9:68, 1953.
171. Sjöstrand, F. S. A new repeat structural element of mitochondrial and certain cytoplasmic membranes. *Nature* (London) 199:1262, 1963.
172. Sjöstrand, F. S. A comparison of plasma membranes, cytomembranes, and mitochondrial membrane elements with respect to ultrastructure features. *J. Ultrastruct. Res.* 9:561, 1963.
173. Skou, J. C. The influence of some cations on the adenosine triphosphatase from peripheral nerves. *Biochim. Biophys. Acta* 23:394, 1957.
174. Skou, J. C. Further investigations on a $Mg^{++}Na^{+}$-activated adenosinetriphosphatase, possibly related to the active, linked transport of Na^{+} and K^{+} across the nerve membrane. *Biochim. Biophys. Acta* 42:6, 1960.
175. Skou, J. C. Preparation from mammalian brain and kidney of the enzyme system involved in active transport of Na^{+} and K^{+}. *Biochim. Biophys. Acta* 58:314, 1962.
176. Skou, J. C. Studies on the $Na^{+} + K^{+}$ activated ATP hydrolysing enzyme system. The role of SH groups. *Biochem. Biophys. Res. Commun.* 10:79, 1963.
177. Skou, J. C. Enzymatic basis for active transport of Na^{+} and K^{+} across cell membrane. *Physiol. Rev.* 45:596, 1965.
178. Skou, J. C., and Hilberg, C. The effect of sulfhydryl-blocking reagents and of urea on the $(Na^{+} + K^{+})$-activated enzyme system. *Biochim. Biophys. Acta* 110:359, 1965.
179. Sneath, J. S., and Sneath, P. H. A. Transformation of the Lewis groups of human red cells. *Nature* (London) 176:172, 1955.
180. Stahl, W. L., Sattin, A., and McIlwain, H. Separation of adenosine diphosphate-adenosine triphosphate-exchange activity from cerebral microsomal sodium-plus-potassium ion-stimulated adenosine triphosphatase. *Biochem. J.* 99:404, 1966.
181. Stasny, J. T., and Crane, F. L. The effect of sonic oscillation on the structure and function of beef heart mitochondria. *J. Cell Biol.* 22:49, 1964.
182. Stoeckenius, W. Some electron microscopical observations on lipid-crystalline phases in lipid-water systems. *J. Cell Biol.* 12:221, 1962.
183. Stoeckenius, W. Some observations on negatively stained mitochondria. *J. Cell Biol.* 17:443, 1963.
184. Stoeckenius, W., and Mahr, S. C. Studies on the reaction of osmium tetroxide with lipids and related compounds. *Lab. Invest.* 14:1196, 1965.
185. Sutherland, R. M., Rothstein, A., and Weed, R. I. Erythrocyte membrane sulfhydryl groups and cation permeability. *J. Cell. Physiol.* 69:185, 1967.
186. Swanson, P. D., and Stahl, W. L. The adenosine diphosphate-adenosine triphosphate-exchange reaction of cerebral microsomes and its relation to the sodium ion-stimulated adenosine-triphosphatase reaction. *Biochem. J.* 99:396, 1966.

187. Szulman, A. E. The histological distribution of blood group substances A and B in man. *J. Exp. Med.* 111:785, 1960.
188. Szulman, A. E. The histological distribution of the blood group substances in man as disclosed by immunofluorescence: II. The H antigen and its relation to A and B antigens. *J. Exp. Med.* 115:977, 1962.
189. Szulman, A. E. The histological distribution of the blood group substances in man as disclosed by immunofluorescence: III. The A, B, and H antigens in embryos and fetuses from 18 mm in length. *J. Exp. Med.* 119:503, 1964.
190. Tanaka, R., and Strickland, K. P. Role of phospholipid in the activation of Na^+, K^+-activated adenosine triphosphatase of beef brain. *Arch. Biochem.* 111:583, 1965.
191. Tobias, J. M. Effects of phospholipases, collagenase and chymotrypsin on impulse conduction and resting potential in the lobster axon with parallel experiments on frog muscle. *J. Cell. Comp. Physiol.* 46:183, 1955.
192. Tobias, J. M. Further studies on the nature of the excitable system in nerve. *J. Gen. Physiol.* 43 (Suppl. 1):57, 1960.
193. Turkington, R. W. Thyrotropin-stimulated adenosine triphosphatase in isolated thyroid cell membranes. *Biochim. Biophys. Acta* 65:386, 1962.
194. VanSteveninck, J., Weed, R. I., and Rothstein, A. Localization of erythrocyte membrane sulfhydryl groups essential for glucose transport. *J. Gen. Physiol.* 48:617, 1965.
195. Vincent, D., Segonzac, C., and Sesque, G. Sur la cholinestérase des érythrocytes du sang humain. *C. R. Soc. Biol.* (Paris) 155:662, 1961.
196. Vogt, P. K. Die Immunologie der Lebermikrosomen. *Z. Naturforsch.* [B] 15:221, 1960.
197. Wallach, D. F. H. Isolation of Plasma Membranes of Animal Cells. In Davis, B. D., and Warren, L. (Eds.), *The Specificity of Cell Surfaces.* Englewood Cliffs, N.J.: Prentice-Hall, 1967. P. 129.
198. Wallach, D. F. H., and Hager, E. B. Association of cell surface antigens with microsomal membrane fraction derived from Ehrlich ascites carcinoma cells. *Nature* (London) 196:1004, 1962.
199. Wallach, D. F. H., and Kamat, V. B. Plasma and cytoplasmic membrane fragments from Ehrlich ascites carcinoma. *Proc. Nat. Acad. Sci. U.S.A.* 52:721, 1964.
200. Wallach, D. F. H., and Ullrey, D. Studies on the surface and cytoplasmic membranes of Ehrlich ascites carcinoma cells: II. Alkali-cation-activated adenosine triphosphate hydrolysis in a microsomal membrane fraction. *Biochim. Biophys. Acta* 88:620, 1964.
201. Wallach, D. F. H., and Zahler, P. H. Protein conformations in cellular membranes. *Proc. Nat. Acad. Sci. U.S.A.* 56:1552, 1966.
202. Walz, F. G., Jr., and Chan, P. C. Adenosine triphosphate-dependent retention of sodium ions by a sodium and potassium-activated adenosine triphosphatase preparation from erythrocyte membranes. *Arch. Biochem.* 113:569, 1966.
203. Warren, L., Glick, M. C., and Nass, M. K. Membranes of animal cells: I. Methods of isolation of the surface membrane. *J. Cell. Comp. Physiol.* 68:269, 1966.

204. Ways, P., and Hanahan, D. J. Characterization and quantification of red cell lipids in normal man. *J. Lipid Res.* 5:318, 1964.
205. Westerman, M. P., Pierce, L. E., and Jensen, W. N. A direct method for quantitative measurement of red cell dimensions. *J. Lab. Clin. Med.* 57:819, 1961.
206. Whittam, R. Potassium movements and ATP in human red cells. *J. Physiol.* (London) 140:479, 1958.
207. Whittam, R. The asymmetrical stimulation of a membrane adenosine triphosphatase in relation to active cation transport. *Biochem. J.* 84:110, 1962.
208. Whittam, R. Directional effects of alkali metal ions on adenosine triphosphate hydrolysis in erythrocyte ghosts. *Nature* (London) 196:134, 1962.
209. Wiener, A. S. *Blood Groups and Blood Transfusion* (3rd ed.). Springfield, Ill.: Thomas, 1943.
210. Wigzell, H. Quantitative titrations of mouse H-2 antibodies using Cr^{51}-labeled target cells. *Transplantation* 3:423, 1965.
211. Winkler, K. C., and Bungenberg de Jong, H. G. Structure of the erythrocyte membrane. *Arch. Neerl. Physiol.* 25:431, 1940.
212. Yamakami, K. The individuality of semen, with reference to its property of inhibiting specifically isohemoagglutination. *J. Immun.* 12:185, 1926.
213. Yamamoto, T. On the thickness of the unit membrane. *J. Cell Biol.* 17:413, 1963.
214. Yoshida, H., Fujisawa, H., and Ohi, Y. Influences of protamine on the Na^+, K^+-dependent ATPase and on the active transport processes of potassium and of L-dopa into brain slices. *Canad. J. Biochem.* 43:841, 1965.

2

Physical Principles of Active Transport and Electrical Excitability

Karl Kornacker

SINCE THE TIME OF GALVANI it has been known that living tissues possess electrical energy, and that nerves can be excited by an electric current. The contemporary view [11] is that electrical energy is generated in living tissues by the active transport of ions across cell membranes, creating and maintaining nonequilibrium electrolyte concentration gradients, and that electrical excitability is due to time-dependent ion permeabilities, allowing the membrane potential to shift back and forth between the equilibrium potentials of different ionic species.

Since merely introducing the concepts of active transport and time-dependent ion permeabilities does not in itself explain the processes involved, considerable experimental and theoretical research has been aimed at characterizing the molecular mechanisms that underlie these phenomena. One major type of membrane research has utilized the voltage clamping technique [4], leading to a characterization of membrane excitability and active transport in terms of empirical electrical parameters. The strongest point in favor of this approach is that it uses a quantitative language which is precisely related both to natural phenomena of interest and to diverse experiments made under controlled artificial conditions. However, these electrical parameters are not related directly to the molecular organization of the membrane.

This work was supported in part under Contract AF 33(615)-3885 from the U.S. Air Force Aerospace Medical Division and Grant GB-5474 from the National Science Foundation.

Another direction of membrane research has utilized chemical techniques, leading mainly to a characterization of the biochemical pathways which provide energy for active transport [16]. There have also been attempts to treat electrical activity as a consequence of macromolecular rearrangements [5, 20]. The chemical approach deals directly with the problem of molecular organization in active membranes and so provides information which is unavailable by electrophysiological methods.

Finally there is a type of approach based on the thermodynamics of irreversible processes [3, 10]. Because active transport and electrical activity are both identified by thermodynamic criteria, the formalism of irreversible thermodynamics is well suited to the study of these phenomena and yields important results unobtainable by other methods.

Clearly it would be desirable to combine the available electrical, chemical, and thermodynamic information into a single complete picture of active biological membranes. This cannot be done, however, without a theoretical framework which includes the basic concepts of all three modes of description. Such a framework is difficult to construct. If an "active channel" is defined as the smallest patch of membrane that can exhibit active transport or electrical excitability, it seems likely that the operation of a single active channel involves only a very few macromolecules. One is confronted, therefore, by the problem of relating the chemical and mechanical behavior of a few membrane molecules to the thermodynamic behavior of the ionic environment.

The general method for relating the mechanical and thermodynamic aspects of a system entails the calculation of averages. Usually these averaging operations are attributed to the experimenter's procedure; the experimenter is supposed to average over randomly varying molecular states by making repeated observations, at least in principle [18]. However, since the actual temporal sequence of macromolecular states in an active channel is certainly not random, the usual random model for the fundamental averaging operation can be only an approximation at best.

In the case of biological membranes, time averages appear in the natural description of the membrane potential because of the transmembrane electrical capacitance (about one microfarad per square centimeter for squid giant axon membrane). The point of view which underlies the development in this chapter[1] is that *the electrical time*

[1] See also Kornacker, K. A theory of inherently macroscopic processes with application to heat and active transport. *Nature* (London), in press.

averaging performed by biological membranes links the molecular mechanics of the active channels to the thermodynamics of the ionic environment. This view is worked out formally in the section on active transport. After consideration of the basic requirements for electrical excitability, the results are combined in the last section to show how electrical excitability may have evolved by specialization of the active transport channels.

MODEL-FREE THEORY OF MEMBRANE POTENTIAL

Model-free theories—that is, theories which involve only rigorous deductions from definitions and universal laws—can often serve as convenient starting points for the construction of model-dependent theories. The following formalism is widely used in modern membrane research and underlies the theoretical developments in this paper.

The conductance, g_γ, and resistance, r_γ, assigned to a charged molecular species γ are defined as

$$g_\gamma = \frac{I_\gamma^{AB}}{V^{AB} - E_\gamma^{AB}} = \frac{1}{r_\gamma} \qquad (1)$$

where V^{AB} is the electrical voltage difference between two distinct homogeneous phases A and B on either side of the membrane, I_γ^{AB} is the net electric current carried between A and B by the γ molecular species, and E_γ^{AB} is the chemical equilibrium potential for the γ species. Since V^{AB} is the same for all ion species, Equation 1 can be summed over all ion species to yield

$$V^{AB} = \frac{\Sigma\, I_\gamma^{AB} + \Sigma\, g_\gamma E_\gamma^{AB}}{\Sigma\, g_\gamma} \qquad (2)$$

Equation 2 may also be written as

$$\Sigma\, g_\gamma = \frac{\Sigma\, I_\gamma^{AB}}{V^{AB} - (\Sigma\, g_\gamma E_\gamma^{AB}/\Sigma\, g_\gamma)} \qquad (3)$$

which closely resembles Equation 1, the main distinction being that $(\Sigma\, g_\gamma E_\gamma^{AB}/\Sigma\, g_\gamma)$ is generally voltage dependent while E_γ^{AB} is not. For squid giant axon membrane, $1/\Sigma\, g_\gamma$ is about 1000 ohms for a square centimeter at rest.

Several points should be made concerning the meaning and use of Equation 2. In the first place this equation says nothing whatever about membrane structure. It does not even refer to membrane physics. The reason is that Equation 1 is a definition, and Equation 2 follows from Equation 1 by algebra alone.

Why then do equilibrium potentials appear in these equations? The reason is that, were the constants E_γ^{AB} in Equation 1 replaced by some other constants (zero for example), the ionic conductances g_γ would

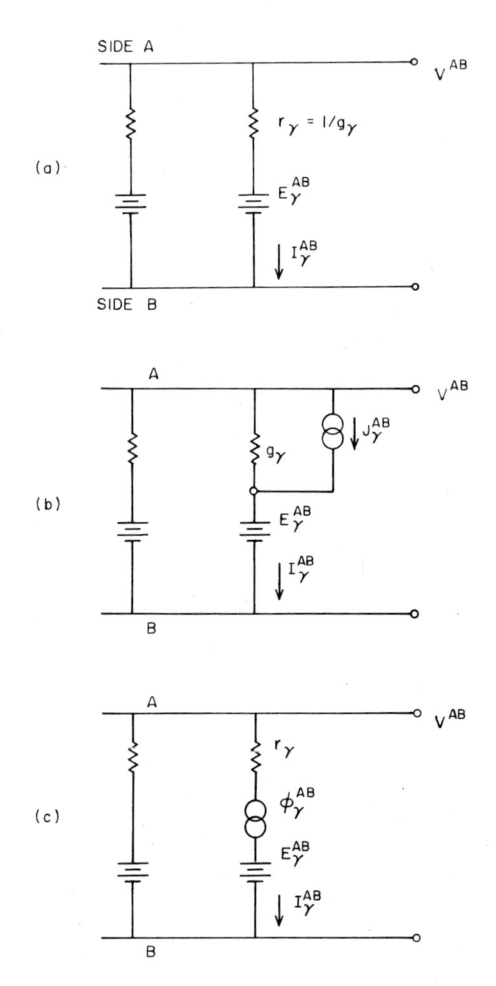

FIG. 2-1. Formal circuit models for the steady-state electrical properties of membrane. (a) Model with only passive current flowing down the electrochemical potential gradients. (b) Model including active transport viewed as a current source. (c) Model including active transport viewed as a voltage source. Note that none of these models gives an accurate physical picture, as is explained in the text.

not remain finite. For zero values of these constants, g_γ would become zero when $I_\gamma^{AB} = 0$ and would become infinite when $V^{AB} = 0$. The presence of E_γ^{AB} in Equation 1 is thus largely one of mathematical convenience.

A problem arises in using Equation 2 when a species is present on only one side of the membrane; in this case the corresponding equilibrium potential becomes infinite. The difficulty can be circumvented by using Equation 1, remembering that I_γ^{AB} and V^{AB} remain finite, to obtain

$$\lim_{E_\gamma^{AB} \to \infty} g_\gamma E_\gamma^{AB} = -I_\gamma^{AB} \tag{4}$$

Equation 4 can generally be used to remove infinite E_γ^{AB} from Equation 2.

The circuit model shown in Figure 2-1a is widely used to represent Equation 2. *In a formal mathematical sense the circuit model is equivalent to Equation 2 and therefore is completely general; it cannot be disproved experimentally. In a physical sense, however, the circuit is misleading because batteries do not always produce the same electrical effects as ionic concentration gradients.* Figure 2-2 compares the effects of a voltage source, a current source, and an ionic concentration gradient on the hypothetical current-voltage relation for a simple rectifying membrane. With a transmembrane concentration gradient the slope of the current-voltage curve can become negative (Fig. 2-2d), a fact of great importance when considering the mechanism of electrical excitability. The reason for this behavior is discussed in the section on electrical excitability (Equation 26).

ACTIVE TRANSPORT

Electrical Aspects

If I_γ^{AB} goes to zero when V^{AB} is not equal to E_γ^{AB} then the conductance g_γ as defined in Equation 1 becomes zero. For squid giant axon membrane at rest the sodium current is zero even though the membrane potential differs from the sodium equilibrium potential by more than a tenth of a volt, suggesting that sodium cannot cross the membrane. However, radioactive sodium crosses the resting membrane in both directions at the rate of approximately one microampere per square centimeter [2].

Such apparently inconsistent facts are formally reconciled by introducing an *active transport current* $\mathbf{J}_\gamma^{AB}$ and writing

$$g_\gamma = \frac{I_\gamma{}^{AB} - \mathbf{J}_\gamma{}^{AB}}{V^{AB} - E_\gamma{}^{AB}} \tag{5}$$

The conductance defined in Equation 5 could be thought of as the conductance of a "passive channel," in which case the "active channel" through which $\mathbf{J}_\gamma{}^{AB}$ passes is viewed as an ideal current generator, with zero conductance (Fig. 2-1b).

Active transport could just as well be represented by a series voltage source, for which the basic relation would be

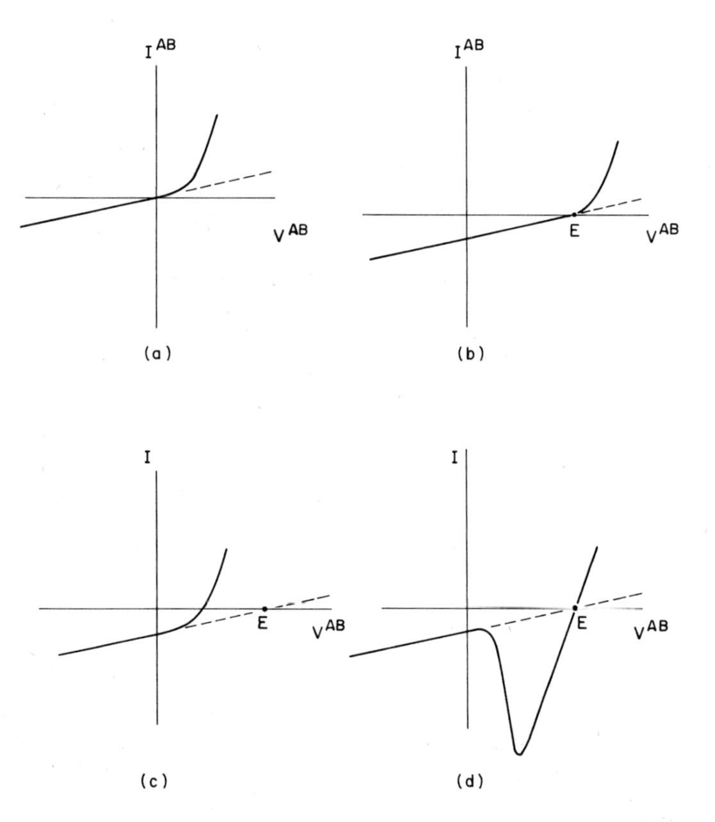

FIG. 2-2. Current-voltage curves for a simple rectifying membrane with different energy sources. Dotted line shows the effect of replacing the rectifier by a fixed resistance. (a) No energy source. (b) Constant voltage source. (c) Constant current source. (d) Ionic concentration gradient. The three energy sources have been adjusted to have identical effects on the fixed resistor.

$$r_\gamma = \frac{V^{AB} - E_\gamma^{AB} - \phi_\gamma^{AB}}{I_\gamma^{AB}} \tag{6}$$

where I_γ^{AB} becomes zero only when $(V^{AB} - E_\gamma^{AB} - \phi_\gamma^{AB})$ is zero. The circuit analog of Equation 6 is shown in Figure 2-1c. It will be shown below that in writing Equations 5 and 6 we can no longer maintain $r_\gamma g_\gamma$ equal to one, and that ohmic relations cannot exist between $\mathbf{J}_\gamma^{AB}$ and ϕ_γ^{AB}.

The modified forms of Equation 2, obtained by summing Equations 5 and 6 over all charged species, are

$$V^{AB} = \frac{\Sigma\, I_\gamma^{AB} - \Sigma\, \mathbf{J}_\gamma^{AB} + \Sigma\, g_\gamma E_\gamma^{AB}}{\Sigma\, g_\gamma}$$

$$= \frac{\Sigma\, I_\gamma^{AB} + \Sigma\, (\phi_\gamma^{AB}/r_\gamma) + \Sigma\, (E_\gamma^{AB}/r_\gamma)}{\Sigma\, (1/r_\gamma)} \tag{7}$$

The additional terms $\Sigma\, \mathbf{J}_\gamma^{AB}/\Sigma\, g_\gamma$ and $\Sigma(\phi_\gamma^{AB}/r_\gamma)/\Sigma(1/r_\gamma)$ are called *electrogenic effects of active transport.*

To evaluate the electrical effects of active transport let us consider the molecular mechanism by which nonzero values for $\mathbf{J}_\gamma^{AB}$ and ϕ_γ^{AB} are generated. The formalism which will be used is attractive because it combines the mechanical and thermodynamic aspects of the problem without requiring artificial assumptions.

The local temporal fluctuation of the electrochemical potential difference across a small patch of membrane (single active transport channel) is defined as

$$\mathfrak{e}_\gamma^{AB} = (\mathscr{V}^{AB} - \mathscr{E}_\gamma^{AB}) - (\overline{\mathscr{V}^{AB} - \mathscr{E}_\gamma^{AB}}) \tag{8}$$

where the bar denotes the time averaging operation performed by the resting membrane (time constant about one millisecond); script quantities include the unobservable high frequency thermal and chemical "noise." Using Equation 8 we can write

$$\begin{aligned} \mathscr{I}_\gamma^{AB} &= \mathfrak{g}_\gamma(\mathscr{V}^{AB} - \mathscr{E}_\gamma^{AB}) \\ &= \mathfrak{g}_\gamma(\overline{\mathscr{V}^{AB} - \mathscr{E}_\gamma^{AB}}) + \mathfrak{e}_\gamma^{AB}\mathfrak{g}_\gamma \end{aligned} \tag{9}$$

Averaging Equation 9 then yields

$$\overline{\mathscr{I}_\gamma^{AB}} = \bar{\mathfrak{g}}_\gamma(\overline{\mathscr{V}^{AB} - \mathscr{E}_\gamma^{AB}}) + \overline{\mathfrak{e}_\gamma^{AB}\mathfrak{g}_\gamma} \tag{10}$$

which may be rewritten

$$\bar{\mathfrak{g}}_\gamma = \frac{\overline{\mathscr{I}_\gamma{}^{AB}} - \overline{\mathfrak{e}_\gamma{}^{AB}\mathfrak{g}_\gamma}}{\overline{\mathscr{V}^{AB}} - \overline{\mathscr{E}_\gamma{}^{AB}}} \tag{11}$$

Comparing Equations 11 and 5, and identifying the average of a script quantity as the corresponding observable quantity, we see that the temporal cross correlation term $\overline{\mathfrak{e}_\gamma{}^{AB}\mathfrak{g}_\gamma}$ plays the role of an active transport current

$$\mathbf{J}_\gamma{}^{AB} = \overline{\mathfrak{e}_\gamma{}^{AB}\mathfrak{g}_\gamma} \tag{12}$$

Note that $\mathbf{J}_\gamma{}^{AB}$ is zero if either $\mathfrak{g}_\gamma$ or $(\mathscr{V}^{AB} - \mathscr{E}_\gamma{}^{AB})$ is strictly constant in time, and differs from zero only if the fluctuations in $\mathfrak{g}_\gamma$ and $(\mathscr{V}^{AB} - \mathscr{E}_\gamma{}^{AB})$ are correlated.

In order to obtain the equivalent voltage source term we look at the current fluctuations

$$\mathfrak{i}_\gamma{}^{AB} = \mathscr{I}_\gamma{}^{AB} - \overline{\mathscr{I}_\gamma{}^{AB}} \tag{13}$$

and write

$$\mathscr{V}^{AB} - \mathscr{E}_\gamma{}^{AB} = \mathfrak{r}_\gamma\overline{\mathscr{I}_\gamma{}^{AB}} + \mathfrak{r}_\gamma\mathfrak{i}_\gamma{}^{AB} \tag{14}$$

whence

$$\overline{\mathscr{V}^{AB} - \mathscr{E}_\gamma{}^{AB}} = \bar{\mathfrak{r}}_\gamma\overline{\mathscr{I}_\gamma{}^{AB}} + \overline{\mathfrak{r}_\gamma\mathfrak{i}_\gamma{}^{AB}} \tag{15}$$

so that

$$\bar{\mathfrak{r}}_\gamma = \frac{\overline{\mathscr{V}^{AB}} - \overline{\mathscr{E}_\gamma{}^{AB}} - \overline{\mathfrak{r}_\gamma\mathfrak{i}_\gamma{}^{AB}}}{\overline{\mathscr{I}_\gamma{}^{AB}}} \tag{16}$$

Comparing Equations 16 and 7, we find that the effective average voltage source term can be identified as another temporal cross correlation:

$$\phi_\gamma{}^{AB} = \overline{\mathfrak{r}_\gamma\mathfrak{i}_\gamma{}^{AB}} \tag{17}$$

Before considering specific physical models for active transport we shall examine the general relation between the two equivalent source terms $\mathbf{J}_\gamma{}^{AB}$ and $\phi_\gamma{}^{AB}$. Multiplying Equation 10 through by $\bar{\mathfrak{r}}_\gamma$, adding and subtracting $(\overline{\mathscr{V}^{AB} - \mathscr{E}_\gamma{}^{AB}})$, and using Equation 15 to identify the voltage source term, we obtain

$$\phi_\gamma{}^{AB} = -\bar{\mathfrak{r}}_\gamma\,\mathbf{J}_\gamma{}^{AB} + (1 - \bar{\mathfrak{r}}_\gamma\bar{\mathfrak{g}}_\gamma)(\overline{\mathscr{V}^{AB} - \mathscr{E}_\gamma{}^{AB}}) \tag{18}$$

A similar treatment of Equation 15 yields

$$\mathbf{J}_\gamma{}^{AB} = -\bar{\mathfrak{g}}_\gamma\,\phi_\gamma{}^{AB} + (1 - \bar{\mathfrak{r}}_\gamma\bar{\mathfrak{g}}_\gamma)\overline{\mathscr{I}_\gamma{}^{AB}} \tag{19}$$

If the factor $(1 - \bar{\mathfrak{r}}_\gamma \bar{\mathfrak{g}}_\gamma)$ were zero, Equations 18 and 19 would express the usual ohmic relations. However, $\mathfrak{r}_\gamma$ and $\mathfrak{g}_\gamma$ are negatively correlated through the relation $\mathfrak{r}_\gamma \mathfrak{g}_\gamma = 1$; an increase of one variable is necessarily accompanied by a decrease of the other. Therefore, the cross correlation $(\overline{\mathfrak{r}_\gamma \mathfrak{g}_\gamma} - \bar{\mathfrak{r}}_\gamma \bar{\mathfrak{g}}_\gamma)$ is negative, and equals zero if and only if $\mathfrak{g}_\gamma$ is strictly constant in time. But if $\mathfrak{g}_\gamma$ were strictly constant in time then $\mathbf{J}_\gamma{}^{AB}$ would vanish according to Equation 12. Therefore, whenever there is active transport, the quantity $(1 - \bar{\mathfrak{r}}_\gamma \bar{\mathfrak{g}}_\gamma)$, which is equal to the cross correlation $(\overline{\mathfrak{r}_\gamma \mathfrak{g}_\gamma} - \bar{\mathfrak{r}}_\gamma \bar{\mathfrak{g}}_\gamma)$, must be negative and not zero. We conclude that ohmic relations between the equivalent source terms $\mathbf{J}_\gamma{}^{AB}$ and $\phi_\gamma{}^{AB}$ can never exist. For this reason the circuits shown in Figures 2-1b and 2-1c, while formally correct, are apt to be misleading.

Metabolic Aspects

In the absence of metabolic energy production, the source terms defined by Equations 12 and 17 have to vanish at equilibrium, because *the thermal fluctuations of current, electrochemical potential, and conductance are all uncorrelated in a system at thermal equilibrium.* Metabolic processes are generally credited with being able to maintain steady nonequilibrium conditions across cell membranes by generating an active transport current. In view of the preceding discussion it should be apparent that *metabolic processes contributing to active transport must generate local nonthermal correlated fluctuations of conductance and electrochemical potential.*

Two basic types of physical mechanism for this activity are possible: (1) the chemical free energy could be coupled *electrically,* directly producing local nonthermal electrical fluctuations, which could then, for example, cause correlated fluctuations in a voltage-dependent conductance [2]; or (2) the energy could be coupled *chemically* (structurally) to some membrane component, producing correlated nonthermal local fluctuations in $\mathfrak{r}_\gamma$ and $\mathscr{I}_\gamma{}^{AB}$.

The first mechanism might, for instance, involve the transient shifting of an electron on the cation-activated adenosine triphosphatase (ATPase) during ATP hydrolysis [16] inside a channel, causing a brief change in the voltage across a rectifying junction in the channel. This mechanism might also involve the simultaneous release of bound ions into the channel [16], causing a correlated change of local chemical potential.

The second mechanism must operate in any active carrier system

[22] because the local membrane resistance becomes a function of the direction in which an ion attempts to cross the membrane, *establishing the required correlation between the local fluctuations of current and resistance.* Metabolic energy might then be used, for example, to raise the affinity of an ion for the carrier on only one side of the membrane. In some cases the actively maintained concentration gradient of one species might maintain the carrier affinity gradient for another species, as has been suggested to explain the mechanism of sodium-dependent metabolite transport [19].

The above active transport mechanisms all entail electrical rectification in the active transport channels. It would be possible, however, for the first mechanism (1) to operate in conjunction with chemically produced fluctuations of conductance, in which case the channel would not necessarily show electrical rectification. The significance of rectification in active transport channels will be discussed in the last section of this chapter.

Experimental Distinctions Between Various Mechanisms

Walker and Eisenman [21] have pointed to a distinguishing feature of transport via mobile carriers, namely, that the instantaneous current-voltage curves should cross the voltage axis at different voltages depending on the preceding steady level of polarization. As yet, however, the experimental data on this point are incomplete. Adelman et al. [1], while studying the mechanism of hyperpolarizing responses in perfused squid giant axons subjected to high external potassium concentrations, obtained a family of instantaneous current-voltage curves with different prior polarizations showing that all the curves cross the voltage axis at the same point. Although this result rules out the possibility that the passive potassium permeability is due to a mobile carrier, it says little about the possibility that active potassium transport is dependent on a mobile carrier.

Usually one cannot study instantaneous sodium currents in squid giant axon because the measurements are dominated by potassium. However, incidental to the study of excitability in squid giant axons that had been depolarized by perfusion with low potassium solutions, Moore et al. [17] found that a negative resistance (see next section) appears in both the instantaneous and steady-state current-voltage curves, due most likely to an observable sodium current. Thus, in this preparation a family of instantaneous current-voltage curves at different prior polarizations may detect a mobile sodium carrier if it exists.

OUTSIDE

INSIDE

CONCENTRATION DROP →

PASSIVE CURRENT →

IONIC RECTIFICATION →

← MEMBRANE RECTIFICATION

← ACTIVE TRANSPORT CURRENT

irections of various vectors associated with sodium transport.

ve discussion provides a basis for explaining the steady-state esistance, and associated hyperpolarizing response, which n squid giant axon under depolarization by a high external n concentration [1, 15]. The steady-state potassium conduc- reases when the membrane is depolarized. Negative resis- uld therefore be expected to appear when the potassium um potential becomes sufficiently more positive than the at which the potassium conductance change occurs and Equa- holds. The resulting steady-state current-voltage relation then have the form shown in Figure 2-2d. An externally ap- olarizing current can be used to shift the current-voltage curve

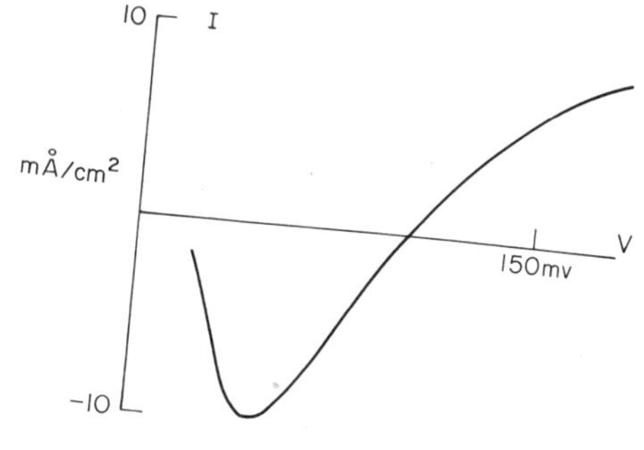

IG. 2-6. Current-voltage relation for maximum sodium current during xcitation of *Xenopus* axon. The rectification due to the sodium concentra- on gradient, most prominent for positive (outward) currents, is opposite o the rectification due to the open, excitable channels, most prominent for negative (inward) currents. (Based on Dodge and Frankenhaeuser [7].)

Conway [6] has observed that in frog sartorius muscles the radioactive sodium efflux is reduced by a factor of about 10 when the membrane is hyperpolarized from 63 to 67 mv by an altered ionic environment. The sharpness of the cutoff seems to argue against a mobile carrier transport system. Therefore, it would be of interest to apply the Walker and Eisenman test to this system.

Hodgkin and Keynes [12] failed to observe any reduction in the radioactive sodium efflux from squid giant axons during apparent electrical hyperpolarizations as great as 40 mv. However, the sodium efflux was significantly reduced in potassium-free bathing solutions, which caused only about 10 mv hyperpolarization. They concluded that the observed reduction in sodium efflux was not an electrical effect but was due to a specific potassium requirement in the active transport system. In view of Conway's striking observations, and the fact that some level of hyperpolarization must eventually decrease the sodium efflux, one might look for possible artifacts in the electrical polarization experiments. Considering the experimental design, one possible artifact would be a slow polarization (within about two minutes) of the silver-silver chloride electrodes, causing a reduction in the transmembrane hyperpolarization. In any case, the experiment should be extended to determine the voltage dependence of the sodium efflux at large transmembrane hyperpolarizations.

ELECTRICAL EXCITABILITY

Negative Resistance

To examine what properties are necessary for electrical excitability, let us consider a general membrane model consisting of a two-terminal device (black box) in parallel with a capacitor, as shown in Figure 2-3. Let us assume for the moment that the black box is characterized by a time-independent current-voltage relation.

The basic open-circuit equations for the above system (no current path bypassing the membrane) are

$$I_1 + I_2 = 0 \tag{20}$$

and

$$\frac{dV}{dt} = I_2/C \tag{21}$$

Substituting Equation 20 into Equation 21 gives

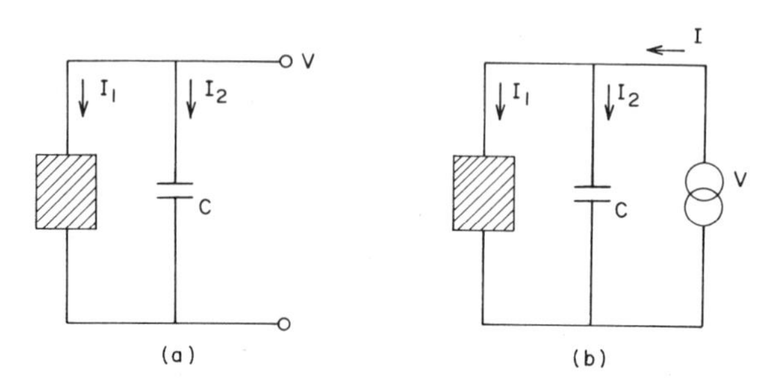

FIG. 2-3. Formal models for transient electrical behavior of membrane. (a) Open circuit condition (no current flowing through an external path). (b) Voltage clamp.

$$\frac{dV}{dt} = -I_1/C \tag{22}$$

and multiplying through by (dI_1/dV) gives

$$\frac{dI_1}{dt} = -I_1 \left(\frac{dI_1}{dV}\right) \Big/ C \tag{23}$$

Now consider a special device for which (dI_1/dV) is negative (negative resistance). Under these circumstances Equation 23 says that the magnitude of I_1 will continually increase with time, so that Equation 22 implies that V will keep changing in one direction at an ever increasing rate. The state $I_1 = 0$ would be an *unstable equilibrium* point in this case because the slightest perturbation would lead to a runaway increase in I_1. Such a device would clearly produce an extreme of electrical excitability.

Any real physical device must eventually show a positive (dI_1/dV) at large I_1 since infinite currents cannot appear in an isolated system. When (dI_1/dV) eventually becomes positive, then, by Equation 23, the magnitude of I_1 will decrease. If I_1 reaches zero while (dI_1/dV) is positive, then both V and I_1 will stop changing and the system will be stable (Fig. 2-4).

The basic requirements for negative (dI_1/dV) can be found by differentiating Equation 1 to obtain

$$\frac{dI_\gamma^{AB}}{dV^{AB}} = (V^{AB} - E_\gamma^{AB}) \frac{dg_\gamma}{dV^{AB}} + g_\gamma \tag{24}$$

Equation 24 is relevant because in our excitable membrane model

showing that $(dI_\gamma^{A}$
order for (dI_1/dV)
One way of writin
using Equation 24,

Since g_γ is always positive
can occur only when $(E_\gamma^{A}$
For a simple rectifier $(dg_\gamma/$
bility of negative $(dI_\gamma^{AB}/dV$
voltages which lie to one si
example, V^{AB} must be less th
The electronic circuit anal
negative (dI_γ^{AB}/dV^{AB}) because
centration gradients simply shi
Equation 26 is never satisfied (F
for a membrane rectifier is strai
concentration gradient giving a
change of ionic environment does
in the membrane. The rectification
itself generally occurs in a different
opposite direction (Figs. 2-5 and 2-6)

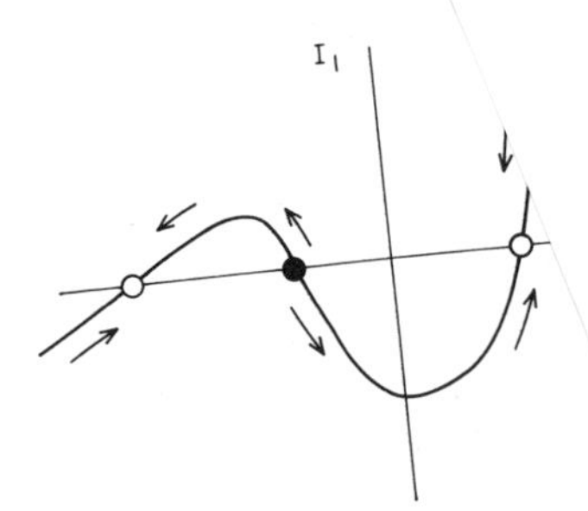

FIG. 2-4. Direction of spontaneous electrical chan
system having the indicated steady-state current-vol
indicates a point of instability (threshold); open dots
bility (resting states).

so that it resembles Figure 2-4; under such conditions the membrane displays two stable states [15].

Reset Mechanisms

A simple transition from one stable state to another via an intervening region of negative (dI/dV) is not yet sufficient to produce electrical excitability as it commonly is observed in biological membranes. Some additional mechanism must be provided to return the system to its initial state (unless a permanent memory unit is contemplated). Three basic types of reset mechanisms come to mind: (1) one in which the current-voltage characteristic of the device changes during excitation; (2) one in which a second negative (dI/dV) region exists near the new $I_1 = 0$ point, moving V back toward its original state; (3) one in which another stimulus drives the system back past threshold. The following considerations underlie an examination of these possibilities.

If an external circuit is connected across the membrane (Fig. 2-3b), Equation 20 must be replaced by

$$I_1 + I_2 = I \tag{27}$$

where I now is the current flowing through the external path. If the external circuit is designed to "voltage clamp" the membrane, meaning that V is temporarily held constant, then from Equation 21

$$I_2 = 0 \tag{28}$$

so that

$$I_1 = I \tag{29}$$

Equation 29 tells us, as expected, that voltage clamp experiments allow a determination of the current-voltage characteristic of the black box. Moreover, a time-dependent black box current-voltage characteristic, as in reset mechanism (1), would cause the measured I to be time dependent following a sudden excitatory change in the clamping voltage.

The most thoroughly studied example of an electrically excitable system which returns to its initial state by means of a time-dependent current-voltage characteristic is the squid giant axon membrane. Although it is generally agreed that the time dependence is associated with the release and readsorption of calcium ions, no definite physical model for this process has emerged [8]. The following considerations treat one aspect of the problem.

The data obtained from voltage clamp studies on squid giant axons which had been perfused with a dilute potassium solution show, among other things, a region of negative (dI^{AB}/dV^{AB}) at "steady state" [17]. Moreover, the voltage range over which the steady-state (dI^{AB}/dV^{AB}) is negative is the same as the voltage range for negative (dI^{AB}/dV^{AB}) during excitation (at the time of maximum inward current). In mathematical terms this relation suggests that, under voltage clamp conditions, the maximum change in conductance during excitation is related to the resting conductance by an approximately voltage-independent multiplicative factor. The content of this description, in physical terms, is that at rest a few excitable channels are "open" while the majority are "closed." The multiplicative factor then describes the opening of closed channels, and the reset mechanism involves the closing of open channels.

Explaining the mechanisms for opening and closing the excitable channels is one of the most important unsolved problems in the field of membrane physiology [5] and lies outside the scope of the present chapter. For our purposes it is sufficient to observe that *open excitable channels possessing a negative resistance may be present in the resting state.*

The preceding considerations suggest that when the membrane is exposed to calcium-free solutions, allowing the excitable channels to remain permanently open, a marked negative resistance should persist in the steady state and the membrane should be able, with proper polarization, to display two stable states as in Figure 2-4. This effect has been observed [15].

PHYSICAL MODEL OF ELECTRICALLY ACTIVE MEMBRANE

Intuitively, or formally from Equation 12, we know that the transport current $\mathbf{J}_\gamma^{AB}$ must flow in the "forward direction" through its rectifier—that is, in the same direction as an externally applied current which would increase g_γ. The actively maintained gradient of $Z_\gamma c_\gamma$ will then tend to point opposite to the forward direction of the rectifier, where Z_γ denotes the charge carried by a γ molecule and c_γ denotes its concentration. For a homogeneous membrane, however, the forward direction for I_γ^{AB} is down the gradient of $Z_\gamma c_\gamma$ [9]. *We conclude that the rectification involved in active transport requires an inherently rectifying membrane component* (cf. Figs. 2-5 and 2-6).

An increase of E_γ^{AB}, produced by active transport of the γ species, will increase the negativity of (dI_γ^{AB}/dV^{AB}) if and only if E_γ^{AB} and (dg_γ/dV^{AB}) have the same sign, as can be seen from Equation 26. However, because E_γ^{AB} and the gradient of $Z_\gamma c_\gamma$ have opposite signs, the condition for increasingly negative (dI_γ^{AB}/dV^{AB}) is therefore equivalent to the condition that (dg_γ/dV^{AB}) and the gradient of $Z_\gamma c_\gamma$ have opposite signs. In other words, the direction of rectification for I_γ^{AB} again must point opposite to the gradient of $Z_\gamma c_\gamma$. Because we have found that this condition tends to hold in the active transport channels, we conclude that *there is a tendency for any rectifying active transport channel to develop negative resistance.* This conclusion leads naturally to the suggestion that, at least in squid giant axon membrane, *the open excitable channels may be the active transport channels.*

In the common-channel model the chemical specificity of the excitation current is determined by the specificity of the channel rectifier. The specificity of the transport currents, however, would include both the rectifier specificity and whatever specificity was imposed by the metabolic energy source. For example, the fact that lithium fails to substitute for sodium in the cation-activated ATPase implicated in active cation transport (see Chap. 1) may explain why lithium is not transported even though it can substitute for sodium in excitatory activity [13].

There is evidence that the above common-channel model may apply to membranes other than squid giant axon. One prerequisite for the common-channel model is that the active transport channel show rectification. An indication that this requirement is met generally would be the demonstration of induced electrical excitability during the artificial increase of some actively maintained concentration gradient. This effect apparently has been observed in a brackish water alga which becomes electrically excitable when transferred to an isotonic sucrose solution [14].

Because electrical excitability is generally independent of metabolism [13], another requirement of the common-channel model is that the electrical rectification mechanism of the active transport channels be passive, as opposed to active rectification based on a carrier affinity gradient mechanism. In the algal preparation one could decide between these alternatives by removing the energy supply for active transport. Induced electrical excitability would remain only if the rectification were passive. Demonstration of passive rectification in the active transport channels of a normally inexcitable alga would strongly support the notion that the electrical excitability channels of neurons evolved from, and are perhaps identical with, electrically coupled active transport channels.

REFERENCES

1. Adelman, W. J., Jr., Dyro, F. M., and Senft, J. Long duration responses obtained from internally perfused axons. *J Gen. Physiol.* 48:1, 1965.
2. Candia, O. A. An alternative interpretation of the sodium transport in frog skin. Abstract, 11th Annual Meeting, Biophysical Society, Houston, 1967.
3. Caplan, S. R. The degree of coupling and its relation to efficiency of energy conversion in multiple-flow systems. *J. Theor. Biol.* 10:209, 1966.
4. Chandler, W. K., and Meves, H. Ionic Selectivity in Perfused Giant Axons. In Adelman, W. J., Jr. (Ed.), *Symposium on Physical and Mathematical Approaches to the Study of the Electronic Behavior of Excitable Membranes.* Philadelphia: Wistar Institute, 1965. P. 65.
5. Changeux, J.-P., Thiery, J., Tung, Y., and Kittel, C. On the cooperativity of biological membranes. *Proc. Nat. Acad. Sci. U.S.A.* 57:335, 1967.
6. Conway, E. J., Kernan, R. P., and Zadunaisky, J. A. The sodium pump in skeletal muscle in relation to energy barriers. *J. Physiol.* (London) 155:263, 1961.
7. Dodge, F. A., and Frankenhaeuser, B. Sodium currents in the myelinated nerve fibre of *Xenopus laevis* investigated with the voltage clamp technique. *J. Physiol.* (London) 148:188, 1959.
8. Frankenhaeuser, B., and Hodgkin, A. L. The action of calcium on the electrical properties of squid axons. *J. Physiol.* (London) 137:218, 1957.
9. Goldman, D. E. Potential impedance and rectification in membranes. *J. Gen. Physiol.* 27:37, 1943.
10. Hill, T. L., and Kedem, O. Studies in irreversible thermodynamics: III. Models for steady state and active transport across membranes. *J. Theor. Biol.* 10:399, 1966.
11. Hodgkin, A. L. The ionic basis of electrical activity in nerve and muscle. *Biol. Rev.* 26:339, 1951.
12. Hodgkin, A. L., and Keynes, R. D. Active transport of cations in giant axons from *Sepia* and *Loligo*. *J. Physiol.* (London) 128:28, 1955.
13. Keynes, R. D. Ion transport in excitable cells. *Protoplasma* 63:12, 1967.
14. Kishimoto, U. Voltage clamp and perfusion studies on *Nitella* internodes. *J. Cell. Comp. Physiol.* 66 (Suppl. 2):43, 1965.
15. Koketsu, I., and Koyama, I. Membrane responses of frog's spinal ganglion cells in calcium-free solutions. *J. Physiol.* (London) 163:1, 1962.
16. Mitchell, P. Translocations through natural membranes. *Advances Enzym.* 29:33, 1967.
17. Moore, J. W., Narahashi, T., and Ulbricht, W. Sodium conductance shift in an axon internally perfused with a sucrose and low-potassium solution. *J. Physiol.* (London) 172:163, 1964.

18. Schrodinger, E. *Statistical Thermodynamics* (2nd ed.). London: Cambridge University Press, 1962.
19. Semenza, G. Rate equations of some cases of enzyme inhibition and activation—their application to sodium-activated membrane transport systems. *J. Theor. Biol.* 15:145, 1967.
20. Tasaki, I., and Singer, I. A macromolecular approach to the excitable membrane. *J. Cell. Comp. Physiol.* 66 (Suppl. 2):137, 1965.
21. Walker, J. L., and Eisenman, G. A test of the theory of the steady state properties of a liquid ion exchange membrane. *Ann. N.Y. Acad. Sci.* 137:777, 1966.
22. Wilbrandt, W., and Rosenberg, T. The concept of carrier transport and its corollaries in pharmacology. *Pharmacol. Rev.* 13:109, 1961.

3

Mechanisms of Absorption

Stanley G. Schultz[1]

CELL-MEMBRANE SYSTEMS may be divided into two broad categories. The first comprises single, nonpolar cells whose membranes have symmetrical properties over their entire extent. These membranes are responsible for the maintenance of steady-state intracellular compositions and for the selective net transport of nutrients and metabolic products into and out of the cell. The second category comprises polar cells organized in the form of epithelial tissues. The membranes that surround these cells, in addition to being responsible for the regulation of intracellular composition, are capable of bringing about net transport of solutes and water across an intact multicellular sheet. This oriented net transport requires at least two membranous boundaries that are arranged in series and that have different or asymmetrical properties. For the many epithelial tissues that consist of a single layer of tightly attached cells, the asymmetry applies to the properties of different parts of the membrane surrounding a single cell. Often there are striking morphological differences between the membranes that constitute the opposing borders of the cell, as is the case for small intestine [70] and renal tubule, and it is not surprising that such membranes should differ functionally as well. However, even when morphological differences cannot be resolved, as in the case of toad bladder [64], net transport across a multicellular sheet would apparently be impossible if the membranes forming the two opposing borders of the sheet were functionally symmetrical.

[1] Established Investigator of the American Heart Association.

Thus, the membranes that surround polar cells have dual functions: (1) in common with all cells, they must maintain an intracellular composition differing markedly from that of the surrounding medium; and (2) they must be capable of bringing about net transepithelial transport of selected solutes and water. Although the precise relation between these two functions is as yet uncertain, there is no compelling evidence that they are mediated by entirely separate mechanisms. Both functions can be accomplished by a series arrangement of two membranes provided they have different specific transport properties and permeabilities. An extensive analysis, along these lines, of model systems capable of transcellular transport has been published recently by Parsons and Prichard [48]. Further, numerous basic similarities (e.g., substrate specificities, drug sensitivities) exist between specific transport mechanisms found in nonpolar cells and those responsible for transport across a variety of epithelia (e.g., see references 73, 76 for reviews). Thus, there is every reason to believe that the same or very similar transport mechanisms are present in nonpolar and polar cells but that the latter have acquired a higher level of transport specialization through an asymmetrical distribution of these properties. The ingenious demonstration by Oxender and Christensen [47] of transcellular transport across an artificial "tissue" formed by layering Ehrlich ascites tumor cells on a membrane filter attests to the fact that transepithelial transport does not presuppose basically different mechanisms from those present in symmetrical cells.

In this chapter we shall examine several mechanisms that appear to be involved in transport across intact cell layers. The transport of solutes and water across specific epithelial tissues such as intestine [57], renal tubule [23], isolated frog skin [73], toad urinary bladder [64], and gallbladder [21] has been extensively reviewed in recent years. For this reason our discussion will focus not on a specific tissue but rather on different types of transport mechanisms that are encountered in the study of absorptive and secretory processes and will draw examples from a variety of tissues. The reader interested in the function of a specific tissue may consult the references cited above.

At the outset, it is of value to clarify the sense in which we employ the term *mechanism.* The ultimate aim of the study of membrane phenomena is to define completely the interaction between the membrane and the transported species at the molecular level, e.g., to specify the physical factors involved in simple diffusion across biological membranes, to identify the membrane components that are operationally referred to as carriers and to characterize their interactions with the transported species, and to elucidate the manner by which energy-yielding chemical reactions are linked to transport processes. Unfortunately, far too little is known in these areas to make a discussion of mechanism at the molecular level anything but highly speculative. A more profitable approach, for the moment, is a phenomenological analysis of the flow of a given species in terms of the thermodynamic forces acting upon the species. At this level a complete description of a given transport process requires the identification of all the macroscopic forces that influence the process. This approach, as we shall see, permits a convenient classification of biological transport processes and provides considerable insight into the nature of several of these phenomena in spite of our ignorance of underlying molecular and biochemical mechanisms. Moreover, by uncovering transport processes for which the responsible driving forces cannot be adequately described, the approach suggests directions for further investigation at the molecular and biochemical levels. Clearly, in the final analysis the thermodynamic and molecular approaches are complementary and must lead to mutually compatible descriptions of transport processes.

GENERAL PRINCIPLES

The formal analysis of transport processes in terms of flows and forces rests heavily on the principles of classic thermodynamics and the extension of these principles to steady-state systems in which irreversible processes are taking place. Space limitations preclude an extensive discussion of these principles. The author has attempted to make the following discussion self-sufficient; however, lengthy derivations and justifications of many of the points raised are omitted. (For further detail consult references 2, 12, 32, 35, 38, and 56.)

It is empirically known that in a system characterized by *only a single* flow (e.g., flow of matter, charge, heat, etc.) the flow is linearly related to its driving force if the system is not too far displaced from equilibrium. For such a general system we may write

$$\mathbf{J} = LX \qquad (1)$$

where $\mathbf{J}$ is the flow rate, X is the thermodynamic driving force responsible for the flow (often referred to as the "conjugate" force), and L is a phenomenological coefficient of proportionality whose value is not dependent upon X. L is the flow per unit driving force and has units of conductance. Ohm's law and Fick's first law of diffusion are familiar examples of Equation 1. In a system characterized by the flow of a single species i across a membrane, the responsible or conjugate driving force is the difference in the electrochemical potential of i across the membrane $\Delta\tilde{\mu}_i$. At constant temperature and pressure

$$\Delta\tilde{\mu}_i = \mathrm{RT}\ \ln \frac{a_i'}{a_i''} + z_i\ \mathbf{F}\ \Delta\Psi \qquad (2)$$

where a_i' and a_i'' are the thermodynamic activities of i on the $'$ and $''$ sides of the membrane and $\Delta\Psi$ is the electrical potential difference across the membrane $(\Psi' - \Psi'')$. T is the absolute temperature, $\mathbf{F}$ is the Faraday, R is the gas constant, and z_i is the valence of i. If the system is not too far displaced from equilibrium[2]

$$\mathbf{J}_i = -\mathrm{L}\ \Delta\tilde{\mu}_i \qquad (3)$$

Clearly, L is related to the permeability of the membrane to i.

If a system is characterized by more than one flow, it is experimentally observed that any single flow may be influenced by all the other flows and, thus, by all the other forces present. Such a system can be described by the following set of "phenomenological equations":

$$\begin{aligned} \mathbf{J}_i &= \mathrm{L}_{ii}\ \mathrm{X}_i + \mathrm{L}_{ij}\ \mathrm{X}_j + \ldots + \mathrm{L}_{in}\ \mathrm{X}_n \\ \mathbf{J}_j &= \mathrm{L}_{ji}\ \mathrm{X}_i + \mathrm{L}_{jj}\ \mathrm{X}_j + \ldots + \mathrm{L}_{jn}\ \mathrm{X}_n \\ &\vdots \\ \mathbf{J}_n &= \mathrm{L}_{ni}\ \mathrm{X}_i + \mathrm{L}_{nj}\ \mathrm{X}_j + \ldots + \mathrm{L}_{nn}\ \mathrm{X}_n \end{aligned} \qquad (4)$$

or

$$\mathbf{J}_i = \mathrm{L}_{ii}\ \mathrm{X}_i + \sum_{\substack{k=1 \\ k \neq i}}^{n} \mathrm{L}_{ik}\ \mathrm{X}_k \qquad (5)$$

Here, L_{ii} is the "straight" phenomenological coefficient that relates $\mathbf{J}_i$ to its conjugate force X_i. The L_{ik} where $k \neq i$ are "cross" phenomeno-

[2] The conjugate driving force for the flow of matter in a continuous system is the *negative* of the electrochemical potential gradient since a positive flow takes place in the direction of decreasing electrochemical potential. Further, it is important to note that, although L is not a function of $\Delta\tilde{\mu}_i$, it is not a constant but rather a function of the state of the system, and, in particular, of the concentration of i. For further discussion of Equation 3, see reference 32, Chap. 10.

logical coefficients, and they reflect the extent to which the flow of *i* is influenced by the other forces in the system.

For our purposes it is sufficient to view the cross coefficients as empirical factors that account for the presence of coupling or interaction between two or more flows; the molecular mechanisms responsible for these interactions need not concern us at this point. Nevertheless, brief consideration of the possible bases of coupled flows in biological transport processes may provide a more intuitive grasp of these important interactions. Coupling between the flows of matter—for example, coupling between the flows of solvent and solute—may be viewed as arising from a frictional interaction between moving particles, and it is possible to express cross coefficients in terms of mutual frictional coefficients [36]. The flows of two chemical reactions may interact if they share a common product or reactant (including heat). The simplest form of such interactions may be called "stoichiometric coupling." A more difficult problem, the subject of extensive discussion in recent years, is the attempt to explain the interaction between the *nondirected* flow of a chemical reaction and the *directed* flow of matter [2, 32]. Suffice it to say that in an anisotropic membrane a chemical reaction could impart direction to the flow of matter by altering the local thermodynamic potential of the transported species or of a mobile membrane component [33] or by effecting local changes in membrane structure [2].

Equations 4 and 5 express flows as functions of forces. These expressions may be transformed into a new set of phenomenological equations in which the forces are functions of the flows, i.e.,

$$
\begin{aligned}
X_i &= R_{ii}\,\mathbf{J}_i + R_{ij}\,\mathbf{J}_j + \ldots + R_{in}\,\mathbf{J}_n \\
X_j &= R_{ji}\,\mathbf{J}_i + R_{jj}\,\mathbf{J}_j + \ldots + R_{jn}\,\mathbf{J}_n \\
&\;\vdots \\
X_n &= R_{ni}\,\mathbf{J}_i + R_{nj}\,\mathbf{J}_j + \ldots + R_{nn}\,\mathbf{J}_n
\end{aligned}
$$

or

$$
X_i = R_{ii}\,\mathbf{J}_i + \sum_{\substack{k=1 \\ k\neq i}}^{n} R_{ik}\,\mathbf{J}_k \tag{6}
$$

Solving for $\mathbf{J}_i$ we obtain

$$
\mathbf{J}_i = \frac{X_i}{R_{ii}} - \sum_{\substack{k=1 \\ k\neq i}}^{n} R_{ik}\,\mathbf{J}_k/R_{ii}
$$

The R_{ik} are resistance coefficients and are related to the L_{ik} of Equations 4 and 5.

Using this approach, Kedem [33] presented the following general equation describing the flow of species i across a membrane within which a chemical reaction takes place.

$$\mathbf{J}_i = -\frac{\Delta\tilde{\mu}_i}{R_{ii}} - \sum_{\substack{j=1 \\ j\neq i}}^{n} R_{ij}\,\mathbf{J}_j/R_{ii} - \frac{R_{ir}}{R_{ii}}\,\mathbf{J}_r \tag{7}$$

The primary value of Equation 7 is that it explicitly subdivides the total or overall driving force responsible for the flow of species i into three general categories. The first term on the right represents the contribution to $\mathbf{J}_i$ resulting from a difference in the electrochemical potential of i across the membrane. The middle term represents the summation of all the contributions to $\mathbf{J}_i$ resulting from interactions, by whatever mechanisms, with flows of other solutes or solvents across the membrane. The final term represents the contribution to $\mathbf{J}_i$ arising from coupling with the flow of a chemical reaction, $\mathbf{J}_r$. Clearly, when all $\mathbf{J}_j$ and $\mathbf{J}_r$ are zero, $1/R_{ii}$ is the permeability of the membrane to i and is equal to L_{ii}. All flows must be to some extent influenced by their conjugate driving force unless the membrane is absolutely impermeable to the substance (i.e., $R_{ii} = \infty$).

This subdivision of overall driving forces permits a convenient and useful categorization of biological transport processes which will serve as the framework for the following discussion. We shall show that many absorptive processes fall into one of the following three categories:

1. Flows that appear to be dependent only on their conjugate driving force.
2. Flows that cannot be attributed solely to the conjugate driving force and that appear to be coupled to metabolic reactions.
3. Flows that cannot be attributed solely to the conjugate driving force and that appear to be coupled to the flows of other solutes or solvents.

It should be apparent that the experimental application of this approach, and in particular Equation 7, to biological transport processes is limited to systems in which all the net flows, the concentrations (or activities) of all species on the two sides of the membrane, and the electrical potential difference across the membrane can be determined unequivocally. For this reason studies of transport across epithelial tissues, where the solutions bathing both sides of the tissue are readily accessible, have made important contributions to the thermodynamic description of transport processes. The application of this approach to transport across single-cell membranes is obviously difficult and fraught with uncertainties.

TRANSPORT DEPENDENT ONLY ON CONJUGATE DRIVING FORCE

We will first consider the case in which the flow of a given species is dependent only on its conjugate driving force and is uninfluenced by other flows of matter or chemical reactions. With reference to Equation 7, this is the condition in which all $R_{ij,j\neq i} = 0$ and $R_{ir} = 0$, so that

$$\mathbf{J}_i = -\Delta\tilde{\mu}_i/R_{ii} \tag{8}$$

Thus, $\mathbf{J}_i = 0$ when $\Delta\tilde{\mu}_i = 0$. If i is an uncharged solute ($z_i = 0$), this condition is realized when the activities of i on the two sides of the membrane are equal. The situation is somewhat more complicated in the case of charged solutes since most epithelial structures display an electrical potential difference between the two sides of the tissue, so that $\Delta\tilde{\mu}_i \neq 0$ when $a_i' = a_i''$ (Equation 2). Under these circumstances the short-circuit technique, introduced by Ussing and Zerahn [74] in their studies of ion transport across frog skin, may be employed to impose the condition $\Delta\tilde{\mu}_i = 0$ for all i regardless of charge. A diagram of a typical short-circuit apparatus is shown in Figure 3-1. The tissue (*S*) is mounted as a flat sheet between the two chambers (*C*), and the electrical potential difference across the tissue is measured by means of two electrolyte bridges (*A* and *A'*) situated close to the sides of the tissue and connected to a high-impedance potentiometer (*P*). Current may be passed through the tissue from an external battery by means of electrodes *B'* and *B* and monitored by a microammeter (*M*). The compositions of the solutions bathing the two surfaces and the electrical potential difference across the tissue may be varied independently, and, with appropriate tracer techniques, $\mathbf{J}_i$ can be evaluated over a wide range of values for $\Delta\tilde{\mu}_i$. A particularly useful set of conditions under which to evaluate $\mathbf{J}_i$ occurs when both sides of the tissue are bathed with identical solutions and the external current is adjusted so that the electrical potential difference across the tissue is zero. Under these conditions the tissue is said to be "short-circuited"

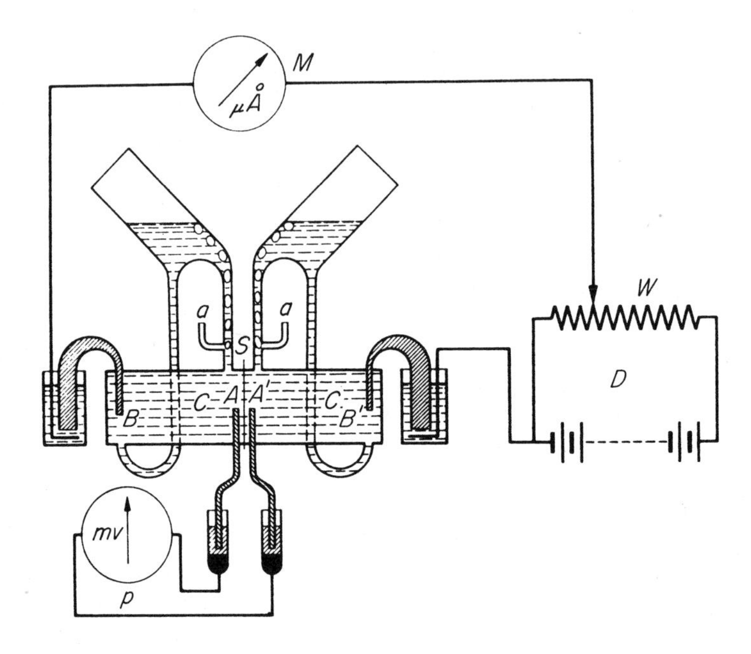

FIG. 3-1. Schematic of apparatus employed by Ussing and Zerahn [74] for the study of ion transport across isolated frog skin.

and $\Delta\tilde{\mu}_i = 0$ for all i. If $\mathbf{J}_i$ is dependent only on its conjugate driving force, $\mathbf{J}_i$ will be zero under short-circuit conditions. The finding $\mathbf{J}_i \neq 0$ when $\Delta\tilde{\mu}_i = 0$ is conclusive evidence that the transport of i cannot be attributed solely to external differences in concentration and electrical potential across the tissue and that additional forces must be involved.

The apparatus and technique described above offer another approach with which one can determine whether the transport of a charged species can be attributed solely to external differences in the electrochemical potential. Ussing [71] has shown that if the flow of an ion across a membrane is due entirely to diffusion the ratio of the unidirectional fluxes across the membrane should be given by

$$\frac{\overrightarrow{\mathbf{J}_i}}{\overleftarrow{\mathbf{J}_i}} = \frac{\mathrm{a}_i'}{\mathrm{a}_i''}\, e^{z_i\mathbf{F}(\Psi'-\Psi'')/RT} \tag{9}$$

so that the net flux is given by

$$\mathbf{J}_i = \overrightarrow{\mathbf{J}_i} - \overleftarrow{\mathbf{J}_i} = \overleftarrow{\mathbf{J}_i}\left[\frac{\mathrm{a}_i'}{\mathrm{a}_i''}\, e^{z_i\mathbf{F}(\Psi_i'-\Psi_i'')/RT} - 1\right]$$

where $\mathbf{J}_i$ is the unidirectional flux from side $'$ to side $''$ and $\overleftarrow{\mathbf{J}_i}$ is the unidirectional flux in the opposite direction. Clearly, under short-circuit conditions $\overrightarrow{\mathbf{J}_i} = \overleftarrow{\mathbf{J}_i}$ and $\mathbf{J}_i = 0$. Because the "flux ratio" analysis permits an examination of $\mathbf{J}_i$ over a wide range of chemical activities and electrical potential differences, it may strengthen the conclusion drawn from the demonstration that $\mathbf{J}_i = 0$ when $\Delta\tilde{\mu} = 0$ by essentially eliminating the possibility that other forces exert an influence on $\mathbf{J}_i$ but that these influences vanish or fortuitously cancel when the tissue is short-circuited.

Although many biological transport processes are adequately described by both Equation 8 and Equation 9, these two equations are based on different considerations and warrant more detailed examination. Equation 9 is based on the kinetics of ionic diffusion, and a number of assumptions are implicit in its derivation.

1. The membrane is considered to separate two well-mixed compartments, and the substance *i* in solution is assumed to be in thermodynamic equilibrium with *i* in the membrane at both membrane-solution interfaces; that is, passage through the membrane must be the rate-limiting step in the transfer of *i* from $'$ to $''$.
2. The chemical state of *i* within the membrane is the same as in the surrounding solutions; that is, *i* is not influenced by chemical reactions, association, binding, or complex-formation within the membrane.
3. There is no interaction between the flow of *i* and the flows of other substances through the membrane.
4. The membrane is in a steady state with respect to *i*, so that the flux in any direction is constant, and is the same at every point within the membrane in the direction of flow.
5. Since unidirectional fluxes can be determined only with the use of appropriate isotopes, the flows of the two isotopes (tracer and abundant species) must be independent. Coupling between these flows will disturb the flux ratio and complicate its interpretation.
6. Both unidirectional fluxes permeate the membrane through pathways having identical properties.

Thus, the flux-ratio equation is dependent upon a number of assumptions related to the kinetics of transport and to the properties of the membrane pathways through which the flow takes place. For these reasons, if the flow of an ion conforms to the predictions of the flux-ratio equation, the mechanism of transport is kinetically indistinguishable from simple diffusion. Conversely, if the unidirectional fluxes of an ion fail to conform to Equation 9, one *cannot* conclude that forces in addition to the external electrochemical potential difference influence the *net* flow of this ion. If the flows of tracer and abundant species are not independent, for whatever reason, the unidirectional fluxes and the flux ratio will be affected. Kedem and Essig [34] have recently examined the influence of isotope interactions on the flux ratio and have demonstrated that two phenomena, (1) exchange diffusion [12, 40] and (2) "single-file" diffusion [27], that lead to deviation from the predictions of Equation 9 can be described formally in terms of coupling between flows of abundant and tracer species. It should be stressed that these phenomena, regardless of their underlying mechanisms (e.g., mobile carriers; long, narrow pores), are in fact tracer *artifacts* that influence the *measurement* of unidirectional fluxes. Their presence has no effect on the determination of net flux either directly, by chemical means, or as the difference between the two unidirectional tracer fluxes.

In contrast with the flux-ratio equation, the derivation of Equation 8 assumes only that thermodynamic equilibrium exists at the membrane-solution interfaces, that the gradient of the electrochemical potential is continuous through the membrane, and that the system is in a steady state. Because it is founded exclusively on thermodynamic principles, it is concerned mainly with the initial and final states of the transported species and is independent of the pathways or kinetics of permeation; therein lie both its strength and its weakness.[3] Failure of a flow to conform to Equation 8 is conclusive evidence that additional forces are required to account for the observed flow regardless of membrane structure, the mechanism of permeation, or the presence of isotope interactions. On the other hand, the finding $\mathbf{J}_i = 0$ when $\Delta\tilde{\mu}_i = 0$ sheds no light on the mechanism of permeation; it does not distinguish between permeation by simple diffusion and permeation by complex formation with membrane components such as mobile carriers (e.g., facilitated transfer [12]).

In summary, if the flow of an ion conforms to the Ussing flux-ratio equation, it must also conform to Equation 8, and diffusion is strongly

[3] In essence, Equation 8 is dependent upon assumptions 1–4 listed above. Kedem and Essig [34] have shown that the flux-ratio equation can be derived from the more general considerations of irreversible thermodynamics providing assumptions 5 and 6 dealing with the absence of isotope interactions and identical permeation pathways are included.

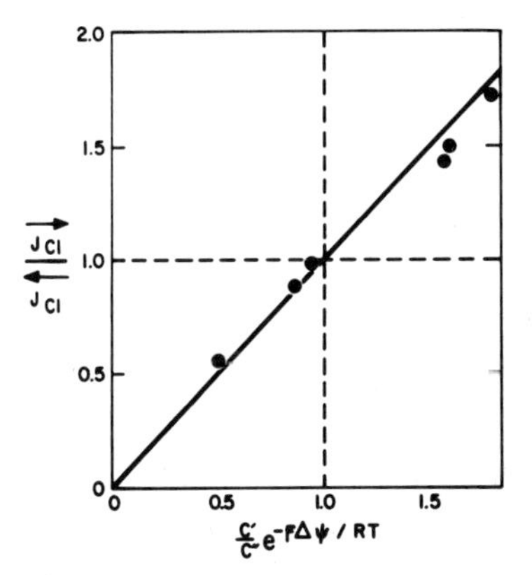

FIG. 3-2. Comparison of the ratio of the unidirectional fluxes of Cl across in vivo rat colon (ordinate) with the predictions of the Ussing flux-ratio equation for simple ionic diffusion (abscissa). C' and C'' represent the Cl concentrations in the solutions bathing the ′ and ″ sides of the tissue; the line is the line of identity. (Data from Curran and Schwartz [14].)

implicated as the mechanism of transport. If the flow fails to conform to Equation 8, it must also fail to conform to Equation 9, and forces in addition to the difference in electrochemical potential across the membrane must be involved. If the flow conforms with Equation 8 but does not agree with the flux-ratio equation, isotope interactions or complex formation within the membrane is strongly suspected.

Chloride transport across isolated frog skin [73], in vitro preparations of rat [4] and rabbit ileum [63], or in vivo rat colon [14] provides a good example of an ionic movement that, under many experimental conditions, appears to be influenced almost exclusively by the differences in the electrochemical potential for Cl between the two bathing solutions. Under short-circuit conditions, $\mathbf{J}_{Cl} = 0$ across all four of these preparations. Further, as illustrated in Figure 3-2, the excellent agreement between the observed ratio of unidirectional Cl fluxes and that predicted by Equation 9 indicates that the mechanism of Cl transport, under the conditions of these experiments, is kinetically indistinguishable from simple diffusion.

TRANSPORT COUPLED TO METABOLIC REACTIONS

When an isolated segment of rabbit ileum is mounted in the apparatus shown in Figure 3-1 and both sides of the tissue are bathed with saline solutions having identical compositions, the serosal surface of the tissue is found to be electrically positive with respect to the mucosal solution by approximately 10 mv. Under these conditions net flows of Na, Cl, and water from the mucosal solution to the serosal solution are observed [57]. As we have just discussed, the net flow of Cl can be completely attributed to the $\Delta\tilde{\mu}_{Cl}$, which, in this case, is due solely to the difference in electrical potential. The net flow of Na, on the other hand, is directed from a region of lower electrochemical potential to a region of higher electrochemical potential—a direction opposite to that which would be observed if Na transport were due to simple diffusion. When the electrical potential difference across the tissue is reduced to zero by passage of current from an external source, net Cl transport is abolished but net Na and water transport from mucosa to serosa persist. Clearly, $\mathbf{J}_{Na}$ cannot be adequately described by Equation 8, and additional forces must be sought to explain this flow; these forces can only derive from coupling between $\mathbf{J}_{Na}$ and other flows in the system. Stated in another way, the increase in free energy of Na resulting from its transfer from a region of lower electrochemical potential to a region of higher electrochemical potential must be compensated for by a process involving a decrease in free energy to which $\mathbf{J}_{Na}$ is coupled. Under short-circuit conditions, two other flows are present that could, singly or in combination, provide the required driving force for $\mathbf{J}_{Na}$: the flow of water and the flow of metabolic reactions.

The dependence of intestinal Na transport on energy-yielding metabolic processes has been recognized for many years. Inhibition of glycolysis or aerobic metabolism with fluoride, iodoacetate, dinitrophenol, cyanide, or anaerobiosis completely abolishes net Na transport across isolated small intestine; that is, in the presence of metabolic inhibitors, $\mathbf{J}_{Na} = 0$ when $\Delta\tilde{\mu}_{Na} = 0$ [57]. As shown in Table 3-1, the effect of inhibitors on $\mathbf{J}_{Na}$ is entirely the result of a decrease in the unidirectional Na flux from mucosa to serosa; the normal serosa-to-mucosa flux is unaffected. These data indicate that the interaction between $\mathbf{J}_{Na}$ and metabolic processes, whether direct or indirect, is one which influences the movement of Na only from mucosa to serosa. The implication of this finding will be discussed further below.

It is important to stress at this point that a dependence of $\mathbf{J}_i$ on metabolic processes does not imply a *direct* link or coupling between the flow of i and the flow of metabolic reactions. Thus, the flow of i could be dependent upon coupling with the flow of another species, say j, which is in turn directly coupled to metabolism. Inhibition of

TABLE 3-1. *Effect of Inhibitors on Na Transport**

Tissue	$\mathbf{J}_{ms}$†	$\mathbf{J}_{sm}$†	$\mathbf{J}_{net}$†
Rat ileum, in vitro [9]			
Control	5.0	2.1	2.9
2,4-Dinitrophenol	2.1	—	0
Iodoacetate	2.2	—	0.1
Anaerobiosis	2.5	—	0.4
Rabbit ileum, in vitro [60]			
Control	9.6	5.7	3.9
Ouabain	6.2	6.4	−0.2

* The unidirectional flux from mucosa to serosa is designated by the subscript *ms*, and the serosa-to-mucosa unidirectional flux is designated by the subscript *sm*; $\mathbf{J}_{net} = \mathbf{J}_{ms} - \mathbf{J}_{sm}$.

† μmoles/cm² hr.

metabolism would result in inhibition of both $\mathbf{J}_j$ and $\mathbf{J}_i$, and, in the absence of further information, one cannot determine whether it is $\mathbf{J}_j$, $\mathbf{J}_i$, or both that are directly linked to metabolism. This point is exceedingly important and will be the focus of much of the succeeding discussion. It can be succinctly expressed using the format of Equation 7. Thus, if there are net flows of i and j under short-circuit conditions when $\Delta\tilde{\mu}_i = 0$ and $\Delta\tilde{\mu}_j = 0$, we may have any of the following relations:

(i)
$$\mathbf{J}_i = -(R_{ij}/R_{ii})\mathbf{J}_j - (R_{ir}/R_{ii})\mathbf{J}_r$$
$$\mathbf{J}_j = -(R_{ji}/R_{jj})\mathbf{J}_i$$

(ii)
$$\mathbf{J}_j = -(R_{ji}/R_{jj})\mathbf{J}_i - (R_{jr}/R_{jj})\mathbf{J}_r$$
$$\mathbf{J}_i = -(R_{ij}/R_{ii})\mathbf{J}_j$$

(iii)
$$\mathbf{J}_i = -(R_{ir}/R_{ii})\mathbf{J}_r$$
$$\mathbf{J}_j = -(R_{jr}/R_{jj})\mathbf{J}_r$$

where all $R \neq 0$.[4] Clearly, in all three instances $\mathbf{J}_i$ and $\mathbf{J}_j$ will equal zero when $\mathbf{J}_r = 0$. Distinction between these three possible explanations for $\mathbf{J}_i$ and $\mathbf{J}_j$ depends upon an identification of the pertinent reaction, $\mathbf{J}_r$, and demonstration of an interaction between i and $\mathbf{J}_r$ (i.e., $R_{ir} \neq 0$), or j and $\mathbf{J}_r$ (i.e., $R_{jr} \neq 0$), or both.

[4] Interactions between $\mathbf{J}_j$ and $\mathbf{J}_i$ have been omitted from relation (iii) above. Although such interactions could be included, they are not essential to the present argument.

Returning to the problem of Na transport across intestine and other epithelia, we may raise the following question: Is $\mathbf{J}_{Na}$ directly linked to metabolic processes or is Na absorption the result of coupling to another flow, say water, that is in turn directly linked to metabolism? Although we will return to this important question later, we can mention here the compelling evidence that Na transport across intestine and other epithelia is brought about by a highly specific mechanism that is directly coupled to energy-yielding chemical reactions within the tissue. This evidence is based, in part, on the similarities between the mechanisms responsible for net Na transport across epithelia and those found in single-cell systems in which Na transport is not accompanied by a net flow of water [73]. In many such systems a membrane fraction, possessing adenosine triphosphatase (ATPase) activity, has been identified. The ATPase is activated by the combined presence of Na and K, it is inhibited by the cardiac glycoside ouabain, and convincing evidence has been compiled that it is intimately involved in Na transport [65]. Thus, in these systems progress has been made toward defining $\mathbf{J}_r$ and the stoichiometric relation between $\mathbf{J}_r$ and $\mathbf{J}_{Na}$. Although similar Na, K–activated ATPase activity has been found in membrane preparations from small intestine, frog skin, toad urinary bladder, and renal tissue [66], the evidence implicating these enzyme activities in Na transport is not as direct as in the case of the human erythrocyte. Nevertheless, there is good reason to believe that they are involved in the coupling of Na transport to exergonic chemical reactions and that the inhibitory effect of ouabain on Na transport across all the epithelia mentioned is mediated through its action on this ATPase.

NaCl Transport and the Transepithelial Electrical Potential Difference

A characteristic of many epithelial tissues capable of transporting ions against electrochemical potential differences is the presence of an electrical potential difference, $\Delta\Psi$, across the tissue even when both surfaces are bathed with identical solutions. Indeed, the finding of a steady-state $\Delta\Psi$ across a membrane that separates two identical solutions is conclusive evidence that the flow of at least one charged species must be due to a mechanism other than simple diffusion. Although the precise origin of this potential difference is still uncertain, its relation to transepithelial ion transport may be clarified by examining a simplified case.

Consider a tissue that has the following properties: (1) Cl transport across the tissue is entirely attributable to $\Delta\tilde{\mu}_{Cl}$, and (2) Na can be transported from side ′ to side ″ against an electrochemical potential difference by means of a special carrier mechanism that is directly coupled to energy-yielding metabolic reactions. Assume further that no other flows significantly influence the movements of Na and Cl. These are essentially the characteristics of NaCl transport across a variety of epithelial tissues, including isolated frog skin [72], and several in vitro preparations of small and large intestine [57]. When the tissue is bathed on both surfaces with identical NaCl solutions, there will be a flow of NaCl from side ′ to side ″, and side ″ will be electrically positive with respect to side ′. According to Equation 7 the flows of Na and Cl are given by

$$\mathbf{J}_{Na} = -\Delta\tilde{\mu}_{Na}/R_{Na} - R_r\,\mathbf{J}_r/R_{Na}$$

$$\mathbf{J}_{Cl} = -\Delta\tilde{\mu}_{Cl}/R_{Cl} \tag{10}$$

Since both solutions have identical compositions, the electrochemical potential differences of Na and Cl are due solely to the $\Delta\Psi$, so that $\Delta\tilde{\mu}_{Na} = -\Delta\tilde{\mu}_{Cl} = \mathbf{F}\,\Delta\Psi$. Further, since bulk electroneutrality must be maintained, the flow of cation across the membrane must be matched by an equivalent flow of anion, so that $\mathbf{J}_{Na} = \mathbf{J}_{Cl}$. The following relation is obtained by combining the above equations:

$$\Delta\Psi = -\frac{R_{Na}\,R_{Cl}}{(R_{Na} + R_{Cl})\mathbf{F}}\left(\frac{R_r}{R_{Na}}\,\mathbf{J}_r\right) \tag{11}$$

Thus, $\Delta\Psi$ is a direct function of the rate of the metabolic reaction and its degree of coupling to the flow of Na. Further, it is clear from Equation 10 that the term $R_r\,\mathbf{J}_r/R_{Na}$ represents the rate of Na transport that would be observed when the tissue was short-circuited (i.e., when $\Delta\tilde{\mu}_{Na} = 0$). Thus, $\Delta\Psi$ is directly related to that portion of $\mathbf{J}_{Na}$ that is not attributable to $\Delta\tilde{\mu}_{Na}$.

The relation between $\Delta\Psi$ and Na transport may be further clarified by viewing Equation 11 as formally analogous to Ohm's law. The term $R_r\,\mathbf{J}_r/\mathbf{F}\,R_{Na}$ has units of coulombs per unit time and is therefore the "Na current" generated by the direct link to metabolic reactions. The term $R_{Na}\,R_{Cl}/(R_{Na} + R_{Cl})$ represents the combined resistance of the tissue to Na and Cl flows assuming that these flows do not interact (i.e., that they take place through independent or parallel pathways). When R_{Na}, R_r, and $\mathbf{J}_r$ are constant, $\Delta\Psi$ increases with increasing R_{Cl}. That is, under these conditions $\Delta\Psi$ is inversely related to the permeability of the tissue to Cl. This relation follows intuitively from the condition of electroneutrality and the stipulation that the driving force for Cl transport is due entirely to $\Delta\Psi$ when the Cl concentrations in the two bathing solutions are equal. For a fixed value of $R_r\, \mathbf{J}_r/R_{Na}$, then, the $\Delta\Psi$ must be of sufficient magnitude to bring about a flow of Cl equal to that of Na. The lower the permeability of the tissue to Cl, the greater must be the value of $\Delta\Psi$ in order to comply with the constraint $\mathbf{J}_{Na} = \mathbf{J}_{Cl}$.

Equation 11 also indicates that the external current required to abolish the spontaneous $\Delta\Psi$, the "short-circuit current," must be equal to $R_r\, \mathbf{J}_r/\mathbf{F}\, R_{Na}$, or, stated explicitly, the current necessary to short-circuit the tissue must be equal to the net flow of Na observed under short-circuit conditions. This equality between short-circuit current and $\mathbf{J}_{Na}$ when $\Delta\tilde{\mu}_{Na} = 0$ has been demonstrated for isolated frog skin, toad urinary bladder, and several preparations of small and large intestine and provides a means by which the flow of Na resulting from forces other than $\Delta\tilde{\mu}_{Na}$ can be readily monitored [57].

Finally, it is instructive to consider the more general case in which both $\mathbf{J}_{Na}$ and $\mathbf{J}_{Cl}$ may be linked to chemical reactions. The flow equations are

$$\mathbf{J}_{Na} = -\frac{\Delta\tilde{\mu}_{Na}}{R_{Na}} - \frac{R_{Na,r}}{R_{Na}}\,\mathbf{J}'_r$$

$$\mathbf{J}_{Cl} = -\frac{\Delta\tilde{\mu}_{Cl}}{R_{Cl}} - \frac{R_{Cl,r}}{R_{Cl}}\,\mathbf{J}''_r$$

where $\mathbf{J}'_r$ and $\mathbf{J}''_r$ represent two, possibly different, chemical reactions. When the two solutions have identical compositions, we obtain

$$\Delta\Psi = \left(\frac{R_{Na}\ R_{Cl}}{R_{Na} + R_{Cl}}\right)\left(\frac{R_{Cl,r}}{R_{Cl}}\,\mathbf{J}''_r - \frac{R_{Na,r}}{R_{Na}}\,\mathbf{J}'_r\right)$$

Clearly, if the metabolically linked flows of Na and Cl have the same direction, $\Delta\Psi$ will reflect the difference between these rates. On the other hand, if these flows are oriented in opposite directions (e.g., Na absorption and Cl secretion) $\Delta\Psi$ is related to the sum of their rates. If, as appears to be the case for rabbit gallbladder [18, 21], Na and Cl transport are mediated by a neutral NaCl pump mechanism.

$$\frac{R_{Cl,r}}{R_{Cl}}\,\mathbf{J}''_r = \frac{R_{Na,r}}{R_{Na}}\,\mathbf{J}'_r = \frac{R_{NaCl,r}}{R_{NaCl}}\,\mathbf{J}_r$$

and $\Delta\Psi = 0$. Thus, although the presence of an electrical potential difference across a tissue separating identical solutions indicates that the flow of at least one ion cannot be attributed solely to $\Delta\tilde{\mu}$, the absence of an electrical potential difference does not exclude the presence of metabolically linked ion flows.

TRANSPORT DEPENDENT UPON COUPLING WITH THE FLOW OF OTHER SPECIES

Volume flow across an artificial membrane is given by the following expression[5]

$$\mathbf{J}_V = L_p(\Delta P - \sigma \text{ RT } \Delta C) \qquad (12)$$

where L_p is the hydraulic conductivity of the membrane and the term in parenthesis is the conjugate driving force for volume flow. The first term in parenthesis, ΔP, is the difference in hydrostatic pressure across the membrane. The second term in the parenthesis is the contribution to the driving force arising from the difference in water activities on the two sides of the membrane due to the presence of different concentrations of dissolved solute; it is the difference in osmotic pressure across the membrane. In 1951 Staverman [68] demonstrated that the effective osmotic pressure across a membrane that is not impermeable to the solute is given by

$$\Delta\pi_{eff} = \sigma \text{ RT } \Delta C \qquad (13)$$

where ΔC is the solute concentration difference and the value of σ, the "reflection coefficient," may vary from 1.0 for a solute to which the membrane is ideally impermeable to zero for a solute that, as far as the membrane is concerned, is indistinguishable from the solvent.[6] Clearly, when $\sigma = 1.0$, we obtain the classic expression of van't Hoff

[5] Volume flow, $\mathbf{J}_V$, is related to water flow, $\mathbf{J}_{H_2O}$, as follows:

$$\mathbf{J}_V = \bar{V}_{H_2O} \mathbf{J}_{H_2O} + \sum \bar{V}_i \mathbf{J}_i$$

where $\bar{V}$ is the partial molar volume and the subscript i denotes solutes. Since biological secretions and absorbates are generally dilute solutions, $\mathbf{J}_V \approx \bar{V}_{H_2O} \mathbf{J}_{H_2O}$ so that the terms *water flow* and *volume flow* will be used interchangeably. Derivations of Equation 12 are given in references 32 (Chap. 10) and 35.

[6] The reflection coefficient may also assume a negative value as in the case of negative anomalous osmosis; however, for present purposes only the range $0 \leq \sigma \leq 1$ need be considered.

for an ideal, semipermeable membrane. Thus, the conjugate driving force for volume flow has units of pressure, and the hydraulic conductivity is often expressed as cm^3/dyne, sec.

In 1892 Reid demonstrated that an isolated segment of rabbit small intestine, bathed on both surfaces with identical solutions, will transport water from the mucosal solution to the serosal solution even *against* a hydrostatic pressure difference as long as the tissue remains viable [52]. This observation, that water transport across intestine cannot be attributed entirely to differences in water activity (or combined differences in osmotic and hydrostatic pressures), has been confirmed repeatedly. The data of Parsons and Wingate [49], as shown in Figure 3-3, indicate that the mucosal solution bathing in vitro rat small intestine must be made 100 mOsm hypertonic to the serosal solution before net water flow from mucosa to serosa is abolished. The dependence of this water transport mechanism on metabolic energy is dramatically demonstrated by the data obtained by Diamond [17] using fish gallbladder. As shown in Figure 3-4 the gallbladder will transport fluid from the lumen to the surrounding medium even when the mucosal solution is 20 mM hypertonic to the serosal solution. However, when energy-yielding metabolic processes are arrested, the direction of flow is reversed and the gallbladder behaves as would an artificial membrane.

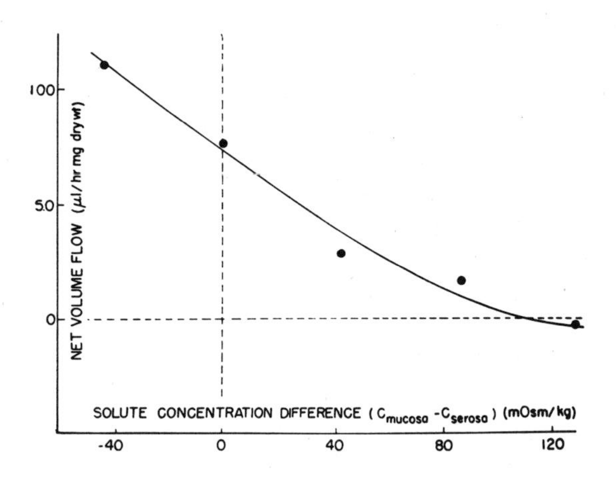

FIG. 3-3. Relation between net volume flow from mucosal solution to serosal solution and solute concentration difference across in vitro rat ileum. (Data from Parsons and Wingate [49].) (From Curran [10].)

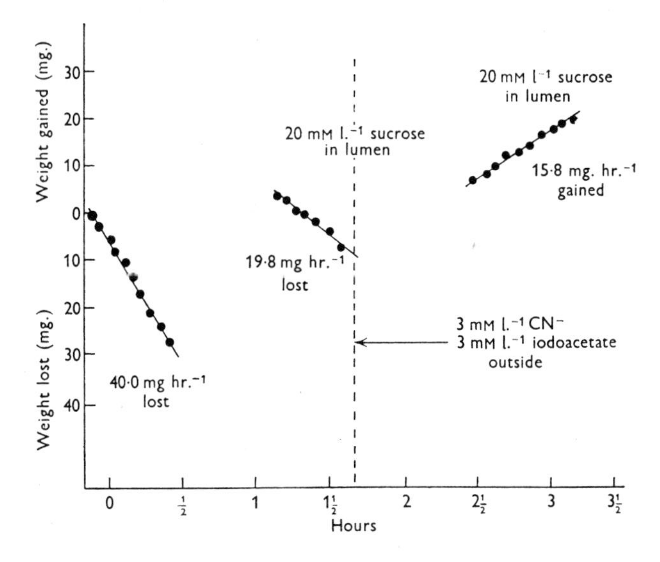

FIG. 3-4. Dependence of water transport across fish gallbladder on metabolic energy. Fluid transported from the lumen of the gallbladder into the serosal bathing solution is designated as "weight lost" and net volume flow from the serosal solution into the lumen is designated as "weight gained." The first line (slope = 40 mg/hr) represents the rate of fluid transported out of the gallbladder when both surfaces of the tissue were bathed with isotonic Ringer's solution. Although the addition of 20 mM sucrose to the luminal fluid slowed this rate (second line), net volume flow was directed against an adverse osmotic pressure difference. In the presence of metabolic inhibitors the direction of net volume flow was reversed so that the flow could be attributed entirely to the difference in water activities on the two sides of the tissue. (From Diamond [17].)

Thus, for the case of water transport across intestine and other epithelia we see that $\mathbf{J}_V \neq 0$ when $\Delta\tilde{\mu}_{H_2O} = 0$, but that $\mathbf{J}_V = 0$ when *both* $\Delta\tilde{\mu}_{H_2O} = 0$ *and* $\mathbf{J}_r = 0$. The immediate question is: Need we postulate a direct link between water transport and metabolic reactions, or can coupling to other flows suffice to explain this process? Two extensively confirmed sets of observations provide the answer. First, there is overwhelming evidence that water transport is dependent upon net solute, and in particular NaCl, transport; in the absence of either net solute transport or a hydrostatic (or osmotic) pressure difference water transport across the tissue has not been demonstrated convincingly for any epithelial tissue. This obligatory dependence of water

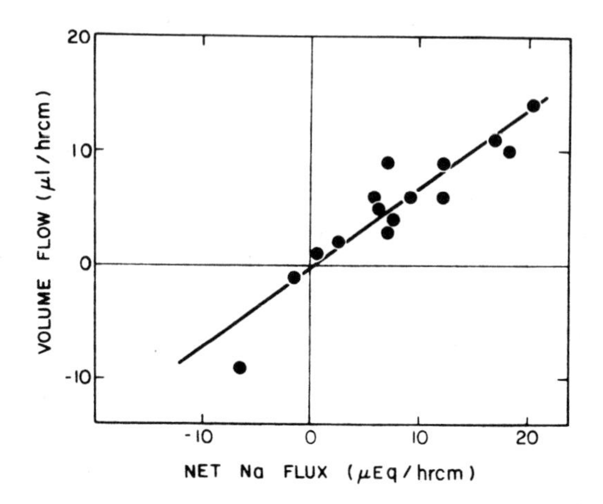

FIG. 3-5. Relation between volume flow and net Na transport across in vitro rat ileum. Positive values indicate flow from mucosa to serosa. (Data from Curran [9].)

movement on net solute transport across small intestine was demonstrated by Curran and Solomon [15], who made use of the observation that there is a close relation between the rate of water absorption and the NaCl concentration in the lumen providing the osmolarity of the luminal fluid is maintained constant by the addition of an indifferent and relatively impermeant solute such as mannitol. As shown in Figure 3-5, water absorption by an in vitro preparation of rat ileum ceases when net Na absorption ceases [9].

The second set of observations is that, in addition to the obligatory dependence of water transport on solute transport, there is a stoichiometric relation between the two rates of transport such that the transported fluid is essentially isotonic, with the solution bathing the mucosal (or luminal) surface of the tissue. The clearest demonstration of this relation is provided by the studies of Diamond [19]. He examined the osmolarity of the fluid transported across the wall of isolated rabbit gallbladder when the osmolarity of the luminal solution was varied over an eightfold range either by changing the NaCl concentration or by adding impermeant solutes. His results, shown in Figure 3-6, indicate that regardless of the composition and osmolarity of the luminal solution the absorbate was essentially an NaCl solution that was isotonic with the luminal solution.

These observations form the basis of the widely held view that the

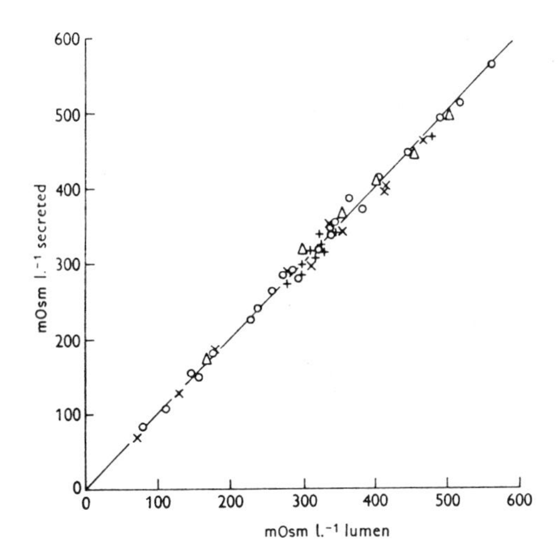

FIG. 3-6. Relation between the osmolarity of the fluid transported across an in vitro preparation of rabbit gallbladder and the osmolarity of the solution in the lumen. (From Diamond [19].)

driving force for the flow of water across epithelial membranes separating identical solutions arises from physical coupling to the net flow of solute. The perplexing problem was to conceive of a system in which the flow of a single molecule of solute would be accompanied by the flow of several hundred water molecules so as to yield an isotonic absorbate. A model system that could satisfy all of the characteristics of biological water transport was proposed by Curran [9–11, 46] several years ago and is now generally referred to as the double-membrane hypothesis. This model involves at least two membranes arranged in series as shown in Figure 3-7. A solute, i, is assumed to be

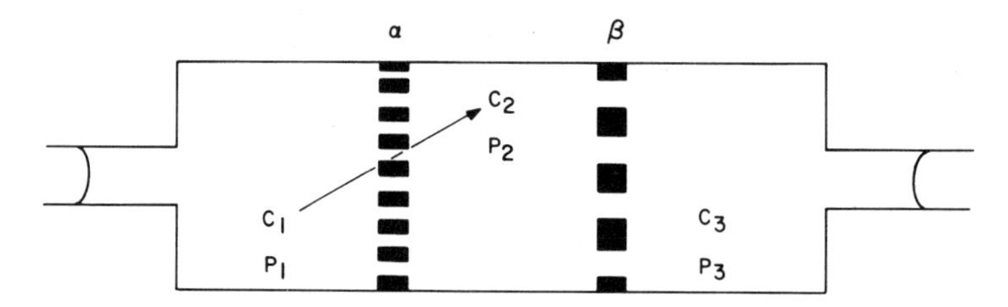

FIG. 3-7. "Double-membrane" model proposed by Curran.

transported from compartment 1 to compartment 2 across membrane α, resulting in an accumulation of i in compartment 2. The reflection coefficient of membrane α for this solute, σ_α, is assumed to be much greater than that of membrane β, σ_β. The effective osmotic pressure across membrane α is $\Delta\pi_\alpha = \sigma_\alpha$ RT $(C_2 - C_1)$, and when the concentration of i in 2 is greater than that in 1 the volume flow across membrane α from compartment 1 to compartment 2 is given by $\mathbf{J}_{V\alpha} = L_{p\alpha}\sigma_\alpha RT\ (C_2 - C_1)$. Since $\sigma_\beta \ll \sigma_\alpha$ the effective osmotic pressure difference across β and the osmotically induced volume flow from compartment 3 to 2 will be small. If the volume of compartment 2 is constrained, the entrance of fluid from compartment 1 will result in an increase in the hydrostatic pressure in 2 which, in turn, will force a flow of fluid from compartment 2 to compartment 3 across the more permeable membrane β.

A formal analysis of this double-membrane model for a simplified case may clarify the operation of the system. We assume that the system contains a single solute and that a steady state has been achieved with respect to the concentrations of the solute in the three compartments and volume flow across both barriers. According to Equation 12

$$\mathbf{J}_{V\alpha} = L_{p\alpha}[(P_1 - P_2) + \sigma_\alpha\ RT\ (C_2 - C_1)]$$

$$\mathbf{J}_{V\beta} = L_{p\beta}[(P_2 - P_3) + \sigma_\beta\ RT\ (C_3 - C_2)]$$

where P and C denote the hydrostatic pressure and solute concentration in each compartment respectively. Since in the steady state $\mathbf{J}_{V\alpha} = \mathbf{J}_{V\beta} = \mathbf{J}_V$, we can solve the above equations for the simplifying condition that $P_1 = P_3$ and obtain

$$\mathbf{J}_V = L_p[\sigma_\alpha\ RT\ (C_2 - C_1) + \sigma_\beta\ RT\ (C_3 - C_2)] \qquad (14)$$

where L_p is equal to $L_{p\alpha}L_{p\beta}/(L_{p\alpha} + L_{p\beta})$. Further, when $C_1 = C_3$ we obtain $\mathbf{J}_V = L_p(\sigma_\alpha - \sigma_\beta)$ RT $(C_2 - C_1)$. Thus, there will be a flow of volume from compartment 1 to compartment 3 ($\mathbf{J}_V > 0$) when $\sigma_\alpha > \sigma_\beta$ and $C_2 > C_1$ even though the hydrostatic pressure and solute concentrations in compartments 1 and 3 are identical. Also, it is clear from Equation 14 that volume flow from 1 to 3 can take place even when $C_1 > C_3$ (e.g., when 1 is hypertonic to 3) providing $C_2 > C_1$ and σ_α is sufficiently greater than σ_β. Further discussion of the general behavior of a series arrangement of two membranes with asymmetrical properties is beyond the scope of this chapter. The interested reader should consult the detailed analysis by Patlak, Goldstein, and Hoffman [50].

Thus, the general requirement for solute-coupled water transport across epithelial membranes is the presence, within the tissue, of a constrained compartment into which solute is "pumped." In addition, this compartment must have boundaries that are functionally different with respect to solute and water permeabilities; this functional asymmetry imparts direction to the net volume flow. The boundaries, it should be noted, need not be discrete membranes; they may include regions of tissue or even several membranous structures, providing their composite properties are consistent with those described above.

Diamond [20, 69] has recently proposed a mechanism for isotonic water transport that incorporates many of the basic principles of the double-membrane hypothesis and extends them to take into account the detailed anatomic structure of several epithelial tissues. According to this model, illustrated in Figure 3-8, the intracellular compartment (analogous to compartment 1 of Figure 3-7) rapidly achieves osmotic equilibrium with the luminal fluid. Solute, primarily NaCl, is pumped out of the cell across the lateral membranes (analogous to membrane α) into the blind end of the cul-de-sac formed by the lateral inter-

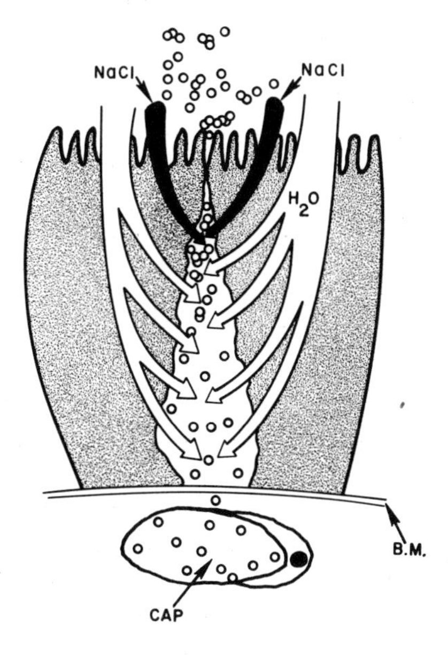

FIG. 3-8. Diamond's "standing osmotic gradient" model for isotonic fluid absorption by gallbladder. B.M., basement membrane; CAP, an underlying capillary.

cellular spaces and the tight junctions that bind the cells together at their apical surfaces. The accumulation of solute in this small volume creates a region of hypertonicity and provides the driving force for the flow of water out of the cell into the cul-de-sac. The resulting increase in the volume of this small region combined with the viscous resistance of the neighboring structures leads to a slight elevation of the local pressure, and the solution is forced down the length of the lateral intercellular space toward the basement membrane and underlying capillary bed.

An important feature of this model is that it provides an explanation for the observation that the absorbate is always isotonic with the luminal fluid that is compatible with the ultrastructure of the tissue. According to Diamond, the solution at the blind end of the cul-de-sac is always hypertonic to the cell interior, but it becomes progressively equilibrated with the intracellular osmolarity as it passes down the intercellular space. This long, slow passage through the tortuous lateral space is what permits the absorbate to achieve osmotic equilibrium with the cell interior.

The major modification of the double-membrane hypothesis introduced by Diamond is that compartment 2, instead of being homogeneous, contains a "standing osmotic gradient" ranging from hypertonic at the blind end to isotonic at the basement membrane. The site at which solute is pumped into the lateral space is restricted to the blind end, whereas water flow is permitted to take place across the entire extent of the lateral membrane. Clearly, in Diamond's model the lateral intercellular space combines the properties of both compartment 2 and membrane β of the double-membrane model, and one may treat the lateral space as if it were comprised of a continuum of infinitely small compartments separated by membranes whose reflection coefficients for the transported solute are essentially zero. Thus, the underlying asymmetry, proposed by Curran, that is responsible for imparting direction to the volume flow is essentially retained.

The double-membrane model introduced by Curran, and its extension and modification by Diamond, can fully account for all the characteristics of isotonic water transport across epithelial membranes and certainly represents one of the major advances of the past decade in our understanding of biological transport processes. The reader interested in this particular problem is strongly urged to examine the brilliant series of papers by these authors. In addition, the theory of solute-coupled water flow represents a high point in the application of irreversible thermodynamics, particularly the concept of the reflection coefficient, to biological processes. Although a direct link between water absorption and metabolic processes cannot be unequivocally ruled out, it is now clear that there is no need to postulate such a link; water transport can be completely accounted for in terms of simple physical forces generated within the tissue by net solute transport.

OTHER SOLUTE-SOLVENT INTERACTIONS

The dependence of water absorption on net solute flow represents one type of solute-solvent interaction in which solute transport generates the driving force for water transport. It is also possible for water flow to provide the driving force for net solute flow, though the mechanism of coupling differs from that described above. This phenomenon, referred to as solvent drag, is generally ascribed to the frictional force exerted on the solute by the flow of solvent and suggests that solvent and solute share, at least in part, a common pathway through the membrane (e.g., the solute may be considered to be entrained in a fluid stream flowing through aqueous channels or pores that traverse the membrane).

Although the possibility of solute entrainment in a solvent stream had been entertained earlier [22], the first explicit demonstration of solvent drag across a biological membrane is attributed to Andersen and Ussing [1]. These investigators showed that net water flow across isolated toad skin in response to an osmotic pressure difference can bring about net movement of an uncharged solute in the absence of a chemical potential difference. They inferred that the drag effect can arise only when the solute moves in a continuous water phase (e.g., through pores) so that molecules diffusing in the direction of net water movement will be accelerated while those diffusing in the opposite direction will be retarded. Solvent drag has also been demonstrated for urea transport across isolated toad urinary bladder [39] in response to osmotically induced water flow. A particularly interesting study of solvent drag across intestine has been published by Hakim and Lifson [26, 41], who demonstrated that water flow across in vivo as well as in vitro canine small intestine can bring about net movement of urea against a concentration difference. Further, as shown in

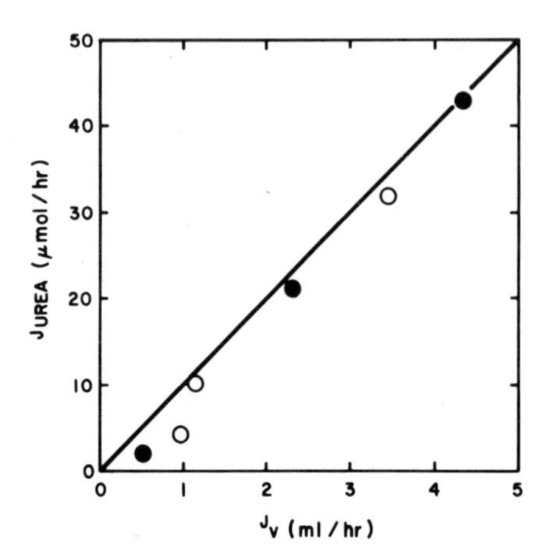

FIG. 3-9. Net movement of urea across in vitro canine small intestine due to solvent drag. The open circles represent flow from the mucosal solution to the serosal solution in response to spontaneous water absorption; the closed circles represent flows from the serosal solution to the mucosal solution produced by application of a hydrostatic pressure to the serosal solution. In both instances net flow of urea was directed against a chemical potential difference from a solution in which its initial concentration was 10 mM to a solution in which its initial concentration was 20 mM. At high flow rates the urea concentration in the transported fluid approaches that of the solution from which the flow originated (10 mM). (Data from Hakim and Lifson [26].)

Figure 3-9, both spontaneous water flow from mucosa to serosa[7] and water flow in the opposite direction induced by the application of a hydrostatic pressure to the serosal solution exerted approximately the same drag effect on urea. These data represent a convincing demonstration of solvent drag arising from spontaneous water flow and suggest that the same or similar channels are involved in spontaneous and pressure-induced water flows.

The effect of solvent drag can be described by the following expression for the flow of an uncharged solute, *s*, across a membrane assuming that the solute may only interact with the flow of water:

$$\mathbf{J}_s = \bar{C}_s \ (1 - \sigma_s) \ \mathbf{J}_V + P_s \ \Delta C_s \tag{15}$$

[7] The term *spontaneous water flow* refers to the normal solute-linked water absorption which, in the experiments of Hakim and Lifson, was stimulated by the presence of glucose in the mucosal solution.

where $\bar{C}_s$ is the average concentration of the solute in the membrane, ΔC_s is the concentration difference across the membrane, and P_s is the permeability coefficient determined when $\mathbf{J}_V = 0$. When $\Delta C_s = 0$, $\mathbf{J}_s = C_s(1 - \sigma_s)\ \mathbf{J}_V$. Thus, the magnitude of the solute flow due *entirely* to solvent drag not only is a function of the solute concentration and the rate of water flow but is also dependent upon the reflection coefficient σ_s. The latter is a function of the combined structure and properties of the membrane and the solute (e.g., size, shape, charge, lipid solubility). Clearly, when $\sigma_s = 1$, as in the case for the ideal semipermeable membrane, net flow of solute due to solvent drag is not possible.

At present it is difficult to evaluate precisely the physiological importance of solvent drag as a transport mechanism. It should be noted that solvent drag across epithelial tissues has been demonstrated convincingly only for relatively small, uncharged solutes. According to the data of Leaf and Hays [39], the reflection coefficient for urea passage across toad urinary bladder is 0.8 and that for thiourea is essentially 1.0. On the other hand, Hakim and Lifson report a value of 0.3 for the reflection coefficient for urea movement across in vivo dog small intestine, and a value close to zero for an in vitro preparation. These low values for σ suggest that spontaneous water flow across small intestine might exert a significant solvent drag effect on larger molecules. Indeed, the earlier observations of Fisher [22] suggest that spontaneous water absorption from the lumen of isolated rat small intestine can provide the driving force for net absorption of the six-carbon alcohol, sorbitol. Yet Lindemann and Solomon [42] have reported that mannitol has a reflection coefficient of 0.99 in rat small intestine, and Smyth and Wright [67], using this value for mannitol, have reported reflection coefficients for urea, erythritol, and lactose in rat small intestine of 0.8, 0.9, and 1.0 respectively. These data indicate that solvent drag should not significantly influence the flow of molecules having six or more carbons and are in apparent conflict with the results of Fisher. The disagreement may be related to the fact that these investigators employed different methods and, in some instances, different segments of the rat intestine. Fisher studied the rates of solute and water flow across the entire thickness of intestinal wall, and his data are best described by means of Equation 15. Lindemann and Solomon, and Smyth and Wright calculated reflection coefficients using Equation 13 by determining the zero-time effective osmotic pressure difference produced by adding known concentrations of various solutes to the mucosal solution.

It is not clear that the different methods employed should give the same results when applied to a composite membrane system containing series as well as parallel inhomogeneities. When a "membrane" consists of a brush border, serosal and lateral membranes, tight junctions, intercellular spaces, basement membrane, connective tissue, muscle layers, etc., as well as denuded areas caused by exfoliation, results derived from simplified relations must be interpreted with caution. Although the reflection coefficient is a well-defined property of a single membrane, the "overall" reflection coefficient of a composite membrane system is a complex function of several independent membrane properties. As shown by Kedem and Katchalsky [37], the overall reflection coefficient of a composite membrane composed merely of a parallel array of elements is not just an average of the single reflection coefficients; it may be smaller than the smallest or larger than the largest elementary σ. Thus "overall" descriptive terms such as *permeability, effective pore radius, hydraulic conductivity,* etc., must be interpreted with care when applied to complex tissues, and, when comparing data obtained using different methods, one must be quite sure that these data reflect the *same elementary components* of the composite system.[8]

We have dwelt on the problem of solvent drag at length because it represents a potentially important transport mechanism that is as yet incompletely explored. Since epithelial tissues such as intestine and renal tubule absorb large quantities of water daily, solvent drag effects could, in principle, significantly influence the absorption of a variety of solutes. Additional studies along the lines of those described by Hakim and Lifson using a variety of probing molecules are required before the physiological importance of solvent drag can be adequately evaluated.

Sodium and Water: Chicken and Egg

We will now return to a problem that has been the subject of considerable controversy for many years. In view of the fact that epithelial

[8] Pidot and Diamond [51] have suggested that osmotically induced water flow and spontaneous water flow across gallbladder utilize separate and different channels, and Smyth and Wright [67] have arrived at the same conclusion for the in vitro rat small intestine. If this is the case, the reflection coefficients determined by Lindemann and Solomon [42] and Smyth and Wright [67] cannot be applied directly to the problem of solvent drag arising from spontaneous water flow, so that their data and those of Fisher [22] need not be incompatible.

tissue can transport both Na and water against their thermodynamic potential differences, several relations among Na transport, water transport, and the flow of metabolic reactions are a priori equally possible. Thus, (1) Na transport may be directly linked to metabolic processes with water transport secondary to Na transport; (2) water transport may be directly coupled to metabolism with Na transport resulting from solvent drag; and (3) both Na transport and water transport may derive their driving force directly from metabolic reactions. As pointed out previously, a definitive choice between these alternatives is dependent upon an unequivocal demonstration of the presence or absence of direct coupling of Na transport and/or water transport with defined chemical reactions. In the absence of this information, although kinetic data may provide a compelling argument in favor of one of these alternatives, such evidence cannot be considered proof. This is the present status of the problem.

The possibility that water transport is directly coupled to metabolic processes and that Na transport is the result of solvent drag forms the basis of a number of theories, the most famous of which is the "fluid circuit" theory advanced by Ingraham, Peters, and Visscher in 1938 [29]. This theory proposed that intestinal salt absorption is the result of two oppositely directed "forced" fluid streams. The first stream is directed from lumen to plasma and contains a high content of entrained solute because it traverses a region of relatively high solute permeability. The stream directed from plasma to lumen, on the other hand, passes through a region that restricts the flow of solute. This differential sieving effect, illustrated in Figure 3-10, results in a net flow of solute from mucosa to serosa. The original "fluid circuit" theory has been discussed in detail in a recent review [57], and the primary reason for dismissing this model will simply be summarized. As formulated above, the model predicts that when both surfaces of the tissue are bathed with identical solutions there will be net solute flow from mucosa to serosa when net volume flow is zero, i.e., when the forward and backward fluid streams are of equal magnitude. This conclusion is contrary to all the experimental observations that net volume flow and net solute flow between identical solutions cease simultaneously. The data of Clarkson and Rothstein [3], shown in Figure 3-11, provide an excellent illustration of this point.

Further, there is no evidence that Na transport is significantly affected by solvent drag. The unidirectional and net fluxes of Na across rat [24] and dog [75] intestine are not influenced when net

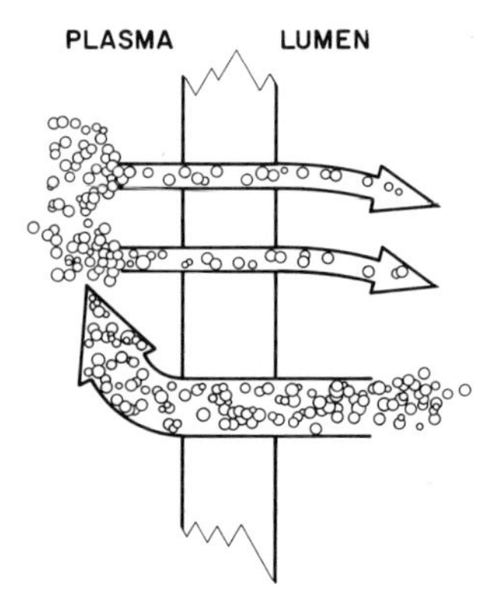

FIG. 3-10. "Fluid circuit" theory of Ingraham, Peters, and Visscher [29].

volume flow is abolished or its direction reversed by making the mucosal solution hypertonic. Similarly, net NaCl movement out of the gallbladder against a combined electrochemical potential difference persists even when there is a significant osmotically induced volume flow into the lumen [18]. Finally, as discussed previously, the serosa-to-mucosa unidirectional fluxes of Na across rat and rabbit small intestine appear to be due to simple diffusion and are unaffected by

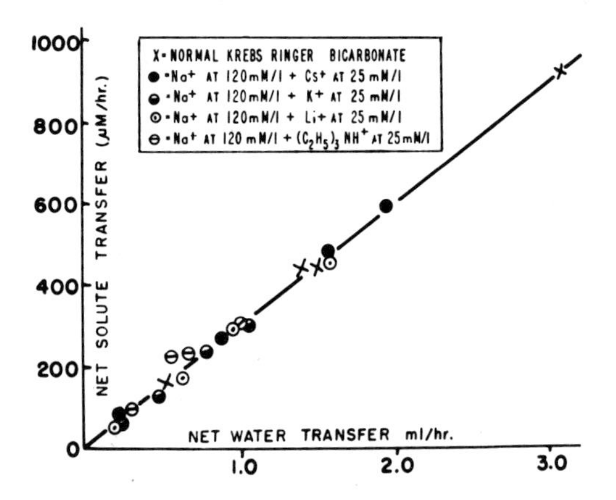

FIG. 3-11. Relation between net flows of solute and volume across rat ileum in vitro. The slope of the line corresponds to a chemical tonicity of 300 mM, in good agreement with that of the bathing solutions (313 mM). (From Clarkson and Rothstein [3].)

metabolic inhibitors, which surely must abolish net volume flow [57]. In contrast to the absence of evidence for a dependence of Na transport on water transport, there is abundant evidence for the converse relation. In all studies on fluid transport between identical solutions, solute transport appears to be the independent variable, water transport being the dependent variable. For example, when net solute transport is inhibited by replacement of Na with poorly transported solutes, water transport is slowed proportionately (Figs. 3-5 and 3-11). In gallbladder, substitution of a wide variety of cations for Na markedly inhibits water absorption. Of particular interest is the observation that fish gallbladder is equally permeable to Na and K, yet when K replaces Na in the luminal solution water transport is inhibited [16, 17]. If water transport were the primary flow, one might expect it to continue unabated and to entrain KCl rather than NaCl. This striking selectivity of fluid transport mechanisms for Na (or NaCl) is characteristic of many solute transport mechanisms but is rather difficult to reconcile with a simple picture of solvent drag or pinocytosis [25].

It should be stressed that although all the above evidence argues against the possibility that solute transport is secondary to water flow (possibility 2) it does not constitute a proof. Certainly, all the observations just cited, and many others, conclusively rule out "classic" solvent drag and pinocytosis as possible mechanisms, and convincing alternative models have not been advanced by the proponents of a primary water transport mechanism. Nevertheless, one could, in principle, construct models, however complicated, that would reconcile some of the above evidence with a primary, metabolically linked water transport mechanism. For example, the failure of osmotically induced water flow to affect Na transport could be explained by the presence of different channels for water flow in response to an external osmotic pressure difference and spontaneous water flow, with only the latter interacting with solute. Indeed, evidence for distinct pathways for osmotically induced and spontaneous, solute-linked water flow has been presented by Pidot and Diamond [51]. Thus, the strongest evidence against the possibility that Na transport is entirely secondary to water flow includes (1) the similarities between Na transport mechanisms in epithelia and the mechanisms found in single cells that do not transport water, and (2) the finding of cation-sensitive ATPase activity in epithelial tissues. Further investigation of this interesting enzyme activity could establish an unequivocal direct link between transepithelial Na transport and exergonic chemical reactions. Although this would not exclude the additional, though remote, possibility that water transport is also directly coupled to metabolism, the model discussed above eliminates the necessity for postulating such a link.

SOLUTE-SOLUTE INTERACTIONS

In recent years condiderable evidence has accumulated that interactions between Na transport and the flow of other solutes may play an important role in a number of vital absorptive processes. In this section we will focus primarily on intestinal absorption of sugars and amino acids, in which the role of Na appears to be best defined. However, there are indications that a fundamentally similar interaction may be involved in the absorption of other essential nutrients [7].

Interactions Between Sodium and Intestinal Transport of Sugars and Amino Acids

The ability of small intestine to transport sugars and amino acids from low concentrations in the mucosal solution to higher concentrations in the serosal solution has been recognized for many years. It is known that these absorptive mechanisms exhibit high degrees of substrate selectivity, including stereospecificity, saturation kinetics, competitive inhibition by structurally similar sugars or amino acids, and, in the case of sugars, specific inhibition by low concentrations of the glycoside phlorizin. These characteristics have suggested that sugar and amino acid transport are mediated by distinct membrane components, possessing enzyme-like properties, located on or near the microvillus border of the epithelial cell. Combination of the sugar or amino acid with these "carrier molecules" is presumed to be prerequisite for passage through the brush border into the absorptive cell. The ability of the intestinal cell to accumulate these solutes, achieving intracellular concentrations that exceed concentrations in the luminal solution, and the dependence of the transport processes on the availability of metabolic energy have naturally prompted the idea that the carrier mechanisms for sugars and amino acids are directly linked to the flow of metabolic reactions [5, 76, 77].

Two independent sets of observations have suggested that direct links between the sugar and amino acid transport mechanisms and metabolic processes are not necessary but that the driving force for the flows of these solutes may be derived from coupling to the flow of Na. First, it has been noted that transepithelial sugar and amino acid transport [8], as well as the accumulation of these solutes within the intestinal cell, is dependent upon the presence of Na in the mucosal solution [6, 59]. The studies of Crane and his associates [6] indicated clearly that Na is required for the *entry* of transported sugars into the epithelial cell, and similar findings have been reported for the case of amino acids [59]. Second, the absorption of sugars and amino acids by small intestine was seen to be accompanied by an increase in the rate of Na absorption [61, 62]. These findings are best illustrated by results obtained using short-circuited segments of isolated rabbit ileum in which the current necessary to abolish the spontaneous electrical potential difference across the tissue is a measure of the rate of net Na transport from mucosa to serosa in the absence of an external driving force [60, 61]. The increase in short-circuit current following the addition of 3-O-methylglucose to the solution bathing the mucosal surface of rabbit ileum and the effect of the subsequent addition of phlorizin are shown in Figure 3-12. Further, the increase in current, as illustrated in Figure 3-13 for the case of L-alanine, is a saturable function of the concentration of sugar or amino acid in the mucosal solution [61, 62]. The fact that nonmetabolized sugars (such as 3-O-

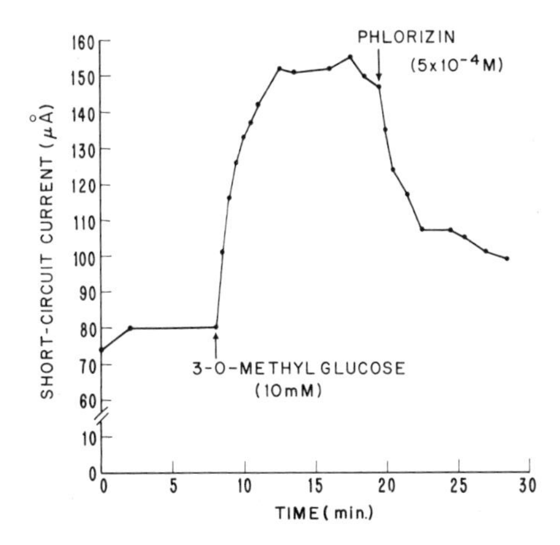

FIG. 3-12. Effect of 3-O-methyl-D-glucose on the short-circuit current across isolated rabbit ileum. (Reproduced from Schultz and Zalusky [61] by permission of the Rockefeller Institute Press.)

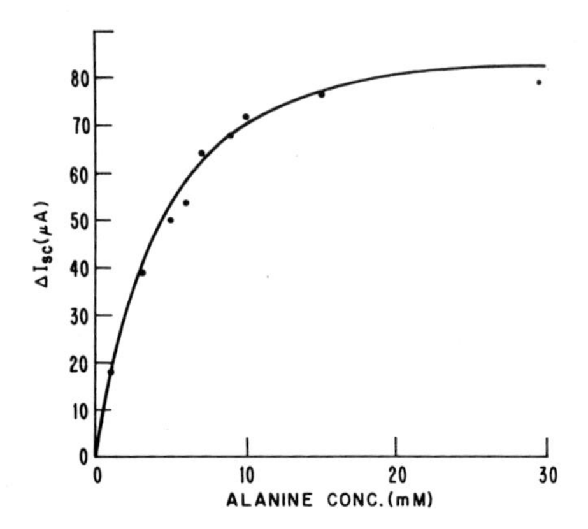

FIG. 3-13. The increase in the short-circuit current (ΔI_{sc}) across isolated rabbit ileum as a function of the concentration of L-alanine in the mucosal bathing solution. (Reproduced from Schultz and Zalusky [62] by permission of *Nature* [London].)

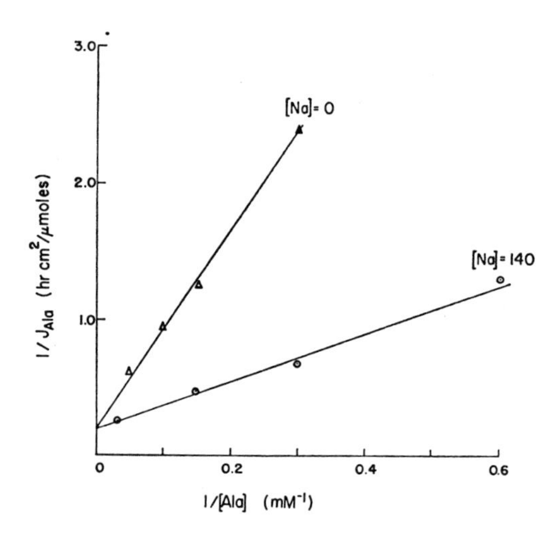

FIG. 3-14. Lineweaver-Burke plots of alanine influx across brush border of isolated rabbit ileum as a function of alanine concentration in the mucosal solution. The linear double-reciprocal plots indicate that alanine influx conforms to Michaelis-Menten kinetics in the absence as well as in the presence of Na. (Reproduced from Curran, Schultz, Chez, and Fuisz [13] by permission of the Rockefeller Institute Press.)

methylglucose) and amino acids increase the rate of net Na transport indicates that the underlying mechanism is dependent only on whether the solute is specifically transported by the small intestine, and is independent of its metabolic fate.

This reciprocal relation (i.e., that the absorption of sugars and amino acids not only requires the presence of Na but also stimulates the absorption of Na) strongly suggested a coupling between Na transport and sugar and/or amino acid transport at the brush border of the intestinal cell [7, 61, 62]. Recent determinations of the simultaneous unidirectional influxes of neutral amino acids (primarily L-alanine) and Na from the mucosal solution across the brush border into the intestinal epithelial cell have provided direct evidence for such coupling [13, 58]. These studies demonstrate that the unidirectional influxes of L-alanine, L-valine, and L-leucine into the cell across the mucosal membrane exhibit saturation kinetics of the classic Michaelis-Menten type. As shown in Figure 3-14, when the Na concentration in the mucosal solution is reduced by replacement with choline, the maximum influx (i.e., that which would be observed in the presence of an infinite amino acid concentration), $\mathbf{J}_A{}^{im}$, is unaffected but the concentration of amino acid necessary to elicit a half-maximal influx, K_t, increases. Further, at each Na concentration studied, there was a linear relation between amino acid and Na influxes such that an in-

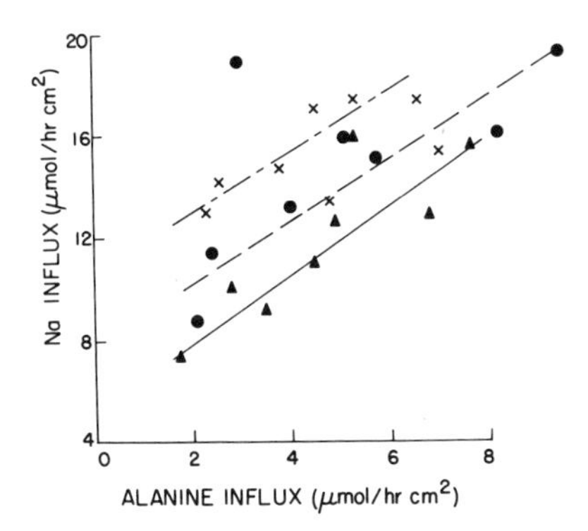

FIG. 3-15. Relation between alanine influx and Na influx across brush border of rabbit ileum. Alanine influx was varied by varying the alanine concentration in the mucosal solution; the Na concentration was maintained constant at 70 mM. The different symbols represent three experiments, each involving eight simultaneous determinations of alanine and Na influxes. (Data from Curran, Schultz, Chez, and Fuisz [13].)

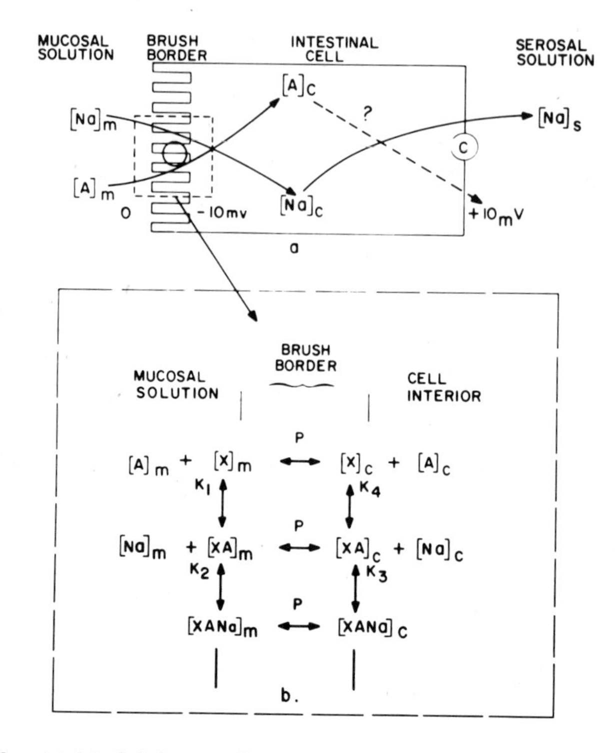

FIG. 3-16. (a) Model for Na-linked amino acid transport across small intestine. The value for the electrical potential difference across the brush border is that reported by Wright [78]; and the transcellular electrical potential difference is that reported by Schultz and Zalusky [60]. *C* represents the ouabain-sensitive Na transport mechanism that is presumed to be directly linked to metabolic reactions and is responsible for extrusion of Na from the cell interior across the serosal and/or lateral membranes into the serosal solution. The dashed line represents the exit process for amino acids, which, as discussed in the text, is poorly understood. (b) Kinetic model of the amino acid-Na interaction at the brush border of small intestine. The subscripts *m* and *c* denote the concentrations at the surfaces of the brush border facing the mucosal medium and cell interior, respectively. *P* is the rate constant for translocation of *X*, *XA*, and *XANa* between the two surfaces of the brush border.

increase in amino acid influx was accompanied by an increase in Na influx (Fig. 3-15).

A kinetic model [13] that satisfies these observations on amino acid transport across intestinal brush border is shown in Figure 3-16b. According to this model, amino acid (A) combines with a component of the brush border (X) to form a binary complex (XA). The reaction is assumed to be governed by the laws of mass action and characterized by a dissociation constant K_1. XA may then cross the membrane, or it may combine with Na to form a ternary complex, $XANa$, which then crosses the membrane; the dissociation constant for the ternary complex is K_2. The rates at which X, XA, and $XANa$ cross the brush border in both directions are assumed to be rate limiting and equal for all three forms. The unidirectional influx of alanine, according to this model, is given by [13]

$$\mathbf{J}_A{}^i = \frac{\mathbf{J}_A{}^{im}[A]_m}{\dfrac{K_1K_2}{K_2 + [Na]_m} + [A]_m} \quad (16)$$

where $[A]_m$ is the alanine concentration in the mucosal solution, $[Na]_m$ is the Na concentration in the mucosal solution, $\mathbf{J}_A{}^{im}$ is the maximum alanine influx, and the experimentally determined values for K_1 and K_2 are 70 mM and 17 mM, respectively. Equation 16 has the form of the Michaelis-Menten equation where K_t, the concentration at which $\mathbf{J}_A{}^i = \mathbf{J}_A{}^{im}/2$, is inversely related to the mucosal Na concentration.

The efflux of alanine out of the cell across the brush border is given by

$$\mathbf{J}_A{}^e = \frac{\mathbf{J}_A{}^{im}[A]_c}{\dfrac{K_3K_4}{K_3 + [Na]_c} + [A]_c}$$

where the subscript c now denotes the intracellular concentrations. If the system is symmetrical—$K_1 = K_4$ and $K_2 = K_3$—the net flux of alanine across the brush border is given by

$$\mathbf{J}_A = \mathbf{J}_A{}^{im}\left[\frac{[A]_m}{\dfrac{K_1K_2}{K_2 + [Na]_m} + [A]_m} - \frac{[A]_c}{\dfrac{K_1K_2}{K_2 + [Na]_c} + [A]_c}\right]$$

Since the intracellular Na concentration is lower than the Na concentration in the mucosal medium [59]—$[Na]_m > [Na]_c$—net flux of alanine into the cell ($\mathbf{J}_A > 0$) will take place even when $[A]_c > [A]_m$; that is, because of the low intracellular Na concentration, alanine can be transported from a region of lower concentration (the mucosal solution) to a region of higher concentration (the cell interior) even though no direct link between the transport process and metabolic energy is included in the model. In kinetic terms, the affinity (or $1/K_t$) of the efflux process for amino acid is much lower than that of the influx process owing to the inverse relation between K_t and Na, so that net transport of amino acid into the cell will persist even when $[A]_c > [A]_m$, providing $[Na]_c < [Na]_m$. Clearly, when $[Na]_c = [Na]_m$ the only asymmetry inherent in this model vanishes so that $\mathbf{J}_A = 0$ when $[A]_c = [A]_m$, and the ability of the cell to concentrate the amino acid is lost. The condition $[Na]_c = [Na]_m$ is realized when the tissue is incubated in the presence of metabolic inhibitors or ouabain, or when the incubation medium is rendered Na free. Under all these conditions maximum experimental values for $[A]_c/[A]_m$ do not differ significantly from unity, and the transport processes resemble carrier-mediated facilitated transfer [59] as predicted by the model [13].

It is of interest to examine the energetics of the model illustrated in Figure 3-16. Since the intracellular Na concentration is lower than the concentration in the mucosal solution, and since the cell interior is electrically negative with respect to the mucosal solution, Na transport across the mucosal border into the cell is energetically a spontaneous process in the direction of the electrochemical potential difference and is associated with a decrease in free energy. In the absence of coupling to other flows this change in free energy is simply dissipated. On the other hand, the coupling of the "downhill" movement of Na to the flow of amino acid permits an increase in the free energy of the amino acid at the expense of the decrease in free energy of Na; the downhill movement of one species thus drives the movement of another species uphill.

We may now ask: Is the electrochemical potential difference for Na across the mucosal membrane adequate to account for the observed uphill transport of alanine? Since the process under consideration is a spontaneous one, it must be accompanied by a decrease in total free energy, G, so that

$$\frac{dG}{dt} = \mathbf{J}'_A \, \Delta\tilde{\mu}_A + \mathbf{J}'_{Na} \, \Delta\tilde{\mu}_{Na} < 0$$

where

$$\Delta\tilde{\mu}_A = RT \ln \left([A]_c/[A]_m\right)$$

$$\Delta\tilde{\mu}_{Na} = RT \ln \left([Na]_c/[Na]_m\right) + \mathbf{F} \, \Delta\Psi$$

and the $\mathbf{J}$'s are the *coupled* net flows of alanine and Na. Because the efficiency of energy conversion for this process is not known, the most we can ascertain is the maximum ratio of intracellular to extracellular alanine concentrations that could result if the efficiency were 100 percent—that is, if *all* the energy available from the downhill movement of Na were converted to the uphill movement of alanine so that dG/dt = 0. Under this idealized condition

$$\left(\frac{[A]_c}{[A]_m}\right)^{J'_A} = \left(\frac{[Na]_m}{[Na]_c}\right)^{J'_{Na}} e^{-J'_{Na}F\Delta\Psi/RT} \qquad (17)$$

When $[Na]_m = 140$ mM, $J'_A = J'_{Na}$ [13] and reasonable values for $[Na]_c$ and $-\Delta\psi$ are 20–50 mM and 10–20 mv, respectively. Substituting these values in Equation 17, we obtain maximum ratios of intracellular to extracellular alanine concentrations that range from 4 to 15. These values are of the same order of magnitude as the experimentally observed ratios of 8–10 [59] and suggest that the proposed model is energetically feasible.

We may summarize our present view of the mechanisms of amino acid (and sugar absorption) with the aid of Figure 3-16a. Amino acid and Na appear to interact at the brush border with a common carrier mechanism that results in the coupled transfer of these solutes into the absorptive cell. By virtue of the coupling, part of the free energy made available by the flow of Na down its electrochemical potential difference is expended in driving the flow of amino acid from a lower concentration in the mucosal solution to a higher concentration in the cell interior. The mechanism by which the amino acid leaves the epithelial cell across the serosal membranes is poorly understood and warrants further direct investigation. Since the intracellular amino acid concentration exceeds the concentration in the serosal solution, the exit process may be attributable to either simple diffusion or, more probably, carrier-mediated facilitated transfer; at present there is no compelling reason to believe that forces other than the concentration difference of the amino acid across the serosal membrane are required to account for this movement. The steady-state electrochemical potential difference for Na across the mucosal membrane is maintained by means of a transport mechanism, located on or near the lateral and/or serosal membranes, that extrudes Na from the cell. Since the extrusion involves net movement of Na against an electrochemical potential difference, it is this process that is presumed to be directly linked to energy-yielding metabolic reactions and subject to the inhibitory action of ouabain. Thus, although the direct driving force for accumulation of amino acid within the cell is the thermodynamic potential difference for Na across the mucosal border, the ultimate source of energy is, of course, cellular metabolism. The Na extrusion mechanism and the resulting low intracellular Na concentration essentially serve as transducers of metabolic energy; the electrochemical potential difference for Na across the brush border is a potential source of energy (an "electrochemical battery") for the transport of any substance whose flow is coupled to the influx of Na. Present evidence suggests that such coupling may provide the driving force for intestinal absorption of sugars and nucleosides [7, 8], and mechanisms similar to the one described for neutral amino acid transport may be involved.

Finally, it is of interest to consider the possible implications of the kinetic model for amino acid transport with regard to molecular mechanism. In essence, according to this model, the effect of Na is to markedly increase the stability of the alanine-carrier complex. This action is reflected in the effect of Na on the overall K_t for alanine influx. When $[Na]_m = 140$ mM the carrier is half-saturated with alanine when the amino acid concentration is 7 mM, whereas in the absence of Na a tenfold greater alanine concentration is required to achieve the same degree of saturation. Also, formation of the amino acid–carrier complex appears to be prerequisite for the binding of Na. This phenomenon is characteristic of a group of enzyme-substrate interactions, referred to as allosteric transitions, in which the affinity of an enzyme for substrate is regulated by other, specific molecules differing chemically and structurally from the substrate. In accordance with the model for allosteric transitions proposed by Monod et al. [44, 45], the interaction between the amino acid carrier and Na may result in conformational changes that increase the affinity of the carrier for amino acids, and binding of amino acid may result in alterations that make possible complex formation with Na (so-called heterotrophic cooperative behavior). The possibilities that allosteric transitions are involved in carrier-mediated transport processes [30], and that they form the molecular basis for the stimulatory action of Na in Na-dependent transport processes [7], have been raised, but further investigation is certainly necessary to test these intriguing speculations.

Nonionic Diffusion and Ion Pair Formation

A type of solute-solute interaction that can lead to transport against a difference in electrochemical potential and that resembles the amino acid transport model is one in which salts cross the membrane in the form of neutral, undissociated ion pairs. This transport mechanism, generally referred to as nonionic diffusion, has been demonstrated in biological tissues and is believed to be involved in the absorption and secretion of weak electrolytes including many drugs [28, 43, 54]. Consider, for example, a poorly dissociated sodium salt, NaA, in equilibrium with its dissociation products in aqueous solution NaA $\rightleftharpoons$ Na + A^- with a dissociation constant, K_A, given by

$$K_A = \frac{[Na][A^-]}{[NaA]} \tag{18}$$

Assume further that the membrane is impermeable to A^-, but that the neutral NaA can permeate the membrane by simple diffusion. Thus, the net flow of A^- across the membrane is determined by the net flow of NaA

$$\mathbf{J}_{A^-} = \mathbf{J}_{NaA} = P([NaA]_m - [NaA]_c) \tag{19}$$

where the subscripts *m* and *c* distinguish between the concentrations of NaA in the extracellular and intracellular compartments, respectively, and P is the permeability coefficient. Combining Equations 18 and 19 we obtain

$$\mathbf{J}_{A^-} = \frac{P}{K_A}([Na]_m[A^-]_m - [Na]_c[A^-]_c)$$

Thus, when $[Na]_m > [Na]_c$ flow of A^- into the cell can take place even though $[A^-]_c > [A^-]_m$. Net flow of A^- into the cell will cease ($\mathbf{J}_{A^-} = 0$) when

$$\frac{[A^-]_c}{[A^-]_m} = \frac{[Na]_m}{[Na]_c}$$

so that in the steady state $[A^-]_c > [A^-]_m$ as long as $[Na]_m > [Na]_c$. If one's methods were sensitive only to the ionized form A^-, the transport of A would be characterized as follows: (1) A^- is transported into the cell against a concentration difference and, if the cell interior is electrically negative with respect to the surrounding medium, against an electrical potential difference as well. (2) Entry of A^- into the cell is dependent upon extracellular Na. (3) Transport against an electrochemical potential difference is abolished by conditions that abolish the difference in Na concentrations across the membrane, i.e., metabolic inhibitors, ouabain, etc. Indeed, the primary distinction between this mechanism and the one described for amino acids and sugars is that the influx of A^- does not display saturation kinetics but is a linear function of A^-; i.e.,

$$\mathbf{J}_{A^-}{}^i = \frac{P[Na]_m}{K_A} [A^-]_m \tag{20}$$

If the analytical methods used are sensitive to the total concentration of an anion, including the dissociated and undissociated form, $A_t = A^- + NaA$, Equation 20 becomes

$$\mathbf{J}_{A^-}{}^i = P\left[\frac{[A_t]_m[Na]_m}{K_A + [Na]_m}\right]$$

Thus, at constant Na, $\mathbf{J}_{A^-}{}^i$ remains a linear function of A_t. However, when A_t is maintained constant, $\mathbf{J}_{A^-}{}^i$ is a saturable function of $[Na]_m$ exhibiting classic Michaelis-Menten form. Further, the steady-state distribution of A_t is given by

$$\frac{[A_t]_c}{[A_t]_m} = \frac{[Na]_m(K_A + [Na]_c)}{[Na]_c(K_A + [Na]_m)}$$

so that one would still find $[A_t]_c > [A_t]_m$ providing $[Na]_c < [Na]_m$.

Thus, transport by nonionic diffusion or ion-pair formation is a type of solute-solute interaction that can bring about flow against an electrochemical potential difference in the absence of a direct link to metabolic reactions. Nonionic diffusion, moreover, does not require special membrane components or carriers; permeation is dependent only on the ability of the nondissociated ion pair to enter and diffuse through the membrane phase. Clearly, the driving force for the uphill movement of one member of the ion pair is derived from stoichiometric coupling to the downhill movement of the other member. In the example used, the immediate driving force for the flow of A^- is dependent upon the steady-state asymmetrical distribution of Na, which, in turn, may be directly dependent upon metabolic processes. Needless to say, the picture can become even more complex if the asymmetrical distribution of the counterion arises from coupling to flows of other solutes rather than from direct coupling to metabolic reactions.

The extent to which nonionic diffusion is involved in the transport of physiologically important solutes is not completely defined.[9] Clearly, it may be the only mode of transport of some solutes, or it may supplement other transport mechanisms and so present a rather complex

[9] In this discussion we have considered nonionic diffusion across a single membrane separating intracellular and extracellular compartments. Clearly, the same arguments can be applied to transport from one bathing solution to another across a multicellular epithelial membrane.

overall picture. In view of the lipoprotein structure and low dielectric properties of biological membranes, nonionic diffusion would be expected to be involved in the transport of lipid-soluble weak acids and bases and could mimic carrier-mediated transport processes particularly if there are asymmetrical distributions of H and Na concentrations.

ACTIVE AND PASSIVE TRANSPORT: A PROBLEM OF DEFINITIONS

It is customary in most discussions of biological transport processes to classify flows in either "active" or "passive" categories. Historically, the purpose of classification was to distinguish between flows that can be attributed entirely to physical forces, such as diffusion and osmosis, and flows that are dependent upon "vital forces" or *Triebkraft* peculiar to viable, biological tissues. In this chapter we have refrained from using the terms *active* and *passive* because they have not been defined in a generally acceptable manner. One can adequately describe transport processes without resorting to such classification, and, in the absence of an unambiguous definition, the use of these labels is a potential source of confusion.

Today the most widely accepted definition of active transport is that proposed by Rosenberg in 1948 [55]. An active transport process, he suggested, is one that results in the net movement of a substance from a region of lower electrochemical potential to a region of higher electrochemical potential. The ease with which this criterion can be unambiguously applied, particularly with respect to transport across epithelial membranes, is certainly one of the reasons for its wide acceptance. On the other hand, because the Rosenberg definition is based solely on classic thermodynamic principles and is concerned only with the initial and final states of the transported substance, it includes in the active transport category many flows that are not directly dependent upon metabolic processes. For example, as we have seen, solvent drag and nonionic diffusion can result in transport against an electrochemical potential difference. These transport processes would be classified as active by the Rosenberg definition even though they can be attributed to simple physical forces and are not directly linked to metabolic reactions. To be sure, the ultimate source of energy for all processes that result in transport against an electrochemical potential difference is cellular metabolism. However, the immediate driving force for these flows may be many steps removed from metabolic processes. Because the Rosenberg definition is concerned not with mechanism but only with initial and final states, it does not distinguish between processes that are directly linked to metabolic energy and those that are not. Thus it can shed no light on the possible molecular or anatomic bases of the flow.

Kedem [33] has recently proposed that the single criterion of an active transport process should be the presence of a direct link between the flow of matter and the flow of metabolic reactions. With reference to Equation 7, a substance i is said to be actively transported if $\mathbf{J}_{ir} \neq 0$. This is a radical departure from the Rosenberg definition because it excludes all flows dependent only on coupling with flows of other substances, and it may include flows that, under the conditions of study, take place in the direction of their electrochemical potential difference; the only requirement is a contribution to that flow arising from a direct link to chemical reactions. A comparison of the definitions of active transport proposed by Rosenberg and Kedem is given in Equation 21.

$$\mathbf{J}_i = \overbrace{-\frac{\Delta\tilde{\mu}_i}{\mathrm{R}_{ii}}}^{\text{Rosenberg: passive}} \overbrace{\underbrace{- \sum_{\substack{j=1 \\ j\neq i}}^{n} \mathrm{R}_{ij}\, \mathbf{J}_j/\mathrm{R}_{ii}}_{\text{Kedem: passive}} \underbrace{- \frac{\mathrm{R}_{ir}}{\mathrm{R}_{ii}}\, \mathbf{J}_r}_{\text{active}}}^{\text{active}} \qquad (21)$$

Clearly, water absorption, uphill transport of urea due to solvent drag, alanine transport, etc., would all be considered active processes by the Rosenberg criterion without equivocation. On the other hand, *if the models we have presented for these transport processes are correct,* they would be classified as passive by the Kedem criterion; of all the processes discussed, only Na transport would be classified as active. Failure to distinguish between these two quite different usages of the term *active transport* can obviously lead to unnecessary confusion and disagreement.

Although, in the opinion of this author, the Kedem definition corresponds more closely to the intuitive concept of "vital forces," *Triebkraft,* or "active transport," its major shortcoming is that it is difficult to apply in practice. It requires an unequivocal demonstration of the presence or absence of a direct link to a defined chemical reaction, and no such evidence is at present available for any transepithelial transport process. In contrast, the Rosenberg criterion is readily applicable, and a determination of whether the flow of a given substance takes place against an electrochemical potential difference is often the first step in the experimental study of the transport of any substance. If the transport of this substance is found to be active by the Rosenberg criterion, it simply means that further study is required to identify the responsible driving forces. In this sense the term merely defines the extent of our understanding of a given flow and serves to designate those processes requiring further investigation. When the forces responsible for a given flow are fully defined, the use of labels such as *active* or *passive* is unnecessary except as a descriptive shorthand.

The problem of rigorously defining the term *active transport* has been recently discussed in detail [12] and will not be further elaborated upon here. The purpose of the present discussion was simply to point out the absence of a generally accepted, unambiguous definition and to stress that two considerably different definitions are in current use. The choice between these definitions is arbitrary and to a large extent a matter of personal preference. Clearly, if one chooses to use the term *active transport,* the sense in which it is employed should be stated explicitly so that unnecessary confusion and apparent disagreement can be avoided.

RECAPITULATION: BIOENERGETIC IMPLICATIONS OF COUPLED TRANSPORT PROCESSES

In this chapter we have examined several general mechanisms that appear to be involved in the transport of specific solutes and water across epithelial tissues. No attempt has been made to be exhaustive; instead, we have focused our attention on processes that are sufficiently well defined for an analysis in terms of flows and driving forces. For example, the role of pinocytosis in transepithelial transport and the important area dealing with hormonal regulation of absorptive and secretory processes have been omitted either because they are not readily subject to this type of analysis or because they do not strictly fall within the scope of a discussion of general mechanisms.

One of the main points emerging from the thermodynamic analysis of biological transport phenomena is that many absorptive processes are not linked directly to metabolic reactions but appear, rather, to be dependent upon interactions with the flow of other solutes, particularly Na. The types of interactions involved are varied. Thus, solute-linked water flow is apparently dependent upon specific ultrastructural features of epithelial tissues. Solvent drag presumably presupposes common pathways or channels for the flows of solvent and solute. In contrast with these physical interactions, the coupling between Na and the transport of sugars and amino acids seems to be mediated by chemical interactions with carrier-like components of the brush border. All these interactions can result in transport against an electrochemical potential difference with the flow of Na serving, ultimately, as a transducer of metabolic energy.

Finally, it would be remiss not to consider briefly the implications of coupled transport processes as far as the energetics of absorptive processes are concerned. As discussed previously, the coupling between the flow of Na and a metabolic reaction means only that part of the energy released by the exergonic reaction is expended in increasing the free energy of Na by directing its flow from a region of lower electrochemical potential to a region of higher electrochemical potential. In the absence of other interactions the difference between the energy released by the flow of the chemical reaction and the energy required for the flow of Na against its electrochemical potential difference is simply dissipated in the form of internal entropy production on the part of the system. The dissipated energy could represent a major fraction of free energy change of the chemical reaction, particularly in tissues (e.g., small intestine) in which differences in the electrochemical potential for Na are small compared to the free energy of hydrolysis of ATP. However, the amount of useful work obtained per mole of Na transported may be significantly increased through the action of coupling mechanisms that enable the flow of Na to provide the driving force for the transport of other solutes and solvent against respective electrochemical potential differences. Under these conditions less of the free energy change of the chemical reaction is dissipated, and the efficiency of energy conversion by the tissue is increased. This situation is obviously of distinct biological advantage in a tissue which, like the small intestine, is responsible for transporting a large number of solutes and a large volume of solvent against relatively small thermodynamic potential differences. Without the types of coupling we have described each transport process would have to be individually linked to exergonic chemical reactions, and the efficiency with which chemical energy is converted into osmotic work (i.e., the work of transport) would almost certainly decrease sharply.

In the words of Professor Aharon Katchalsky [31], "Life is a constant struggle against the tendency to produce entropy by irreversible processes." It is intriguing to speculate that the evolution of coupled biological transport processes was guided by these principles of energy conversion.

REFERENCES

1. Andersen, B., and Ussing, H. H. Solvent-drag on non-electrolytes during osmotic flow through isolated toad skin and its response to antidiuretic hormone. *Acta Physiol. Scand.* 39:228, 1957.
2. Caplan, S. R., and Mikulecky, D. C. Transport Processes in Membranes. In Marinsky, J. A. (Ed.), *Ion Exchange: A Series of Advances.* New York: Dekker, 1966. Vol. 1.
3. Clarkson, T. W., and Rothstein, A. Transport of monovalent cations by the isolated small intestine of the rat. *Amer. J. Physiol.* 199:898, 1960.
4. Clarkson, T. W., and Toole, S. R. Measurement of short-circuit current and ion transport across the ileum. *Amer. J. Physiol.* 206:658, 1964.
5. Crane, R. K. Intestinal absorption of sugars. *Physiol. Rev.* 40:789, 1960.
6. Crane, R. K. Hypothesis for mechanism of intestinal active transport of sugars. *Fed. Proc.* 21:891, 1962.
7. Crane, R. K. Na-dependent transport in the intestine and other animal tissues. *Fed. Proc.* 24:1000, 1965.
8. Csáky, T. Z. Significance of sodium ions in active intestinal transport of nonelectrolytes. *Amer. J. Physiol.* 201:999, 1961.
9. Curran, P. F. Na, Cl, and water transport by rat ileum *in vitro. J. Gen. Physiol.* 43:1137, 1960.
10. Curran, P. F. Ion transport in intestine and its coupling to other transport processes. *Fed. Proc.* 24:993, 1965.
11. Curran, P. F., and McIntosh, J. R. A model system for biological water transport. *Nature* (London) 193:347, 1962.
12. Curran, P. F., and Schultz, S. G. Transport Across Membranes: General Principles. In *Handbook of Physiology. Section 6: Alimentary Canal.* Washington: American Physiological Society, 1968. Vol. 3.
13. Curran, P. F., Schultz, S. G., Chez, R. A., and Fuisz, R. E. Kinetic relations of the Na–amino acid interaction at the mucosal border of intestine. *J. Gen. Physiol.* 50:1261, 1967.
14. Curran, P. F., and Schwartz, G. F. Na, Cl, and water transport by rat colon. *J. Gen. Physiol.* 43:555, 1960.
15. Curran, P. F., and Solomon, A. K. Ion and water fluxes in the ileum of rats. *J. Gen. Physiol.* 41:143, 1957.
16. Diamond, J. M. The reabsorptive function of the gall bladder. *J. Physiol.* (London) 161:442, 1962.
17. Diamond, J. M. The mechanism of water transport by the gall bladder. *J. Physiol.* (London) 161:503, 1962.
18. Diamond, J. M. Transport of salt and water in rabbit and guinea pig gall bladder. *J. Gen. Physiol.* 48:1, 1964.

19. Diamond, J. M. The mechanism of isotonic water transport. *J. Gen. Physiol.* 48:15, 1964.
20. Diamond, J. M., and Bossert, W. H. Standing-gradient osmotic flow: A mechanism for solute-linked water transport in epithelia. *J. Gen. Physiol.* 50:2061, 1967.
21. Dietschy, J. M. Recent developments in solute and water transport across the gall bladder epithelium. *Gastroenterology* 50:692, 1966.
22. Fisher, R. B. The absorption of water and of some small solute molecules from the isolated small intestine of the rat. *J. Physiol.* (London) 130:655, 1955.
23. Giebisch, G., and Windhager, E. E. Renal tubular transfer of sodium, chloride and potassium. *Amer. J. Med.* 36:643, 1964.
24. Green, K., Seshadri, B., and Matty, A. J. Independence of transfer of solute and solvent across the rat ileum. *Nature* (London) 196:1322, 1962.
25. Grim, E. A mechanism for absorption of sodium chloride solutions from the canine gall bladder. *Amer. J. Physiol.* 205:247, 1963.
26. Hakim, A. A., and Lifson, N. Urea transport across dog intestinal mucosa *in vitro*. *Amer. J. Physiol.* 206:1315, 1964.
27. Hodgkin, A. L., and Keynes, R. D. The potassium permeability of a giant nerve fibre. *J. Physiol.* (London) 128:61, 1955.
28. Hogben, C. A. M., Tocco, D. J., Brodie, B. B., and Schanker, L. S. On the mechanism of intestinal absorption of drugs. *J. Pharmacol. Exp. Ther.* 125:275, 1959.
29. Ingraham, R. C., Peters, H. C., and Visscher, M. B. On the movement of materials across living membranes against concentration gradients. *J. Phys. Chem.* 42:141, 1938.
30. Jardetzky, O. Simple allosteric model for membrane pumps. *Science* 211:969, 1966.
31. Katchalsky, A. Non-equilibrium thermodynamics. *Intt. Rev. Sci. Technol.* (Quoted from Lehninger, A. L., *Bioenergetics*. New York: Benjamin, 1965, chap. 12.)
32. Katchalsky, A., and Curran, P. F. *Nonequilibrium Thermodynamics in Biophysics*. Cambridge, Mass.: Harvard University Press, 1965.
33. Kedem, O. Criteria of Active Transport. In Kleinzeller, A., and Kotyk, A. (Eds.), *Membrane Transport and Metabolism*. Prague: Czechoslovak Academy of Science, 1961. P. 87.
34. Kedem, O., and Essig, A. Isotope flows and flux ratios in biological membranes. *J. Gen. Physiol.* 48:1047, 1965.
35. Kedem, O., and Katchalsky, A. Thermodynamic analysis of the permeability of biological membranes to non-electrolytes. *Biochim. Biophys. Acta* 27:229, 1958.
36. Kedem, O., and Katchalsky, A. A physical interpretation of the phenomenological coefficients of membrane permeability. *J. Gen. Physiol.* 45:143, 1961.
37. Kedem, O., and Katchalsky, A. Permeability of composite membranes. *Trans. Faraday Soc.* 59:1918, 1963.
38. Lakshminarayanaiah, N. Transport phenomena in artificial membranes. *Chem. Rev.* 65:491, 1965.

39. Leaf, A., and Hays, R. M. Permeability of the isolated toad bladder to solutes and its modification by vasopressin. *J. Gen. Physiol.* 45:921, 1962.
40. Levi, H., and Ussing, H. H. The exchange of sodium and chloride ions across the fibre membrane of isolated frog sartorius. *Acta Physiol. Scand.* 16:232, 1948.
41. Lifson, N., and Hakim, A. A. Simple diffusive-convective model for intestinal absorption of a non-electrolyte (urea). *Amer. J. Physiol.* 211:1137, 1966.
42. Lindemann, B., and Solomon, A. K. Permeability of luminal surface of intestinal mucosal cells. *J. Gen. Physiol.* 45:801, 1962.
43. Milne, M. D., Scribner, B. H., and Crawford, M. A. Non-ionic diffusion and the excretion of weak acids and bases. *Amer. J. Med.* 24:709, 1958.
44. Monod, J., Changeux, J., and Jacob, F. Allosteric proteins and cellular control systems. *J. Molec. Biol.* 6:306, 1963.
45. Monod, J., Wyman, J., and Changeux, J. On the nature of allosteric transitions: A plausible model. *J. Molec. Biol.* 12:88, 1965.
46. Ogilvie, J. T., McIntosh, J. R., and Curran, P. F. Volume flow in a series-membrane system. *Biochim. Biophys. Acta* 66:441, 1963.
47. Oxender, D. L., and Christensen, H. N. Transcellular concentration as a consequence of intracellular accumulation. *J. Biol. Chem.* 234:2321, 1959.
48. Parsons, D. S., and Prichard, J. S. Properties of some model systems for transcellular active transport. *Biochim. Biophys. Acta* 126:471, 1966.
49. Parsons, D. S., and Wingate, D. L. The effect of osmotic gradients on fluid transfer across rat intestine *in vitro. Biochim. Biophys. Acta* 46:170, 1961.
50. Patlak, C. S., Goldstein, D. A., and Hoffman, J. F. The flow of solute and solvent across a two-membrane system. *J. Theor. Biol.* 5:426, 1963.
51. Pidot, A. L., and Diamond, J. M. Streaming potentials in a biological membrane. *Nature* (London) 201:701, 1964.
52. Reid, E. W. Preliminary report on experiments upon intestinal absorption without osmosis. *Brit. Med. J.*, p. 1133, 1892.
53. Riklis, E., and Quastel, J. H. Effects of cations on sugar absorption by isolated surviving guinea pig intestine. *Canad. J. Biochem. Physiol.* 36:347, 1958.
54. Rosen, H., Leaf, A., and Schwartz, W. B. Diffusion of weak acids across the toad bladder. Influence of pH on non-ionic permeability coefficients. *J. Gen. Physiol.* 48:379, 1964.
55. Rosenberg, T. On accumulation and active transport in biological systems: I. Thermodynamic considerations. *Acta Chem. Scand.* 2:14, 1948.
56. Schlögl, R. *Stofftransport durch Membranen.* Darmstadt: Steinkopff, 1964.
57. Schultz, S. G., and Curran, P. F. Intestinal Absorption of Sodium Chloride and Water. In *Handbook of Physiology. Section 6: Alimentary Canal.* Washington: American Physiological Society, 1968. Vol. 3.
58. Schultz, S. G., Curran, P. F., Chez, R. A., and Fuisz, R. E. Alanine and sodium fluxes across mucosal border of rabbit ileum. *J. Gen. Physiol.* 50:1241, 1967.

59. Schultz, S. G., Fuisz, R. E., and Curran, P. F. Amino acid and sugar transport in rabbit ileum. *J. Gen. Physiol.* 49:849, 1966.
60. Schultz, S. G., and Zalusky, R. Ion transport in isolated rabbit ileum: I. Short-circuit current and Na fluxes. *J. Gen. Physiol.* 47:567, 1964.
61. Schultz, S. G., and Zalusky, R. Ion transport in isolated rabbit ileum: II. The interaction between active sodium and active sugar transport. *J. Gen. Physiol.* 47:1043, 1964.
62. Schultz, S. G., and Zalusky, R. Interactions between active sodium transport and active amino acid transport in isolated rabbit ileum. *Nature* (London) 205:292, 1965.
63. Schultz, S. G., Zalusky, R., and Gass, A. E., Jr. Ion transport in isolated rabbit ileum: III. Chloride fluxes. *J. Gen. Physiol.* 48:375, 1964.
64. Sharp, G. W. G., and Leaf, A. Mechanism of action of aldosterone. *Physiol. Rev.* 46:593, 1966.
65. Skou, J. C. Enzymatic Aspects of Active Linked Transport of Na and K Through the Cell Membrane. In Butler, J. A. V., and Huxley, H. E. (Eds.), *Progress in Biophysics and Molecular Biology*. New York: Macmillan, 1964. Vol. 14.
66. Skou, J. C. Enzymatic basis for active transport of Na and K across cell membrane. *Physiol. Rev.* 45:596, 1965.
67. Smyth, D. H., and Wright, E. M. Streaming potentials in the rat small intestine. *J. Physiol.* (London) 182:591, 1966.
68. Staverman, A. J. The theory of measurement of osmotic pressure. *Rec. Trav. Chim.* 70:344, 1951.
69. Tormey, J. M., and Diamond, J. M. The ultrastructural route of fluid transport in rabbit gall-bladder. *J. Gen. Physiol.* 50:2032, 1967.
70. Trier, J. S., and Rubin, C. E. Electron microscopy of the small intestine: A review. *Gastroenterology* 49:574, 1965.
71. Ussing, H. H. The distinction by means of tracers between active transport and diffusion. *Acta Physiol. Scand.* 19:43, 1949.
72. Ussing, H. H. Transport of Electrolytes and Water Across Epithelia. In *The Harvey Lectures,* Series 59. New York: Academic, 1965. P. 1.
73. Ussing, H. H., Kruhoffer, P., Thaysen, J. H., and Thorn, N. A. *The Alkali Metals in Biology*. Berlin: Springer, 1960.
74. Ussing, H. H., and Zerahn, K. Active transport of sodium as the source of electric current in the short-circuited isolated frog skin. *Acta Physiol. Scand.* 23:110, 1951.
75. Vaughan, B. E. Intestinal electrolyte absorption by parallel determination of unidirectional sodium and water transfers. *Amer. J. Physiol.* 198: 1235, 1960.
76. Wilbrandt, W., and Rosenberg, T. The concept of carrier transport and its corollaries in pharmacology. *Pharmacol. Rev.* 13:109, 1961.
77. Wilson, T. H. *Intestinal Absorption.* Philadelphia: Saunders, 1962.
78. Wright, E. M. The origin of the glucose dependent increase in the potential difference across the tortoise small intestine. *J. Physiol.* (London) 185:486, 1966.

4
Biochemical and Metabolic Aspects of Transport

Robert P. Davis

GENERAL CONSIDERATIONS

All cells engage in a variety of physiological and biochemical processes which result in relative homeostasis of the internal cellular environment. This constant dynamic state of intermediary metabolism is essential to proper cell function. The intracellular phase represents but one compartment in the open thermodynamic system characterizing most living organisms. The physiological efficiency of this system and the regulation of its metabolic work are highly dependent on the accession and the removal of various substrates, intermediates, and products from the intracellular compartment. The mechanisms by which such regulation takes place include, particularly, the processes of cellular absorption and excretion. These processes represent aspects of the general phenomenon of transport.

Processes of cellular transport are carried on at three levels of organization:

1. Primary order of transport: the translocation of a substance across a single integrated biological membrane.
2. Secondary order of transport: transcellular transport.

Supported in part by grants AM-04004, AM-04699, and HD-00674 from the National Institutes of Health and by a grant from the Life Insurance Medical Research Fund. The author was formerly a Career Scientist of the Health Research Council of the City of New York (I-252).

3. Tertiary order of transport: transport resulting from integrated organ function.

Varieties of Transport

Primary transport, whereby a substance moves across a biological membrane, generally involves a single mechanism for the transport of the substance. This process is likely to be characterized in terms of (1) whether the net movement is along or against an electrochemical potential gradient and (2) whether the process requires the expenditure of cellular energy. The two commonest forms of primary transport are diffusion and active transport. In the case of diffusion the substance moves along its electrochemical potential gradient with no energy cost to the cell. In active transport the substance moves against its electrochemical potential gradient at the expense of cellular energy. The thermodynamic criteria for distinguishing between diffusion and other forms of transport have been presented by Ussing [179] and are discussed below.

Other processes of primary transport encountered in biological systems are "downhill active transport" and facilitated diffusion. In downhill active transport a substance moves down its electrochemical potential gradient *at the expense of cellular energy*. In facilitated diffusion the net thermodynamic result is equivalent to the process of diffusion, with the substance moving in a direction determined by the electrochemical potential gradient. However, the kinetic description of this transport process implies the presence of a mediating carrier for the substance, as in the case of glucose transport into erythrocytes [118]. The unitary processes of primary transport are subject to subtle variations, but they share the gross biochemical, kinetic, and thermodynamic properties described above.

In addition to these unitary molecular processes for the movement of substances *through* biological membranes, there are further mechanisms for the translocation of substances *across* membranes. The latter include the translocation of a solution or suspension enveloped in a membrane similar in composition to that of the plasma membrane through which the substance is moving. This is the process of vesicular transport. In vesicular secretion a vesicle in contact with a major cellular membrane becomes, through conformational changes of the membrane, contiguous, then continuous with the membrane, subsequently discharging its contents to the other side of the membrane. Vesicular absorption is the reverse: A vesicle is formed by a fold of membrane enveloping a bulk of solution; the vesicle is pinched off through a conformational change of the membrane; the vesicle moves to the side of the membrane opposite to where the vesicular contents originated. Our understanding of these processes is limited to electron microscopic description of the phenomena.

Secondary transport, or the translocation of a substance across cells, generally involves more than one unitary process operating in sequence. Depending upon the polarity of the cell and the structure of the organ of which the cell is a part, this process may be referred to as either absorption or excretion.

Tertiary transport may involve a multiplicity of unitary mechanisms as well as several forms of secondary transport. In the kidney, tertiary transport leads to modification of a glomerular filtrate and either to net reabsorption of a substance or to net secretion. The net *excretion* of a substance by a complex organ composed of several cell types represents the summation of sequential processes of secondary transport. Special modifications of transport processes may be introduced: in the case of the kidney, for example, by the countercurrent system whereby the kidney concentrates the urine.

There has been great interest in primary and secondary transport occasioned by several recent technological developments. In particular, these developments include availability of isotopic tracers for the measurement of bidirectional fluxes, electrophysiological methods for the measurement of cell potentials and net rates of ion transport, and electron microscopic techniques for the study of cell structure. Since the utilization of these techniques has provided much new information, especially on the processes of primary and secondary transport, the latter will occupy much of our attention in this discussion.

Furthermore, the transport of the sodium ion has been studied in detail and appears to be fundamental to the biology of the cell. Sodium transport is closely integrated with processes affecting the net balance of energy in the cell and is coupled to processes whereby many other substances are transported. We shall therefore devote most of our discussion to cellular mechanisms utilized in the primary and secondary orders of transport of sodium. Several excellent and detailed reviews of active transport have been published recently [3, 85, 131]. We shall focus on only a few of the important problems in our understanding of the molecular mechanisms of transport and the integrative aspects of transport and cellular function.

CLASSIC THERMODYNAMIC CRITERIA FOR ACTIVE TRANSPORT

According to the criteria of Ussing [179], for simple net passive diffusion in the absence of a net flow of solvent, the ratio of fluxes across a membrane will be

$$\frac{M_{in}}{M_{out}} = \frac{C_o}{C_i} e^{\frac{zFE}{RT}}$$

where M_{in} = inward flux
M_{out} = outward flux
C_o = concentration of the substance in the outside solution
C_i = its concentration in the inside solution
E = potential difference between the outside and the inside solution
z = charge borne by the substance
F = Faraday constant
R = gas constant
T = absolute temperature

Under circumstances in which the resting potential of the cell is zero, *or* under short-circuited conditions, *or* when the transported species is uncharged and therefore uninfluenced by electrical field, the flux equation reduces to

$$\frac{M_{in}}{M_{out}} = \frac{C_o}{C_i}$$

as required by simple laws of diffusion for an uncharged substance in the absence of solvent drag.

When this flux equation holds for transport through any membrane, the process of transport is generally diffusion in accord with the simplest of physical forces. The transport is, moreover, uninfluenced by interactions between solute and solvent or by chemical reactions in the membrane. The forces then operating upon the transport process include the diffusion force and the electrical field. When there is bulk flow of solution through pores in the membrane, friction is exerted upon the solutes. This third force is termed *solvent drag* [7].

Active transport is most clearly demonstrated when the transport overcomes the sum of the forces of diffusion, electrical field, and solvent drag [180]. The flux ratio, calculated according to the above formula, will then differ from the observed ratio of fluxes. However, not every case of inapplicability of the flux equation of Ussing should be considered active transport. By itself, such a case implies that, in the transport process, the transported substance takes part in chemical reactions in the membrane phase. Active transport may be best characterized as that form of transport not obeying the flux equation of Ussing *and* requiring the expenditure of cellular energy.

We should note that in applying the Ussing equation to a transport process it is necessary to know the activities rather than the simple molar concentration of the transported species on each side of the membrane. In many physiological circumstances it is impossible to have these data precisely. Some apparent failures of the equation may in fact be due to uncertainty concerning the activity coefficients of the solute species [123]. The covalent or coulombic binding of solutes may result in profound differences between the activity and the measured molar concentration of a solute, including simple ions like sodium [125].

More general analysis of the thermodynamic aspects of transport is presented elsewhere in this volume (Chap. 3), utilizing principles derived from the thermodynamics of irreversible processes. The forces operating upon a transport process, including the cross coefficients representing the interactions between these forces, are explicitly considered, as are the assumptions implicit in the classic and irreversible thermodynamic equations of transport.

THE PARTICIPATION OF CELL MEMBRANES IN ACTIVE TRANSPORT

The Structure of Cell Membranes in Relation to Transport

From kinetic studies it is clear for many substances that the cell membrane must participate in one or another stage of the transport process. The structure of cell membranes imposes restrictions on possible mechanisms for this interaction. A model for the structure of the plasma membrane has been proposed by Davson and Danielli [54]. Electron microscopic and chemical evidence favors a chemically and structurally homogeneous lipoprotein sandwich as making up the unit membrane. This model is not supported by recent infrared spectroscopic studies of erythrocyte membranes (Chap. 1). Moreover, there is a striking heterogeneity of specific chemical functions evident in the diversity of transport functions ascribable to the cell membrane. Undoubtedly a variety of specific proteins with identifiable transport functions will be found. For sodium transport, at least one of these potential mediators has been described in some detail already. The chemical makeup of the plasma membrane requires that lipid solubility be a primary characteristic of the molecular transporters inasmuch as most biological substances undergoing transport have little lipid solubility themselves. Perhaps for this reason, simple diffusion processes are of limited importance for biological systems.

Despite its apparent chemical homogeneity, the unit membrane appears to behave as an anisotropic structure, contributing a vectorial quality to scalar chemical reactions. This may result from chemical participation of membrane molecules in the transport process. The exact disposition of the molecular carriers within plasma membrane is not known. Conventional electron photomicrographs and the Davson-Danielli model imply a lamellar cell membrane. Special electron microscopic methods have demonstrated a ladder-like structure within unit membrane of mitochondria and endoplasmic reticulum [165, 166], interpreted as showing lipid micelles within a protein matrix. As temperature is varied, these membranes undergo transformation between lamellar and micellar structures. Globular structures 50–60 Å in width have been found [14] extensively distributed over isolated plasma membranes of liver cells. The relationships of the lamellar-micellar transformation and of the globular structures to transport properties of cell membrane are unknown.

In addition to these structural alterations within the bimolecular leaflet of plasma membrane, there appear to be modifications of the plasma membrane in different regions of the cell. The modifications possibly bear some relationship to the transport process carried out by the specialized regions of the cell membrane. For example, there are various coatings on the external surface of epithelial cells engaged in transport. These external coatings on plasma membrane may appear as hairlike processes composed of protein-polysaccharide macromolecules [22, 67] or particulate arrays of unknown composition [82]. The hairlike processes and particulate arrays, related generally to the external part of cells, may modify diffusion processes since the lateral aspects of cells, which are probably more concerned with active transport, are devoid of them.

The plasma membrane is further modified as regards active transport by variations in surface area through multiple infoldings. These infoldings are in close relationship to mitochondria in kidney tubular cells [153], the nasal salt gland of marine birds [59], fish gills [143], and other organs. The degree of infolding may vary according to the adaptation of the species to its environment [154].

Within many cells, furthermore, there is an elaborate membrane-bound tubular system close to the external plasma membrane. It has been suggested that this tubular system has an important function in the secretion of hydrochloric acid by the gastric parietal cells [97, 164] and in the transport of calcium by muscle [144]. What role these channels may have in the transport of water through cells has not been studied.

The Carrier Function of Cell Membrane

Kinetic analysis of many primary transport processes shows that a constituent of cell membrane participates in the transport process as a "carrier." The precise identity of the carrier molecule which facilitates the transport through membranes has not been made clear in most instances. As we shall discuss below, some information is available concerning the carrier mediating active sodium transport.

Carrier-mediated transport, whether it be facilitated diffusion or active transport, requires as a transitional stage the association of the transported substance with the carrier molecule within the plasma membrane. In all probability the transport process is carried out through conformational changes of the carrier molecule or of the membrane itself. Data are insufficient to clarify at this time whether carrier molecules move through the membrane [94, 99, 139] or bring about the translocation of the substance through conformational changes of a molecule fixed within the plasma membrane [191].

Great effort has been expended in an attempt to justify one or another model for carrier-mediated transport. The phenomenological data, on the other hand, are compatible with the concept that exchange diffusion, facilitated diffusion, and active transport may be variations of a single process. In this process the carrier molecule within the membrane undergoes appropriate conformational change under a variety of stimuli. With the conformational changes, the affinity of the carrier for the transport substrate changes. For active transport the stimulus for conformational change may be promotion of some ligand within the carrier molecule to an energy-rich state, in some cases through phosphorylation. The energy-rich carrier can then bind the molecule being transported. With other appropriate stimuli, the carrier-solute complex may undergo further conformational change resulting in translocation of the transported material. In facilitated diffusion and in exchange diffusion, activation from energy sources of the cell may not be essential. Here the conformational change may be induced by either of two processes: (1) simple chemical addition as a ligand of the transported substance; (2) activation of the carrier through a combination of the carrier with various secondary substrates. The latter, for example, may be the mechanism of sodium and potassium–induced sugar and amino acid transport [86, 88, 100, 107, 120, 126, 127, 128, 156]. Better insight into these processes will be gained only when we have better identified the specific compartments in and out of which transport takes place and when we have better characterized the chemical composition of these various compartments and the membranes bounding them.

The presence of a carrier, as noted above, has been inferred from the kinetic behavior of a variety of transport processes. These processes demonstrate saturation kinetics, characterized by variations of the Michaelis-Menten relationship. Substances related chemically to the transported material frequently show mutual inhibition in kinetic studies of transport. Moreover, conformational changes of the carrier molecule [191] would be sufficient to explain most interactions of unidirectional fluxes in a bidirectional transport system. Almost all studies of transport, therefore, are currently consistent with the concept of the fixed carrier integral to the Davson-Danielli membrane.

The primary processes of transport, including active transport [94, 99, 139], are consistent with cyclic configurational changes of membrane-bound carriers. The energy requirement for cellular processes of transport will be dictated by the free energy changes undergone by both the solute and the carrier molecule during the transport process. This energy requirement will depend specifically on (1) the difference in electrochemical potential of the solute across the membrane, (2) the energy required for the conformational changes of the carrier molecule, (3) the change in free energy of the solute particle in going from an aqueous solution to a component of a bimolecular leaflet which is primarily lipid in character, (4) the consequent change in hydration of the solute, and (5) possible specific chemical interactions between the solute and the appropriate binding groups of the carrier molecule. From these considerations it is conceivable, for example, that downhill active transport is that form of carrier-mediated transport down an electrochemical potential gradient wherein an external source of energy is required for the conformational changes of the carrier molecule and the change in solvation characteristics of the solute on entering and leaving the membrane. A generalized phenomenological view of carrier mediated transport, whether active or passive, permits a simpler interpretation of data than that predicted by the multiplicity of models so prevalent in discussions of transport processes.

ACTIVE TRANSPORT OF IONS

Cellular Functions of Active Ion Transport

The transport of the sodium ion is essential for a variety of cellular functions. Not only is this transport the mechanism for regulating the sodium and potassium content of cells, but it plays a particular role in several physiological functions including excitation processes in nerve, cardiac function, muscular contraction, and renal secretory and reabsorptive process. The transport of sodium, moreover, is a strongly conditioning factor in the cellular action of insulin and in the transport of amino acids and sugars by intestinal epithelium, muscle cells, erythrocytes, etc.

In addition, water movement and the regulation of cell volume are intimately coupled to the movement of solutes across cells and tissues [85]. An active transport mechanism for water appears unlikely [142]. Neither hydrostatic pressure differences nor differences in the activity of water across animal cell membranes can account for water movement and cell volume regulation in the steady state [175]. Accordingly, the net movement of water against osmotic or hydrostatic gradients must be related to the movement of solutes. Active sodium and potassium transport systems take part in regulating the volume of cells and, secondarily, the spatial relations and interactions of macromolecules within cells [175].

The active transport systems participating in the movement of sodium render the cell functionally impermeable to this ion. They establish concentration gradients for sodium and potassium between cytoplasm and environment. This potential energy, in the form of ion gradients, confers upon cell membranes their excitable properties, so important in physiological functions like nerve impulse conduction and muscular contraction. The role of the ionic gradients in changes in cell configuration is not yet clear, but the regulation of cell volume is an important cellular function depending upon active transport of ions rather than active water movement.

Gross Features of Cation Transport

Ussing has described the overall features of cation transport in epithelial and other cells [180]. Epithelial cells such as amphibian skin and urinary bladder engage in net ion transport. Nerve fibers and other cells, while actively exchanging ions with the extracellular medium, do not engage in net ion transport. Regardless of whether the cell carries on net ion movement, the elements of the cellular transport process across membranes (the primary transport processes) are similar. They include a sodium selective passive diffusion element, an active sodium-potassium pump, and a potassium selective diffusion element. In tissues where there is net transport, the sodium and potassium selective diffusion elements are spatially separated, lending polarity to the cell. In nerve fiber and other excitable cells the two elements are found in the same membrane, although the sodium selective element may be functional over short time periods. These three primary processes of transport have been combined into a phenomenological model of sodium transport for epithelial cells (Fig. 4-1) by Koefoed-Johnsen and Ussing [111].

In the original conception of this model one surface of the cell (the outer one of the frog skin) is presumed to be selectively permeable to sodium whereas the other surface of the cell (the inner one for the frog skin and presumably the surface in contact with the basement membrane of the epithelium) is selectively permeable to potassium. At some surface of the cell, again presumed (probably incorrectly) to be the inner surface for the frog skin, there functioned a sodium-potassium exchange pump coupled to energy-yielding reactions of the cell. The sodium-potassium exchange pump was thought to function in an electrically neutral manner. The outer membrane of the frog skin behaved in this model as a simple sodium electrode and the inner one as a potassium electrode.

Experimental results are forcing considerable modification of this simple model. Sodium does indeed appear to enter the cell by a diffusion process, although not freely so, since the kinetics of this process show a high degree of saturability [30, 71]. Through an appropriate range of sodium concentration in the medium external to the frog's skin, however, this surface of the cell behaves like a sodium electrode.

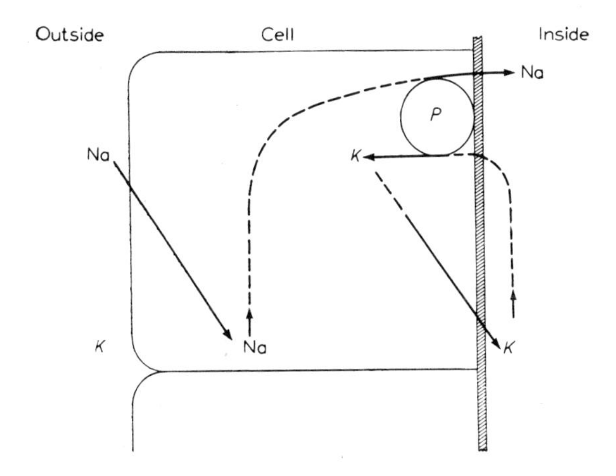

FIG. 4-1. The transepithelial movement of cations according to the model proposed by Koefoed-Johnsen and Ussing [111]. Sodium enters the cell by diffusion across the outer cell membrane, down an electrochemical potential gradient. At the inner cell membrane, an active pump (P) results in the extrusion of sodium from the cell in exchange for potassium.

Microelectrode studies have demonstrated a more complicated physical situation than that predicted by the model [182]. The inner membrane of the cell does not behave like a simple potassium electrode [50, 105, 170], nor does the energy-requiring pump for sodium function as a simple sodium-potassium exchange pump in either the frog skin [23], toad bladder [64, 65], frog skeletal muscle [48, 95, 96, 134], or cardiac muscle [78]. The sodium pump in epithelial cells appears to be electrogenic. Indeed, the fact that sodium-potassium fluxes were observed to be tightly coupled in a variety of transport systems is not proof that these ions utilized a common carrier [68]. In cardiac muscle, moreover, the movement of potassium appears to be passive and driven by the potential generated by the active transport of sodium [140, 141]. The effect of potassium on sodium flux may in fact be a consequence of a depolarization of the cell membrane by potassium [95, 96].

From the model for sodium transport proposed by Koefoed-Johnsen and Ussing it can be inferred, too, that the potential difference across the cell is the sum of two potential differences associated, one each, with the inner and outer cell membranes. This has been confirmed repeatedly [29], but other studies [37] show that these potential gradients may be distributed nearly continuously across the entire cell. Furthermore, there appear to be several intracellular compartments and boundaries, possibly related to membrane-bound cytoplasmic structures, taking part in the transcellular transport process.

In addition to the electrical evidence, microscopic and chemical studies suggest that the Koefoed-Johnsen and Ussing model is oversimplified. In the frog skin [66, 184] there are shunt pathways for ions through intercellular spaces, and intercellular bridges for the diffusion of ions from one cell to the adjacent seem to be important pathways of ion movement. From these studies it has been inferred that the transepithelial pathway for sodium may not be directly transcellular but a more circuitous path. Moreover, a sodium and potassium–sensitive adenosine triphosphatase (ATPase), an enzyme closely associated with active ion transport, has been found [66, 104] to be distributed along the portion of the cell membrane facing intercellular spaces, rather than along the basement membrane. There must be further clarification of the specific pathway of sodium through the cell, with detailed identification and chemical characterization of the subcellular structures through which sodium moves, before the exact nature of the transport process for sodium and the relationship of electrical to physiological phenomena become clear.

Regardless of these difficulties, the overall phenomena of the transport of sodium are consistent with the observations of a passive leak of sodium into the cell and an active transport step outward. The active step is limited in rate by the availability of intracellular sodium, which is, in turn, limited by the rate of the passive sodium entry [30]. This pump and leak mechanism holds both for cells which are polarized and engaged in a net transcellular movement of sodium and for single-cell systems such as the erythrocyte. It is currently impossible to clarify the exact relationship of the sodium pump to potassium movements; the details of this relationship appear to differ from cell type to cell type. The active transport of sodium out of cells in general requires potassium in the external medium. Moreover, this active transport of sodium is generally inhibited by cardiac glycosides [69, 89, 95], although the active transport of sodium in red cells has been shown [112] to have a ouabain-insensitive component.

Cell Volume and Ion Discrimination

The net transport of cations determines the distribution of these cations in the steady state. Net cation transport is the resultant of the active transport of sodium out of cells and potassium into cells and the diffusion of these cations along their electrochemical potential gradients. Because of the free permeability of the cell membrane to water, a change in the rate of active transport of sodium and potassium or in the passive leak permeabilities for these cations will lead to a change in the steady-state concentration of these ions within the cell. Secondarily there will be swelling or shrinkage of the cell. Swelling of cells has been shown to take place under conditions leading to preferential inhibition of active sodium transport [58, 115, 178, 192].

This is well shown in studies carried out on sheep erythrocytes [175, 178]. There are two genetically and chemically distinct classes of sheep erythrocytes: one has a high potassium content (HK cells), the other a low potassium content (LK cells). HK cells have a higher capacity for active sodium-potassium transport and a lower passive inward leak permeability coefficient for potassium than LK cells. The passive inward leak permeability coefficient for potassium is only slightly higher than that for sodium in HK cells, whereas LK cells are considerably more leaky to potassium than to sodium.

Inhibition of active ion transport by strophanthidin results in swelling of incubated erythrocytes of both classes. The observed increase in cell volume is proportioned to the increase in cell cation content in a manner predicted for colloid osmotic swelling. The same relationship holds for both HK and LK sheep cells, although LK cells, having a higher passive permeability to cations, swell more. Moreover, HK cells swell at nearly equivalent rates in a high-potassium and a high-sodium medium. LK cells, on the other hand, swell much more rapidly in a high-potassium than in a high-sodium medium. These rates parallel closely the selective passive permeability coefficients of the HK and LK cells for the different ions. Such studies underline the importance of both active transport and passive leaks of ions in the regulation of cell volume.

Colloid osmotic swelling of cells can also be brought about by preferential increases in passive ion leaks into cells. As discussed by Tosteson [175], the studies by Cook [40, 41] on the hemolytic effect of ultraviolet light on human erythrocytes provide a striking example since the ultraviolet irradiation can be stopped at any time with consequent complete control over the duration of toxic irradiation and with full knowledge of the total dose of ultraviolet irradiation. Irradiated red cells swell as a result of increased passive ion leak permeabilities. The swelling is dependent on the total dose of irradiation and independent of the intensity of the light. The rate of hemolysis and the increase in ion permeabilities for sodium and potassium are proportional to the square of the dose of ultraviolet light, indicating a two-event reaction of the erythrocyte membrane to ultraviolet light. When active transport of ions is inhibited, the rates of water movement and erythrocyte swelling are determined by the increase in passive ion permeabilities and are proportional to the rate of ion movements as predicted for colloid osmotic swelling. Thus the total content of ions as well as the volume of cells is regulated by the balance between passive leaks and active transport systems for sodium and potassium. The quantitative and kinetic aspects of swelling of cells due to changes in the balance between passive leak permeabilities and active pumps for ions have been studied by Tosteson and Hoffman [178] and by Tosteson [175].

The differential permeability of the cell membrane to sodium and potassium is an important factor in the regulation of cell size and ion composition. Not all cells discriminate sensitively between sodium and potassium. For example, microorganisms in the stationary phase are permeable to sodium and potassium to approximately the same degree [163]. Growing bacteria regain the ability to discriminate between potassium and sodium. The balance of the pump and leak system described above provides discrimination between cations in animal cells. In yeast the active transport system for sodium and potassium is the discriminatory mechanism [10, 39] which is very sensitive to the concentration of hydrogen ion; the relative permeability of sodium and potassium may be reversed by changes in pH [10]. Undoubtedly conformational changes of macromolecules within the cell membrane have important effects on the ability of the cell to discriminate between cations and on the relative leakage permeabilities for these cations.

The Pathway for Sodium in Polar Cells

In polar epithelial cells such as the skin of the frog, the urinary bladder of the toad, or the renal tubule the discriminating mechanisms of passive ion leaks for sodium and potassium and the active ion pumps for sodium and potassium are located in different portions of the cell. This spatial separation permits net physical translocation of ions. In the frog skin and the toad urinary bladder it is the outer or environmental surface of the epithelium which harbors the passive leak permeability properties for the sodium ion. This surface functions as a sodium electrode. Sodium passively leaks into the cell down its electrochemical potential gradient [70, 71]. As discussed above, this passive permeability may be conditioned by the concentration of potassium in the external medium [23, 72]. The barriers to diffusion on the outer surface of the epithelial cell may consist of two layers [7, 116], the outer and less porous being the barrier to sodium entry. The permeability of the barrier to sodium may be increased by amphotericin B [122]. Under most physiological circumstances the rate-limiting step in sodium transport is the passive permeability of sodium into the cell via a carrier mechanism [71].

It is difficult to resolve many of the important aspects of sodium transport because of our inability to identify the several compartments of sodium within the cell and the exact localization of the active transport step in relation to anatomic structures in the cell or tissue. Many of the studies of these problems make use of isotopes to measure various intracellular pools of sodium. There is considerable controversy and disagreement [8, 71] as to whether the isotopically measured pool is sodium that has already been transported or sodium awaiting transport.

The resolution of this problem is crucial to our understanding of the effects of the antidiuretic hormone, vasopressin, on sodium transport. The net influence of vasopressin on sodium transport, an increase in transcellular sodium movement observed by Fuhrman and Ussing [73], may result from an effect of vasopressin on either of the primary orders of sodium transport involved in transcellular movement: the passive inward leak of sodium or the active outward sodium pump. Ussing [180] has suggested that vasopressin administration causes an increase in the passive leak permeability for sodium permitting more sodium to reach the active pump mechanism, without an increase in the driving force for sodium. The effect of vasopressin in increasing the sodium pool has been taken by Frazier et al. [71] to imply just such an increase in the passive leak permeability for sodium. On the other hand, the evidence presented by Andersen and Zerahn [8] that this pool consists of sodium that has already been transported would require a reinterpretation of the data, since a change in size of the sodium pool cannot be a feature which distinguishes critically between an increase in passive leak permeability for sodium and an increase in active sodium transport as a consequence of vasopressin administration.

Changes in passive leak permeability for sodium can influence the net rate of ion transport by polar cells. An increase in permeability of the cell membrane for sodium at its outer barrier has been shown [183] to produce an increase in sodium transport and the short-circuit current associated with this transport while a shrinkage of the membrane decreases active ion transport. At more extreme degrees of cell shrinkage in the frog skin, intercellular shunt pathways are opened and the apparent electrogenicity of the sodium pump declines [181]. These changes in net sodium movement may occur in the absence of specific changes in the active sodium-potassium pump.

CHEMICAL MEDIATORS OF SODIUM TRANSPORT

Sodium, Potassium–Activated ATPase

There is mounting evidence that one of the chemical mediators of sodium transport in a large variety of tissues is an enzyme which can function as an ATPase. This enzyme was first demonstrated ten years ago by Skou [167] in a preparation of membrane particles derived from crab nerve. The enzyme required the presence of both sodium and potassium and, like other ATPases, was stimulated by magnesium. The distribution of this enzyme and its relationship to transport have been studied in a variety of tissues [15, 20, 21, 60, 63, 66, 103, 104, 110, 146, 161]. These aspects of the function of the Na,K–activated ATPase in transport have been discussed extensively elsewhere in this volume (Chap. 1) and will not be reviewed here. Not yet resolved is the important question of the homogeneity of preparations of this particulate enzyme system and the relationship of the Na,K–activated ATPase activity to Mg–activated ATPase activity with which it is associated. We will consider, however, the possible molecular mechanism of the Na,K–activated ATPase in relation to transport.

Molecular Mechanisms of the Na,K–Activated ATPase

The Na,K–activated ATPase has been assumed to play an important role in the transport of sodium because of (1) its ubiquitous association with tissues transporting sodium at high rates, (2) its competitive stimulation and inhibition by sodium and potassium, (3) its localization in cell membranes, and (4) the fact that the quantitative aspects of the inhibition of this enzyme by ouabain are similar to the effect of ouabain on sodium transport. Though it has not been universally acknowledged [161], the ouabain inhibition of sodium transport and of Na,K–activated ATPase can be counteracted by potassium ions [60, 146]. Despite the fact that the enzyme has been measured operationally by its capacity to split ATP and despite the attractive hypothesis that ATP may support transport directly, there is no evidence that the Na,K–activated ATPase functions in this manner during physiological ion transport. On the other hand, because of the simplicity and attractiveness of these hypotheses, many attempts have been made to clarify the molecular mechanism of the Na,K–activated ATPase.

When first described by Skou [167, 168], the Na,K–activated ATPase was demonstrated to carry on an ATP-ADP exchange reaction. The possibility that the reaction mechanism included energy-rich phosphorylated intermediates has been appreciated since then; indeed, several such compounds have been identified. Their role in ion transport as either ion carriers or ion-sensitive energy-rich intermediates participating in the mechanisms of transformation of ion carriers has been difficult to establish. This subject has been reviewed critically [3, 85], and it seems justified at this time not to accept any of the phosphorylated intermediates so far described as the specific carrier for ions during active sodium and potassium transport.

The ADP-ATP exchange activity can be separated from the Na,K–activated ATPase activity of brain cortex [171, 172]. Phosphorylated intermediates capable of carrying out the exchange reaction between ATP and ADP need not function as intermediates of the Na,K–activated ATPase, as postulated by Post and Sen [147, 148]. The simplest concept of the mechanism of action of the Na,K–activated ATPase, a theory which admittedly [3, 85] faces serious criticism, suggests the following reaction sequence: (1) The enzyme reacts with ATP to form a phosphorylated intermediate. This reaction is stimulated by sodium. (2) The phosphorylated intermediate may undergo a number of conformational and other transformations, perhaps with the release of energy. (3) The intermediate ultimately regenerates the enzyme and releases inorganic phosphate. This final breakdown of the intermediate is stimulated by potassium and inhibited by ouabain. The first step of the sequence may result in the translocation of sodium ions whereas the last step causes the translocation of potassium and accounts for the mutual antagonism between potassium and ouabain generally seen in both overall transport reactions and kinetic studies of the Na,K–activated ATPase.

Incorporation of the terminal phosphate of ATP in the presence of sodium and magnesium into the protein portion and not in the phospholipid portion of the Na,K–activated ATPase has been demonstrated [4, 13, 93, 148]. This sodium-dependent labeling with ^{32}P has been shown [148] to turn over at a rate consistent with the assumed mechanism for the Na,K–activated ATPase. Unfortunately, in some preparations [87] of Na,K–activated ATPase the labeled intermediate has been shown not to turn over at all, suggesting that the phosphorylation may be nonspecific. While it has been suggested [2] that a serine residue becomes phosphorylated, the evidence favors an acyl phosphate as the acid-stable phosphorylated intermediate [13, 93, 135]. A membrane-associated phosphatase of wide substrate specificity, activated by potassium and inhibited by ouabain, has been found to catalyze a reaction sequence like that of the Na,K–activated ATPase [12]. On the other hand, at least one such phosphatase activity, the *p*-nitrophenylphosphatase, may be different from the Na,K–activated ATPase activity [5, 74].

A reaction sequence for the Na,K–activated ATPase proposed by Albers [3] is as follows:

$$E_1 + \text{Mg ATP} \overset{Na^+}{\rightleftharpoons} E_1 \sim P + \text{Mg ADP} \quad (1)$$

$$E_1 \sim P \overset{Mg^{++}}{\rightleftharpoons} E_2 - P \quad (2)$$

$$E_2 - P \xrightarrow{K^+} E_2 + P \quad (3)$$

$$E_2 \rightleftharpoons E_1 \quad (4)$$

Reactions (2) and (4) represent allosteric conversions which may become rate limiting and are postulated to account for various inhibition reactions. For sodium transport the various forms of E_1 are enzyme molecules with inwardly directed binding sites for sodium; E_2 represents the forms with outwardly directed binding sites for potassium. The physical translocation of ions achieves its vectorial force through reactions (2) and (4). Were the Na,K–activated ATPase proved to be the mediator of sodium-potassium ion translocations and, despite the current objections, shown to pass through a phosphorylated step, this scheme would be most consistent with currently available data.

THE SOURCE OF ENERGY FOR ACTIVE ION TRANSPORT

The Energy Donor

It has been assumed in many studies that the direct source of energy for active ion transport is generally adenosine triphosphate (ATP). This assumption was derived originally from the fact that ATP is (1) a relatively mobile compound of high energy, (2) a participant in both respiring and glycolytic systems, and (3) the principal high-energy product of intermediary metabolism. More decisive and less circumstantial evidence has been difficult to obtain, although the theory has been bolstered by the demonstration that the Na,K–activated ATPase as described above, can split ATP under kinetic circumstances similar to those characteristic of sodium-transporting systems.

The best evidence in favor of ATP's being the direct source of energy for cation transport is the finding that ATP can support transport directly in several systems. Sodium transport is stimulated in the giant axon of the squid following the direct injection of ATP [27]. On the other hand, phosphorylarginine and phospho*enol*pyruvate support transport in the giant axon of the squid even better than ATP. Although the latter compounds may be rapidly converted to ATP, evidence that conversion occurs in the system studied is lacking. Red blood cell ghosts transport sodium and potassium when ATP is introduced into the intracellular medium [75, 90]. Inosine triphosphate, which in many systems including mitochondria can replace ATP, does not support sodium-potassium transport in erythrocyte ghosts. While most studies show a correlation between levels of available ATP and the rate of active ion transport, they cannot be taken to imply that ATP is the direct source of energy for the transport. It is conceivable that high-energy phosphate of several forms may support transport rather than that ATP is the necessary intermediate.

For some years Conway has supported a hypothesis that an oxidation-reduction pump supports ion transport directly [38]. There is much evidence against this hypothesis in a variety of organ systems. Indirect evidence comes from the finding that in the frog skin [202], the toad bladder [117], and the kidney [113] approximately 20 cation equivalents are moved per mole of oxygen consumed by the tissue in support of transport. This amount is far in excess of that predicted by an oxidation-reduction pump as envisioned by Conway. In a redox pump a maximum of four ion equivalents can be transported per mole of oxygen consumed. A redox pump, furthermore, could be satisfactory only for respiring systems and not for glycolytic systems. Dinitrophenol, which stimulates oxidation-reduction reactions of mitochondria, inhibits ion transport in a number of systems including the urinary bladder of the toad [52, 53]. This dissociation of mitochondrial oxidation-reduction reactions from ion transport is strong evidence against a generalized redox pump. Although the original redox pump theory has been modified considerably and such a pump may possibly play a role in gastric acid secretion or in mitochondrial ion transport, the redox pump is unsatisfactory as a general hypothesis for the source of energy for ion transport.

Much more evidence favors a role for compounds containing a high-energy phosphate bond as the direct energy sources for ion transport. A direct role for ATP, indeed, may not be necessary. In liver [185, 186] and in the urinary bladder of the toad [52, 53], intermediates of oxidative phosphorylation, most probably phosphorylated intermediates, appear to donate energy directly to the ion transporter. In the toad bladder, sodium transport is inhibited by inhibition of electron transfer by anaerobiosis, Amytal, or cyanide. Sodium transport is also inhibited by uncoupling or inhibition of oxidative phosphorylation by dinitrophenol, guanidine derivatives, or oligomycin. ADP administration to the toad bladder results in stimulation of oxidative phosphorylation, a decline in levels of intermediates of oxidative phosphorylation, and an increase in ATP level. Despite increased availability of ATP, sodium transport declines as a consequence of ADP administration. The levels of intermediates of oxidative phosphorylation are better correlated than levels of ATP with rates of sodium transport. Since oligomycin is inhibitory to sodium transport at concentrations which do not inhibit the Na,K–activated ATPase but do inhibit the formation of an active phosphorylated intermediate of mitochondrial oxidative phosphorylation, it is likely that this phosphorylated intermediate of oxidative phosphorylation is the energy donor for sodium transport. Similar findings have been reported by van Rossum for sodium transport by liver. The evidence in favor of intermediates of oxidative phosphorylation acting as the direct energy donors for ion transport is, of course, much more positive for mitochondria engaged in ion transport (see below).

In anaerobic systems as well, there is evidence that high-energy phosphorylated intermediates other than ATP play a role as the donor of energy for ion transport, despite the absence of oxidative phosphorylation coupled to respiration. In anaerobic muscle [61] sodium ion transport is supported by ATP, creatine phosphate, and glycolytic phosphorylation. Under all conditions of inhibition of either source of high-energy phosphate the total high-energy phosphate disappearing is constant for a given level of sodium transport. Studies along these lines are in conflict with the hypothesis that sodium transport is directly linked to electron transfer. Moreover, in the turtle bladder, where ion transport is carried on anaerobically [108, 109], sodium transport is inhibited strongly by dinitrophenol whereas glycolysis is stimulated [24]. It is conceivable that in this system a high-energy intermediate of glycolytic systems, analogous to that shown for oxidative phosphorylation, is the direct donor for ion transport. All the evidence in favor of ATP as the direct energy donor for translocations of sodium, and other ions as well, would be equally compatible with the hypothesis that high-energy phosphorylated intermediates are the donors of energy for ion transport. Another view of this important subject has been presented by Whittam [194].

The Interaction of Metabolism and Ion Transport

It is a natural thermodynamic consequence that, if metabolism can control ion transport, so also can ion transport play a role in regulating metabolism. Both respiration and glycolysis may indeed be partly controlled by the rate of ion transport, perhaps through the accumulation of ADP or inorganic phosphate [151], which are important regulators of the rates of respiration and glycolysis, or through the mechanism of the reversibility of electron transfer and oxidative phosphorylation [33].

Mitochondrial respiration in the kidney [18, 19] and in the brain [197] is increased by sodium and potassium in the presence of the Na,K–activated ATPase. Both potassium transport and respiration are inhibited proportionately by ouabain in brain and kidney cortex [198]. Ions affect glycolysis in erythrocytes directly as shown by the finding that lactate production in human erythrocytes is stimulated by sodium and potassium [62, 196]. Moreover, the effects of inorganic phosphate and ouabain on glycolysis are antagonistic [129]; this inhibition of lactate production in erythrocytes by ouabain is dependent upon the availability of external potassium [195]. The stimulatory effect of ions on glycolysis and on respiration is probably due to a change in the phosphate potential of the cell, the ratio $[ADP]\cdot[P_i]/[ATP]$, which is a controlling feature of the rates of glycolysis and respiration in many systems.

While it is attractive to implicate the Na,K–activated ATPase directly as a regulator of metabolism and transport, the phenomenology of this regulation would be the same regardless of whether ATP or a high-energy intermediate of phosphorylation reactions is the direct donor of energy for sodium transport, as demonstrated for mitochondrial ion transport. Furthermore, studies showing intermediates of oxidative phosphorylation as potential energy donors for transport in intact cells [52, 53, 185, 186] provide a mechanism for the sensitive mutual regulation of ion transport and metabolism. Ion transport phenomena are sensitively geared to the production of useful physiological work and the control imposed upon cellular metabolism by the demands for physiological work.

MITOCHONDRIAL ION TRANSPORT

The relationship between ion transport and respiration has long been known, perhaps from the time of Lundergårdh's studies on anionic respiration in plants [124]. Recent work in this field has been stimulated by the growth of interest in active transport processes generally and by the realization that in mitochondria specifically ion transport affects oxidative phosphorylation. The relationship between mitochondrial ion transport and transport phenomena as a reflection of integrated cell function is not yet clear. On the other hand, cells engaged in active ion transport are often rich in mitochondria. These mitochondria appear in fixed tissue sections to be located in the basal portion of the cell at some distance from the lateral cell margins, where it is currently presumed that active transport may take place. It is conceivable that mitochondria in the unfixed living cell may spend their time in close proximity to an interface important for active transport processes of the cell.

An understanding of the mechanism of oxidative phosphorylation and mitochondrial ion transport may provide important clues to the mechanism of ion transport within whole cells containing mitochondria. This understanding will have substantial bearing on transport in cells such as erythrocytes which do not contain mitochondria, since there may be residues of enzymatic function within the plasma membrane of these cells which resemble—at least for active transport processes—the enzymatic components of the membranous material of mitochondria participating in the reactions of oxidative phosphorylation and ion transport.

Two major facets of ion-accumulating systems in mitochondria have been studied recently: a potassium-transporting system, better described as a potassium-hydrogen exchange reaction, and a calcium-accumulating system. Studies of these transport systems of mitochondria have stimulated new examination and new interpretations of the mechanism of oxidative phosphorylation. Moreover, studies of calcium accumulation have important implications for the understanding of calcium metabolism in cells and whole organisms and for the mechanisms of action of hormones and vitamins [57].

General Aspects of Ion Transport in Mitochondria

The uptake of monovalent and divalent cations by mitochondria in association with an increase in respiration and release of hydrogen ions has been recognized for a long time [32, 169]. Because of current difficulties in identifying the specific intramitochondrial compartment into which ions are transported, the degree of intramitochondrial cation binding, the activity coefficient of the unbound intramitochondrial cation, and the membrane potential of mitochondria, it is uncertain whether the process of mitochondrial ion accumulation is active transport. On the other hand, the transport of cations into mitochondria is clearly energy dependent and proceeds at the expense of either substrate utilization or ATP consumption [25, 56, 187].

When ions are accumulated at the expense of substrate utilization through the electron transfer chain, the energy derived from substrate oxidation plainly does not have to be converted into ATP prior to its utilization in ion transport. Inhibitors of the terminal steps of oxidative phosphorylation like oligomycin have no effect on substrate-supported energy-dependent ion movements in mitochondria. However, if substrates are not provided, ATP is necessary and the energy-dependent accumulation of ions by mitochondria is inhibited by oligomycin. These results suggest that the energy may be provided in either of two ways: (1) through electron transfer reactions or (2) through ATP by reversal of oxidative phosphorylation. Inhibitors of electron transfer and uncoupling agents do inhibit substrate-supported ion transport. Accordingly the source of energy for the ion accumulation by mitochondria must be an intermediate of the mitochondrial energy-yielding reactions prior to those involved in phosphorylation. The most prevalent view of ion accumulation linked to oxidative phosphorylation considers a nonphosphorylated high-energy intermediate of oxidative phosphorylation to be the source of energy [34, 150].

An alternative view of mitochondrial ion accumulation has been proposed by Mitchell [130] and is gaining adherents. This chemiosmotic hypothesis relates to the observation that the accumulation of ions by mitochondria is linked to hydrogen ion extrusion by these mitochondria [36]. According to Mitchell [132], for each pair of electrons transferred in mitochondrial respiration up to six protons may be produced. Only two of the protons arise from the oxidation of substrate. The additional four come from the splitting of water coupled to oxidation-reduction reactions. Since two protons are required for each mole of ATP synthesized from phosphate, with a P/O ratio of 3, the balance of hydrogen ions formed and consumed during oxidative phosphorylation is stoichiometrically appropriate. According to Mitchell, the extrusion of protons from mitochondria provides the driving force for ion accumulation. The energy of redox reactions which would ordinarily be converted into ATP formation is siphoned off through the process of extrusion of protons into the accumulation of cations. Under these circumstances the proton extrusion, rather than the cation accumulation, is the primary and the energy-consuming process.

Potassium Transport by Mitochondria

The low basal rate of potassium accumulation by mitochondria can be increased markedly by the treatment of mitochondria with the toxic antibiotic valinomycin [133]. As is typical for many physiological reactions involving potassium ions, rubidium and cesium ions may be substituted for potassium, but sodium and lithium are not equivalent. In the presence of valinomycin there is a notable increase in the rate of potassium accumulation by mitochondria and an increase in the rate of hydrogen ion extrusion. Hydrogen ion extrusion as the primary process would lead to alkalinity of the intramitochondrial medium, in conformity with actual experimental measurements of the pH of the intramitochondrial crystal space during active ion accumulation. According to Mitchell's hypothesis, the movement of potassium into mitochondria then occurs as a passive phenomenon, secondary to the electrostatic forces generated by the primary hydrogen ion pump.

It is difficult to rationalize the stoichiometry of ion accumulation with this hypothesis. The uptake of potassium and concomitant extrusion of hydrogen ion are dependent upon the permeability of mitochondria to anions of the medium. In the presence of nonpermeant anions such as chloride, there is a greater potassium-hydrogen ion exchange. With permeant anions, like phosphate, arsenate, and acetate, there is less hydrogen ion exchange associated with potassium ion accumulation. Anions enter along with potassium ion, and simultaneous water movement and swelling of mitochondria take place. Anionic entry into mitochondria is a passive phenomenon. If the hydrogen ion transport process were the primary energy-linked process, one would not expect anionic entry. The cation entry would be an exchange reaction dictated by the electrochemical potential generated by hydrogen ion extrusion. None of these forces would be operating on anions, especially since hydroxyl ion is generated within the mitochondrion for each hydrogen ion pumped out. On the other hand, with a primary cation pump directed at potassium movement, the stoichiometric relationship between hydrogen ion extruded and anionic entry would hold, as has been verified repeatedly. At this time, then, there is little direct evidence for a primary hydrogen ion pump in mitochondria.

Experimental phenomena of mitochondrial ion transport are better explained by the conventional chemical coupling hypothesis, as proposed by Lehninger [119] and by Chance [34]. Potassium transport by mitochondria, induced by valinomycin (at a concentration of about $10^{-2}\mu g/mg$ mitochondrial protein), can be supported either by substrate oxidation or by ATP. When ATP is the source of energy, oligomycin inhibits potassium uptake by blocking the reversal of the reactions of oxidative phosphorylation and lowering the energy available for transport in the form of a high-energy nonphosphorylated intermediate. When energy is supplied by substrate oxidation, oligomycin is ineffective as an inhibitor of transport, since the energy comes to the transport process from a source higher up in the sequence of reactions of oxidative phosphorylation than the point of action of oligomycin. Substrate-supported transport can be inhibited by anaerobiosis, by inhibitors of electron transfer (rotenone, antimycin A), or by uncouplers of oxidative phosphorylation (dinitrophenol, etc.). Under such circumstances the transport of potassium ion would be energy dependent and linked to substrate or ATP utilization through a nonphosphorylated intermediate of oxidative phosphorylation.

Since the potassium pump is in all likelihood electrogenic, two phenomena operate to maintain electrical neutrality: (1) the extrusion of hydrogen ion and (2) the passive entry of anions. The contribution of anionic movement depends primarily on the permeability of the mitochondrion to the anion. This hypothesis most readily explains the reciprocal hydrogen-potassium pump and the associated rise in intramitochondrial pH, anionic movement with potassium, and the concomitant swelling of mitochondria resulting from water entry and osmotic equilibration. Following the administration of valinomycin or other antibiotics, the permeability of the mitochondrial envelope is increased and these phenomena are exaggerated.

Because of the reversibility of oxidative phosphorylation, both substrate-dependent electron transfer reactions and ATP breakdown may provide the energy for the production of the intermediate of oxidative phosphorylation which is the donor of the energy necessary for the ion accumulation. The effects of inhibitors of electron transfer and of oxidative phosphorylation depend upon the physiological and biochemical circumstances under which the transport is being supported. There is evidence for the formation of phosphorylated protein intermediates of oxidative phosphorylation during the course of ATP breakdown in support of ion transport in mitochondria [149]. However, evidence is lacking that these phosphorylated derivatives [1, 137], including phosphohistidine, have a specific role in the ion accumulation mechanism. In fact, more than one phosphorylated protein has been identified. It is uncertain whether these phosphoproteins are the identical high-energy phosphorylated intermediates of oxidative phosphorylation which participate in the ADP-ATP exchange and ATPase reactions of mitochondria, reactions that are integral to oxidative phosphorylation.

Divalent Cation Transport by Mitochondria

Divalent cations are also known to influence concomitant respiration and hydrogen ion exchange in mitochondria [32, 152, 169]. Very large amounts of calcium are transported by mitochondria in the presence of phosphate, with either ATP or a substrate for electron transfer acting as energy donor [56, 187]. Studies with inhibitors of electron transfer and substrate utilization and inhibitors and uncouplers of oxidative phosphorylation indicate that, just as for potassium transport, a nonphosphorylated intermediate of oxidative phosphorylation provides the energy for the accumulation of divalent cations in mitochondria.

With massive calcium accumulation calcium phosphate is deposited in large accretions within mitochondria [81, 158]. At lower rates of calcium accumulation, without deposition of calcium phosphate, calcium-hydrogen exchange reactions can be demonstrated just as for monovalent cations. The stoichiometry of calcium and hydrogen ion exchange remains relatively constant at 1 to 2 equivalents of hydrogen ion for each mole of divalent cation in the presence or absence of permeant anions. Calcium resembles ADP in exerting respiratory control over mitochondria [35]. The degree of stimulation of respiration by the accumulation of divalent cation varies with the rate at which calcium is presented to the mitochondria and the permeability of the anion. For permeant anions like acetate or phosphate, one mole of calcium is accumulated per mole of oxygen consumed. With nonpermeant anions like chloride, higher calcium to oxygen ratios are obtained [28, 157].

The inequality in stoichiometric relationship between calcium, hydrogen ion, and oxygen has been taken to support the chemi-osmotic coupling hypothesis of Mitchell as the mechanism for the interaction between ion accumulation and energy metabolism in mitochondria. For potassium accumulation the potassium ion to hydrogen ion ratio varies with the penetrability of the anion. The chemi-osmotic coupling hypothesis is unnecessary to explain the phenomena observed. The stoichiometry of calcium transport is more difficult to interpret since there is uncertainty as to the physical state of the transported calcium. When calcium is transported in the presence of nonpermeant anions, it is conceivable that some of the changes in Ca^{++}/O ratio may be due to uncoupling of oxidative phosphorylation, perhaps by calcium itself or by fatty acids which are liberated during divalent cation accumulation [200]. Uncoupling may account for some of the discrepancy in the quantitative aspects of hydrogen ion extrusion, oxygen consumption, and calcium ion uptake.

It is difficult at this time to clarify completely the relationship between mitochondrial transport of cations and energy metabolism. No hypothesis for the coupling between ion accumulation and energy-yielding reactions of mitochondria is completely satisfactory, but since most of the experimental evidence supports the now conventional view of the chemical coupling, it does not seem appropriate to discard this one yet in favor of the chemi-osmotic coupling hypothesis. Much remains to be worked out in support of the latter, including the mechanism by which redox reactions split water and the mechanism of the primary hydrogen ion pump.

The relationship between mitochondrial transport of ions and integrated transcellular cation transport is also unclear. Similarities exist between the phenomena, including the possibility of related energy donors. Differences between cells and mitochondria in the selectivity of cations for transport may be related more to the selectivity of the external plasma membrane of the cell and the mitochondrial membrane than to differences in the fundamental mechanism of ion transport.

There is almost certainly an important interrelationship of transcellular transport, mitochondrial ion accumulation and exchange, and energy metabolism of the cell. It is already apparent that ion transport and cellular metabolism exert mutual regulation which participates in physiological functions. For understanding of both transcellular ion transport in epithelial tissues and mitochondrial transport, we await information on the physical state of cations, their binding, and their pathway in order to resolve questions concerning the "active" nature of transport and the mechanism of transduction of energy to ion accumulation.

THE EFFECT OF SODIUM ON TRANSPORT OF AMINO ACIDS AND SUGARS

Several systems for the transport of amino acids and sugars are simultaneously at work in many cells. Unfortunately for the understanding of the transport mechanisms, a given amino acid may be handled by more than one transport system. Although there has been much work on the kinetics of amino acid transport, a subject which has been well reviewed recently [43, 49, 106, 201], the chemical aspects of the transport system have not been well characterized. Active amino acid transport is carried on by saturable mechanisms obeying the Michaelis-Menten relationship. A catalogue of amino acids transported by a given mechanism has been determined by the demonstration of mutual inhibition, either by competitive or by noncompetitive mechanism, of transport for the several amino acids involved.

While inhibition of transport of one amino acid or sugar by another has been explained on occasion as evidence for competition for common energy sources, most evidence favors the view that they are competing for a common carrier. Regardless of whether a carrier with multiple binding sites or a variety of carriers operate in parallel, the membrane does impart some chemical specificity to the transport system. The ability of *Escherichia coli* to concentrate proline, a specific function of the bacterial membrane, determines the external proline concentration required for growth [102]. Both the active transport of amino acids and sugars and the exchange diffusion processes for these compounds show kinetic and chemical characteristics of carrier-mediated transport. Exchange diffusion reactions are generally insensitive to metabolic inhibitors and are not inhibited by the removal of sodium ions. For some amino acids it may be necessary to inhibit active transport by the removal of sodium ions in order to demonstrate exchange diffusion [106]. The multiple amino acid transport systems existing in parallel may in fact represent a single carrier with several binding sites for the transported species [98]. Similarly, the effect of sugars in inhibiting transport of amino acids in kidney cortex [173] and in intestinal mucosa [6, 9, 49, 136, 159] may also represent competition for binding sites on a carrier common to the two systems of sugar and amino acid transport [6].

The transport of sugars and amino acids in a variety of cells is dependent upon the presence of sodium ions. The sodium-dependent active transport systems operate in a direction parallel to a downhill flux of sodium [43, 49, 106]. The effect of sodium ions on sugar and amino acid transport is best explained by a coupling between the substrate transport and the passive movement of sodium ions through a carrier, as suggested for sugar transport in the intestine [43]. The direction of sugar translocation is determined by the direction of the sodium gradient [42]. Sodium decreases the Michaelis constant of the carrier for sugars on the external surface of the cell while potassium ions increase the Michaelis constant on the inside of the cell. The apparently uphill transport of sugar could then arise from a difference in Michaelis constant (K_m) on either side of the cellular membrane resulting from this interaction between cations and the carrier.

Studies of rat jejunal sacs reviewed by Crane [43] show that the K_m of the transport carrier for 6-deoxyglucose rises 200-fold when the sodium concentration is changed from 145 mEq/liter to zero. The presence of potassium has an added effect in further increasing the K_m of the carrier for 6-deoxyglucose. The lower the sodium concentration and the higher the potassium concentration, the lower the affinity of the carrier for sugar. For glucose transport in rat jejunum, the results are the same with a 20-fold increase in K_m as sodium concentration is reduced from 96 mEq/liter to 24 mEq/liter. Under these circumstances cellular energy would be utilized to maintain the ionic gradients. The active transport of sugar would not depend directly on metabolism for its energy [26, 44].

Amino acid transport activated by sodium ions may be explained on the same basis [26, 155, 188–190]. Studies of the transport of the nonmetabolizable amino acid analog α-aminoisobutyric acid (AIB) in isolated rabbit lymph node cells [106] show an increase in K_m from 1.1×10^{-4} M at $[Na^+] = 140$ mEq/liter to 4×10^{-3} M at $[Na^+] = 50$ mEq/liter. In such studies as these K_m means the concentration of AIB giving half-maximal entry rate and is probably related to the affinity of the carrier for the amino acid analog. The relationship between AIB entry and sodium concentration, furthermore, is compatible with the participation of a single sodium ion as "cosubstrate" with AIB in the transport process. AIB transport in the rat diaphragm is markedly depressed by high potassium concentration in a manner to suggest that the effect of potassium on the K_m of the carrier for AIB is the inverse of the effect of sodium. The asymmetry of the carrier characteristics of the cell membrane may be based upon the asymmetrical distribution of sodium and potassium across the cell wall.

In lowering the Michaelis constant of the transport system for amino acids and sugars without affecting the maximum velocity of this system, sodium ions mimic the effect of insulin on the facilitated diffusion of sugars into fat cells [17, 45–47]. Reversal of the gradient for sodium ion results in a reversal of the direction of active transport of amino acids. Identical conformational changes may be brought about from either side of the cell membrane. With this finding in mind, it appears that the carrier for the transport process is either highly mobile or, if fixed in the membrane, highly symmetrical.

These interesting effects of cations on amino acid and sugar transport have not yet been fully clarified. While the relationship of active transport of cations to the active uptake of amino acids and sugars is not yet understood in detail, important interactions of transport must exist within cell membranes. Apparently active transport may take place through conformational changes of carriers within cell membranes without direct coupling of the translocation mechanism to energy-yielding processes of the cell. Sodium transport has a unique role in this coupling process, as it does in many other physiological control systems. The transport of sodium appears to be central to the coupling of a variety of specialized cell functions to the more general processes of cell biology relating to intermediary metabolism and structure-function relationships.

TERTIARY TRANSPORT

It would be impossible in a review of this scope to discuss in detail the innumerable aspects of integrated organ function which we have characterized as tertiary transport. On the other hand, there is a particular example of tertiary transport which demonstrates forcefully the integration of several varieties of transport phenomena, including active sodium transport, passive water transport, and urea transport, into a unique anatomic-physiological system for the conservation of water through the concentrating mechanism of the mammalian and avian kidney. This is the countercurrent multiplier system of the kidney, composed anatomically of the loops of Henle, the collecting ducts, and the vasa recta. The physiological components of the system are the hydrodynamic aspects of blood and tubular fluid flow, the mechanisms of active ion transport, selective permeabilities of membranes to water, and the diffusibility of urea. The resultant integrated system for the conservation of water is attuned closely to the requirements of the organism and is sensitive to endocrine regulation.

The Countercurrent Multiplier System

The countercurrent multiplier system of the kidney was postulated originally by Werner Kuhn. The history and mechanism of the countercurrent multiplier system for the concentration and dilution of the urine have been well reviewed in the past [80]. According to Kuhn's original hypothesis [114], if one considers fluid flowing in opposite directions in two parallel tubes connected by a hairpin turn and separated by a semipermeable membrane, a small difference in concentration of osmotic constituents at any given point is multiplied continuously along the length of the tubes by the superposition of a pressure difference between the tubes. A detailed theoretical treatment of the countercurrent multiplier system was presented by Hargitay and Kuhn [83]. Although a model was proposed based upon a pressure difference between the tubes, Hargitay and Kuhn recognized that an electrochemical potential difference, rather than a hydrostatic pressure, could supply the driving force in the kidney. A large body of evidence since then has firmly supported their proposal of a countercurrent multiplier system in the kidney, perhaps the strongest support coming from the repeated demonstration of a progressive increase in osmolarity of the tubular fluid, the renal medullary interstitium, and the blood in the vasa recta throughout the renal medulla, from the corticomedullary junction toward the tip of the renal papilla.

In considering a simple countercurrent multiplier system in which the driving force is a hydrostatic pressure difference (Fig. 4-2), Hargitay and Kuhn point out that

$$\frac{C_x}{C_o} = \frac{1}{1 - \frac{\gamma p}{a u_o} x}$$

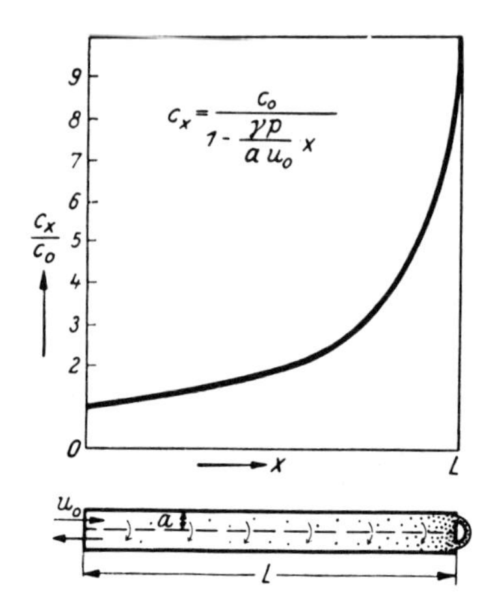

FIG. 4-2. The model for a countercurrent multiplier system proposed by Hargitay and Kuhn [83]. See text.

where C_x = concentration of the solution at point x
C_0 = concentration of the fluid entering the countercurrent multiplier loop
a = diameter of the tubes
u_0 = linear velocity of the fluid entering the system
γ = permeability of the membrane to water
x = distance along the tube
p = difference in hydrostatic pressure across the semipermeable membrane separating the limbs of the countercurrent multiplier loop

Under these circumstances the concentration of a solute will rise hyperbolically along the length of the limb of the tubular loop. The concentration will be maximal at the hairpin turn and minimal at the point of entry and departure from the tubes. Because of the hydrostatic pressure difference there will be a small gradient in solute concentration between the inflow limb and the outflow limb at any given point. This difference in osmotic concentration is called the "single effect." The actual magnitude of the longitudinal gradient is dependent upon the magnitude of the single effect, the length of the loop, and the velocity of inflow. If the diameter of the limbs remains constant, the velocity is a measure of the volume flowing into the system.

A steady state will ultimately be achieved whereby the small concentration gradient between the two limbs is multiplied manyfold along the length of the loop. The details of the confirmation of the countercurrent multiplier system have been reviewed by Gottschalk [79] and by Wirz [199].

The Renal Concentrating Mechanism

The operation of the system in the kidney is depicted in Figure 4-3. The driving force for the concentrating mechanisms is the active transport of sodium out of the ascending limb of the loop of Henle into the interstitium of the renal medulla. The ascending limb is considered impermeable to water, but water diffuses out of the descending limb into the interstitium. The concentration of solutes within the descending limb rises. Accordingly, the single effect is a concentration gradient for sodium chloride between the descending and the ascending limb. Fluid enters the loop of Henle from the proximal tubule,

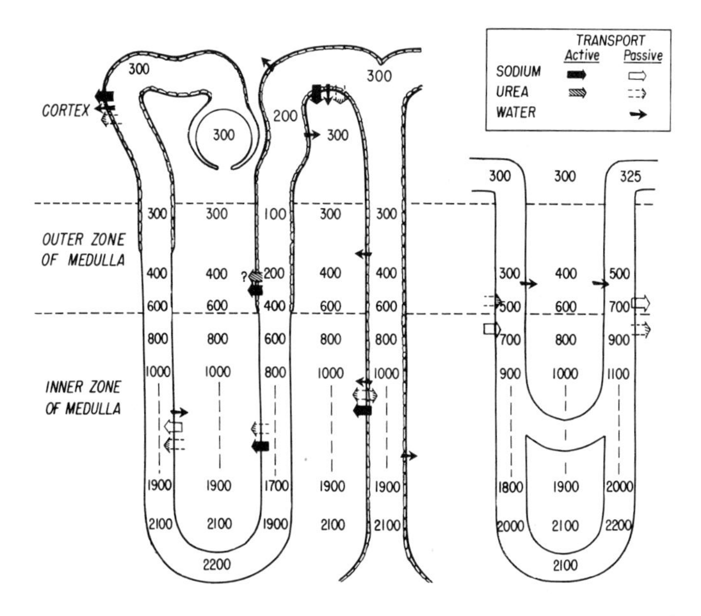

FIG. 4-3. The countercurrent mechanism in a juxta-medullary nephron and adjacent vasa recta according to Gottschalk and Mylle. (From Gottschalk, C. W., and Mylle, M. *Amer. J. Physiol.* 196:927, 1959.)

isosmotic with plasma and at a volume flow of approximately 15 to 25 percent of the glomerular filtration rate. Through continuous sodium chloride transport in the ascending limb of the loop of Henle, a steady state is established in which the small single effect at any given point along the loop of Henle is multiplied longitudinally. This establishes a longitudinal gradient of osmolarity of considerable magnitude, despite the small transverse gradient at any given point.

In the human kidney the longitudinal osmolar gradient achieved can be over 700 to 800 mOsm/liter, while in rodents it may be over 2000 to 3000 mOsm/liter. Yet the magnitude of the single effect is manyfold smaller. The demands on the kidney for energy to establish and maintain the large gradient in osmolarity between the fluid entering the loop of Henle and the fluid at the tip of the hairpin turn are simply the energy requirement for active transport of sodium against a small concentration gradient at each level.

The total sodium reabsorbed by the human kidney can be 25,000 mEq/day, or 18 mEq/minute. Of this amount 14 mEq/minute is reabsorbed actively and isosmotically in the proximal tubule. Only 4 mEq/minute enters the countercurrent multiplier system. As a consequence of the active transport of sodium out of the ascending limb of the loop of Henle into the interstitium of the medulla, the intratubular fluid leaving the loop of Henle to enter the distal tubule becomes hypotonic to plasma and the glomerular filtrate.

The Medullary Interstitium

Hypertonicity of the medullary interstitium is a critical intermediate phase in the effectiveness of the countercurrent multiplier system of the kidney. It removes the necessity for intimate contact between adjacent loops of Henle and the vasa recta. The postglomerular blood vessels of the kidney, before rejoining the venous circulation, plunge in loops of varying length into the medullary tissue adjacent to the loops of Henle. These vasa recta function as countercurrent exchangers, helping to maintain the hypertonicity of the medullary interstitium and contributing further to the renal concentrating process. Because the vasa recta are relatively permeable to water and to solutes like urea and sodium chloride, water diffuses out of the vasa recta as they descend into the medullary pyramid. In the ascending limb of the vasa recta, water is returned to the circulation. The intravascular osmotic concentration throughout the length of the vasa recta is only very slightly less than that of the adjacent interstitium and the loop of Henle fluid at any given point in the longitudinal dimension of the loop. As water leaves the descending limbs of the vasa recta, solutes (especially urea) diffuse into the vasa recta. In the ascending limb, solutes are returned to the medullary interstitium as water reenters and is carried away. Despite the removal of water, there is a continuous recirculation of solute within the medulla. This countercurrent exchange of water and solutes diminishes the loss of sodium chloride and urea from the medulla and contributes to the maintenance of the hypertonicity of the medullary interstitium, a phenomenon necessary to the continuous efficient operation of the countercurrent multiplier system.

The collecting tubule, which passes through the hypertonic medullary interstitium, is relatively permeable to urea and water under conditions of maximal activity of the renal concentrating mechanism (antidiuresis). The countercurrent exchange function of the vasa recta aids in trapping this urea in the interstitial space, further contributing to the efficient function of the renal concentrating mechanism [121]. The diffusion of urea (possibly abetted by active transport of urea [162]) from the collecting duct contributes to the total osmolarity of both the urine and the medullary interstitium. The contribution of urea to the urine osmolarity is independent of the maximal achievable concentration of sodium chloride in the medullary interstitium [16]. An increase in urea presented to the kidney and ultimately to the collecting tubule will contribute significantly to urine osmolarity by the trapping mechanism of the vasa recta and by reentry of urea into the tubular urine via the descending limb of the loop of Henle.

The Action of Antidiuretic Hormone

The countercurrent multiplier system of the loop of Henle and the countercurrent exchange system of the vasa recta function nearly identically during maximal antidiuresis and maximal diuresis. Differences that occur during diuresis are a consequence chiefly of the absence of antidiuretic hormone. The action of antidiuretic hormone is primarily on the distal tubule and on the collecting duct. During antidiuresis the hypotonic fluid entering the distal tubule becomes nearly isotonic before reaching the collecting duct as a result of the reabsorption of most of the solute from the distal tubular fluid and even more of the water. In the absence of antidiuretic hormone, solute and water reabsorption in the distal tubule are considerably less; the fluid remains hypotonic throughout the length of this tubule. Moreover, in the absence of antidiuretic hormone, the collecting tubule is less permeable to urea and relatively impermeable to water. A very dilute urine is formed, therefore, and less urea is recycled in the medullary interstitium. The degree of hyperosmolarity of the intratubular fluid within the loop of Henle and of the medullary interstitium is slightly reduced from that observed during antidiuresis. The effect of antidiuretic hormone on the distal tubule is of considerably greater magnitude than its effect on the collecting duct: threefold more water is reabsorbed in the distal tubule than in the collecting duct during antidiuresis. On the other hand, the relative impermeability of the collecting duct membrane to water and urea during diuresis undoubtedly plays a significant role in reducing the tonicity of the medullary interstitium and facilitating the dilution of the urine during water diuresis.

This mechanism for the concentration of the urine in the avian and mammalian kidney represents a remarkable integration of a few physiological transport processes and a specialized anatomic phenomenon. The processes of active transport of sodium, passive diffusion of water and urea, and possibly active transport of urea are oriented by the countercurrent multiplier system in the loops of Henle and the countercurrent exchange system of the vasa recta of the renal medulla. The resultant system is directed at the conservation of water under a variety of physiological stresses.

CONCLUSION

We have not tried to present a review of all the significant factors affecting transport in living organisms. To discuss the mechanism of hormone action on transport processes, for example, the effects of insulin, aldosterone, antidiuretic hormone, etc., in too little space would have been unfair to the remarkable physiological and chemical information available on these important functions. We have attempted to examine a variety of transport processes and to unify them into important phenomenological groups. We have emphasized the role of the membrane dividing biological compartments, the role of carriers within the membranes, the mutual interaction of transport systems, the importance of sodium transport for the regulation of the composition and volume of aqueous compartments, and the important relationship between transport processes and energy metabolism in cells. In discussing the interaction between energy metabolism and transport we have hoped to demonstrate the sensitive regulation that each phenomenon can impose upon the other. We have emphasized, moreover, the phenomenology of transport and have avoided where possible the introduction of specific models. It should be clear from our discussion that transport processes participate centrally in physiological and biochemical control systems.

REFERENCES

1. Ahmed, K., and Judah, J. D. Mitochondrial phosphoprotein metabolism. *Biochim. Biophys. Acta* 71:295, 1963.
2. Ahmed, K., and Judah, J. D. Identification of active phosphoprotein in a cation-activated adenosine triphosphatase. *Biochim. Biophys. Acta* 104:112, 1965.
3. Albers, R. W. Biochemical aspects of transport. *Ann. Rev. Biochem.* 36:727, 1967
4. Albers, R. W., Fahn, S., and Koval, G. J. The role of sodium ions in the activation of *Electrophorus* electric organ adenosine triphosphatase. *Proc. Nat. Acad. Sci. U.S.A.* 50:474, 1963.
5. Albers, R. W., and Koval, G. J. Sodium-potassium-activated adenosine triphosphatase of *Electrophorus* electric organ: 3. An associated potassium-activated neutral phosphatase. *J. Biol. Chem.* 241:1896, 1966.
6. Alvarado, F. Transport of sugars and amino acids in the intestine: Evidence for a common carrier. *Science* 151:1010, 1966.
7. Andersen, B., and Ussing, H. H. Solvent drag on non-electrolytes during osmotic flow through isolated toad skin and its response to antidiuretic hormone. *Acta Physiol. Scand.* 39:228, 1957.
8. Andersen, B., and Zerahn, K. Method for non-destructive determination of the sodium transport pool in frog skin with radiosodium. *Acta Physiol. Scand.* 59:319, 1963.
9. Annegers, J. H. Some effects of hexoses on the absorption of amino acids. *Amer. J. Physiol.* 210:701, 1966.
10. Armstrong, W. M., and Rothstein, A. Discrimination between alkali metal cations by yeast: I. Effect of pH on uptake. *J. Gen. Physiol.* 48:61, 1964.
11. Askari, A., and Fratantoni, J. C. Effect of monovalent cations on the adenosine triphosphatase of sonicated erythrocyte membrane. *Biochim. Biophys. Acta* 92:132, 1964.
12. Bader, H., and Sen, A. K. (K^+)-dependent acyl phosphatase as part of the ($Na^+ + K^+$)-dependent ATPase of cell membranes. *Biochim. Biophys. Acta* 118:116, 1966.

13. Bader, H., Sen, A. K., and Post, R. L. Isolation and characterization of a phosphorylated intermediate in the ($Na^+ + K^+$) system-dependent ATPase. *Biochim. Biophys. Acta* 118:106, 1966.
14. Benedetti, E. L., and Emmelot, P. Electron microscopic observations on negatively stained plasma membranes isolated from rat liver. *J. Cell Biol.* 26:299, 1965.
15. Berg, G. G., and Chapman, B. The sodium and potassium activated ATPase of intestinal epithelium: I. Location of enzymatic activity in the cell. *J. Cell. Comp. Physiol.* 65:361, 1965.
16. Berliner, R. W., Levinsky, N. G., Davidson, D. G., and Eden, M. Dilution and concentration of the urine and the action of antidiuretic hormone. *Amer. J. Med.* 24:730, 1958.
17. Blecher, M. Phospholipase C and mechanisms of action of insulin and cortisol on glucose entry into free adipose cells. *Biochem. Biophys. Res. Commun.* 21:202, 1965.
18. Blond, D. M., and Whittam, R. The regulation of kidney respiration by sodium and potassium ions. *Biochem. J.* 92:158, 1964.
19. Blond, D. M., and Whittam, R. Effects of sodium and potassium ions on oxidative phosphorylation in relation to respiratory control by a cell membrane adenosine triphosphatase. *Biochem. J.* 97:523, 1965.
20. Bonting, S. L., and Caravaggio, L. L. Studies on sodium-potassium-activated adenosine triphosphatase: V. Correlation of enzyme activity with cation flux in six tissues. *Arch. Biochem.* 101:37, 1963.
21. Bonting, S. L., and Caravaggio, L. L. Studies on Na^+–K^+ activated adenosine triphosphatase: XVI. Its absence from the cation transport system of *Ulva lactuca. Biochim. Biophys. Acta* 112:519, 1966.
22. Brandt, P. W. A consideration of the extraneous coats of the plasma membrane. *Circulation* 26:1075, 1962.
23. Bricker, N. S., Biber, T., and Ussing, H. H. Exposure of the isolated frog skin to high potassium concentrations at the internal surface: I. Bioelectric phenomena and sodium transport. *J. Clin. Invest.* 42:88, 1963.
24. Bricker, N. S., and Klahr, S. Effects of dinitrophenol and oligomycin on the coupling between anaerobic metabolism and anaerobic sodium transport by the isolated turtle bladder. *J. Gen. Physiol.* 49:483, 1966.
25. Brierley, G. P., Bachmann, E., and Green, D. E. Active transport of inorganic phosphate and Mg^{++} by beef heart mitochondria. *Proc. Nat. Acad. Sci. U.S.A.* 48:1928, 1962.
26. Bronk, J. R., and Parsons, D. S. Accumulation and incorporation of amino-acid in rat intestine in vitro. *Nature* (London) 208:785, 1965.
27. Caldwell, P. C., Hodgkin, A. L., Keynes, R. D., and Shaw, T. I. The effects of injecting energy-rich phosphate compounds on the active transport of ions in the giant axons of *Loligo. J. Physiol.* (London) 152:561, 1960.
28. Carafoli, E., Gamble, R. L., Rossi, C. S., and Lehninger, A. L. The effect of salt concentration and pH on the relationship between H^+ ejection and electron transport in isolated mitochondria. *Biochem. Biophys. Res. Commun.* 22:431, 1966.

29. Cereijido, M., and Curran, P. F. Intracellular electrical potentials in frog skin. *J. Gen. Physiol.* 48:543, 1965.
30. Cereijido, M., Herrera, F. C., and Flanigan, W. V. The influence of Na concentration on Na transport across frog skin. *J. Gen. Physiol.* 47:879, 1964.
31. Chan, P. C., and Oblas, B. H. Reversible "uncoupling" of Na, K-activated erythrocyte membrane ATPase by sodium dodecyl sulfate. *Fed. Proc.* 25:655, 1966.
32. Chance, B. On Possible Mechanisms of the Control of Electron Transport in the Respiratory Chain. In *Proceedings of the Third International Congress on Biochemistry*, Brussels. New York: Academic, 1956.
33. Chance, B. The interaction of energy and electron transfer reactions in mitochondria: V. The energy transfer pathway. *J. Biol. Chem.* 236:1569, 1961.
34. Chance, B. Calcium-Stimulated Respiration in Mitochondria. In Chance, B. (Ed.), *Energy-Linked Functions of Mitochondria.* New York: Academic, 1963.
35. Chance, B. The energy-linked reaction of calcium with mitochondria. *J. Biol. Chem.* 240:2729, 1965.
36. Chappell, J. B., and Crofts, A. R. Ion Transport and Reversible Volume Changes of Isolated Mitochondria. In Tager, J. M., Papa, S., Quagliariello, E., and Slater, E. C. (Eds.), *Regulation of Metabolic Processes in Mitochondria.* Amsterdam: Elsevier, 1966.
37. Chowdhury, T. K., and Snell, F. M. A microelectrode study of electrical potentials in frog skin and toad bladder. *Biochim. Biophys. Acta* 94:461, 1965.
38. Conway, E. J. New light on the active transport of sodium ions from skeletal muscle. *Fed. Proc.* 23:680, 1964.
39. Conway, E. J., and Duggan, F. A cation carrier in the yeast cell wall. *Biochem. J.* 69:265, 1958.
40. Cook, J. S. Hemolysis by Ultraviolet Radiation. In Christensen, B. C., and Buchmann, B. (Eds.), *Progress in Photobiology, Proceedings of the Third International Congress on Photobiology.* Amsterdam: Elsevier, 1961.
41. Cook, J. S. Some characteristics of hemolysis by ultraviolet light. *J. Cell. Comp. Physiol.* 47:55, 1956.
42. Crane, R. K. Uphill outflow of sugar from intestinal epithelial cells induced by reversal of the Na^+ gradient: Its significance for the mechanism of Na^+ dependent active transport. *Biochem. Biophys. Res. Commun.* 17:481, 1964.
43. Crane, R. K. Na^+-dependent transport in the intestine and other animal tissues. *Fed. Proc.* 24:1000, 1965.
44. Crane, R. K., Forstner, G., and Eicholz, A. Studies on the mechanism of the intestinal absorption of sugars: X. An effect of Na^+ concentration on the apparent Michaelis constants for intestinal sugar transport, in vitro. *Biochim. Biophys. Acta* 109:467, 1965.
45. Crofford, O. B., Jeanrenaud, B., and Renold, A. E. Effect of insulin on the transport and metabolism of sorbitol by incubated rat epididymal adipose tissue. *Biochim. Biophys. Acta* 111:429, 1965.

46. Crofford, O. B., and Renold, A. E. Glucose uptake by incubated rat epididymal tissue. Rate-limiting steps and site of insulin action. *J. Biol. Chem.* 240:14, 1965.
47. Crofford, O. B., and Renold, A. E. Glucose uptake by incubated rat epididymal adipose tissue. Characteristics of the glucose transport system and action of insulin. *J. Biol. Chem.* 240:3237, 1965.
48. Cross, S. B., Keynes, R. D., and Rybová, R. The coupling of sodium efflux and potassium influx in frog muscle. *J. Physiol.* (London) 181:865, 1965.
49. Curran, P. F. Ion transport in intestine and its coupling to other transport processes. *Fed. Proc.* 24:993, 1965.
50. Curran, P. F., and Cereijido, M. K fluxes in frog skin. *J. Gen. Physiol.* 48:1011, 1965.
51. Davis, R. P. Unpublished data, 1967.
52. Davis, R. P., Canessa-Fischer, M., Edelmann, C. M., Jr., and Hoffman, L. The source of energy for active sodium transport in the toad urinary bladder. *J. Clin. Invest.* 43:1292, 1964.
53. Davis, R. P., Hoffman, L., Canessa-Fischer, M., and Edelmann, C. M., Jr. The source of energy for active sodium transport in the urinary bladder of the toad. Submitted for publication.
54. Davson, H. Growth of the concept of the paucimolecular membrane. *Circulation* 26:1022, 1962.
55. De Graeff, J., Dempsey, E. F., LaMeyer, L. D. F., and Leaf, A. Phospholipids and sodium transport in toad bladder. *Biochim. Biophys. Acta* 106:155, 1965.
56. DeLuca, H. F., and Engstrom, G. W. Calcium uptake by rat kidney mitochondria. *Proc. Nat. Acad. Sci. U.S.A.* 47:1744, 1961.
57. DeLuca, H. F., Engstrom, G. W., and Rasmussen, H. The action of vitamin D and parathyroid hormone *in vitro* on calcium uptake and release by kidney mitochondria. *Proc. Nat. Acad. Sci. U.S.A.* 48:1604, 1962.
58. Deyrup, I. A study of the fluid uptake of rat kidney slices in vitro. *J. Gen. Physiol.* 36:739, 1953.
59. Doyle, W. L. The principal cells of the salt-gland of marine birds. *Exp. Cell Res.* 21:386, 1960.
60. Dunham, E. T., and Glynn, I. M. Adenosinetriphosphatase activity and the active movements of alkali metal ions. *J. Physiol.* (London) 156:274, 1961.
61. Dydynska, M., and Harris, E. J. Consumption of high-energy phosphates during active sodium and potassium interchange in frog muscle. *J. Physiol.* (London) 182:92, 1966.
62. Eckel, R. E., Rizzo, S. C., and Lodish, H. Potassium transport and control of glycolysis in human erythrocytes. *Amer. J. Physiol.* 210:737, 1966.
63. Emmelot, P., Bos, C. J., Benedetti, E. I., and Rumke, P. H. Studies on plasma membranes: I. Chemical composition and enzyme content of plasma membranes isolated from rat liver. *Biochim. Biophys. Acta* 90:126, 1964.

64. Essig, A. Active sodium transport in the toad bladder despite removal of serosal potassium. *Amer. J. Physiol.* 208:401, 1965.
65. Essig, A., and Leaf, A. The role of potassium in active transport of sodium by the toad bladder. *J. Gen. Physiol.* 46:505, 1963.
66. Farquhar, M. G., and Palade, G. E. Functional organization of amphibian skin. *Proc. Nat. Acad. Sci. U.S.A.* 51:569, 1964.
67. Fawcett, D. W. Surface specializations of absorbing cells. *J. Histochem. Cytochem.* 13:75, 1965.
68. Finkelstein, A. Carrier model for active transport of ions across a mosaic membrane. *Biophys. J.* 4:421, 1964
69. Flynn, F. V., and Maizels, M. Cation control in human erythrocytes. *J. Physiol.* (London) 110:301, 1949.
70. Frazier, H. S. The electrical potential profile of the isolated toad bladder. *J. Gen. Physiol.* 45:515, 1962.
71. Frazier, H. S., Dempsey, E. F., and Leaf, A. Movement of sodium across the mucosal surface of the isolated toad bladder and its modification by vasopressin. *J. Gen. Physiol.* 45:529, 1962.
72. Frazier, H. S., and Leaf, A. Electrical characteristics of active sodium transport in the toad bladder. *J. Gen. Physiol.* 46:491, 1963.
73. Fuhrman, F. A., and Ussing, H. H. A characteristic response of the isolated frog skin potential to neurohypophysial principles and its relation to the transfer of sodium and water. *J. Cell. Comp. Physiol.* 38:109, 1951.
74. Fujita, M., Nakao, T., Tashima, Y., Mizuno, N., Nagano, K., and Nakao, M. Potassium-ion stimulated *p*-nitrophenylphosphatase activity occurring in a highly specific adenosine triphosphatase preparation from rabbit brain. *Biochim. Biophys. Acta* 117:42, 1966.
75. Gardos, G. Akkumulation der Kalciumionen durch menschliche Blutkoperchen. *Acta Physiol. Acad. Sci. Hung.* 6:191, 1954.
76. Glynn, I. M. Activation of adenosinetriphosphatase activity in cell membrane by external potassium and internal sodium. *J. Physiol.* (London) 160:18P, 1962.
77. Glynn, I. M., Slayman, C. W., Eichberg, J., and Dawson, R. M. C. The adenosine-triphosphatase system responsible for cation transport in electric organ: Exclusion of phospholipids as intermediates. *Biochem. J.* 94:692, 1965.
78. Goerke, J., and Page, E. Cat heart muscle in vitro: VI. Potassium exchange in papillary muscles. *J. Gen. Physiol.* 48:933, 1965.
79. Gottschalk, C. W. Micropuncture studies of tubular function in the mammalian kidney. *Physiologist* 4:45, 1961.
80. Gottschalk, C. W. Osmotic concentration and dilution of the urine. *Amer. J. Med.* 36:670, 1964.
81. Greenawalt, J. W., Rossi, C. S., and Lehninger, A. L. Effect of active accumulation of calcium and phosphate ions on the structure of rat liver mitochondria. *J. Cell Biol.* 23:21, 1964.
82. Gupta, B. L., and Berridge, M. J. A coat of repeating subunits on the cytoplasmic surface of the plasma membrane in the rectal papillae of

the blowfly, *Calliphora erythrocephala* (Meig.), studied in situ by electron microscopy. *J. Cell Biol.* 29:376, 1966.

83. Hargitay, B., and Kuhn, W. Das Multiplikationsprinzip als Grundlage der Harnkonzentrierung in der Niere. *Z. Elektrochem.* 55:539, 1951.
84. Heald, P. J. Phosphoprotein metabolism and ion transport in nervous tissue: A suggested connexion. *Nature* (London) 193:451, 1962.
85. Heinz, E. Transport through biological membranes. *Ann. Rev. Physiol.* 29:21, 1967.
86. Heinz, E., and Durbin, R. P. Studies of the chloride transport in the gastric mucosa of the frog. *J. Gen. Physiol.* 41:101, 1957.
87. Heinz, E., and Hoffman, J. F. Phosphate incorporation and Na, K-ATPase activity in human red blood cell ghosts. *J. Cell. Comp. Physiol.* 65:31, 1965.
88. Heinz, E., and Walsh, P. M. Exchange diffusion, transport, and intracellular level of amino acids in Ehrlich carcinoma cells. *J. Biol. Chem.* 233:1488, 1958.
89. Hodgkin, A. L., and Keynes, R. D. Active transport of cations in giant axons from *Sepia* and *Loligo*. *J. Physiol.* (London) 128:28, 1955.
90. Hoffman, J. F. The link between metabolism and the active transport of Na in human red cell ghosts. *Fed. Proc.* 19:127, 1960.
91. Hoffman, J. F., Tosteson, D. C., and Whittam, R. Retention of potassium by human erythrocyte ghosts. *Nature* (London) 185:186, 1960.
92. Hokin, L. E., and Hokin, M. R. Biological transport. *Ann. Rev. Biochem.* 32:553, 1963.
93. Hokin, L. E., Sastry, P. S., Galsworthy, P. R., and Yoda, A. Evidence that a phosphorylated intermediate in a brain transport adenosine triphosphatase is an acyl phosphate. *Proc. Nat. Acad. Sci. U.S.A.* 54:177, 1965.
94. Hokin, L. E., and Yoda, A. Inhibition by diisopropylfluorophosphate of a kidney transport adenosine triphosphatase by phosphorylation of a serine residue. *Proc. Nat. Acad. Sci. U.S.A.* 52:454, 1964.
95. Horowicz, P., and Gerber, C. J. Effects of external potassium and strophanthidin on sodium fluxes in frog striated muscle. *J. Gen. Physiol.* 48:489, 1965.
96. Horowicz, P., and Gerber, C. J. Effects of sodium azide on sodium fluxes in frog striated muscle. *J. Gen. Physiol.* 48:515, 1965.
97. Ito, S., and Winchester, R. J. The fine structure of the gastric mucosa in the bat. *J. Cell Biol.* 16:541, 1963.
98. Jaquey, J. A., and Sherman, J. H. The effect of metabolic inhibitors on transport and exchange of amino acids in Ehrlich ascites cells. *Biochim. Biophys. Acta* 109:128, 1965.
99. Jardetzky, O. Simple allosteric model for membrane pumps. *Nature* (London) 211:969, 1966.
100. Johnstone, R. M., and Scholefield, P. G. The need for ions during transport and exchange diffusion of amino acids into Ehrlich ascites carcinoma cells. *Biochim. Biophys. Acta* 94:130, 1965.

101. Judah, J. D., Ahmed, K., and McLean, A. F. M. Ion transport and phosphoproteins of human red cells. *Biochim. Biophys. Acta* 65:472, 1962.
102. Kaback, H. R., and Stadtman, E. R. Protein uptake by an isolated cytoplasmic membrane preparation of *Escherichia* cell. *Proc. Nat. Acad. Sci. U.S.A.* 55:920, 1966.
103. Kamat, V. B., and Wallach, D. F. H. Separation and partial purification of plasma membrane fragments from Ehrlich ascites carcinoma microsomes. *Science* 148:1343, 1965.
104. Kaye, G. I., and Pappas, G. D. Studies on the ciliary epithelium and zonule: III. The fine structure of the rabbit epithelium in relation to the localization of ATPase activity. *J. Microscopie* 4:497, 1965.
105. Kidder, G. W., Cereijido, M., and Curran, P. F. Transient changes in electrical potential differences across frog skin. *Amer. J. Physiol.* 207:935, 1964.
106. Kipnis, D. M., and Parrish, J. E. Role of Na^+ and K^+ on sugar (2-deoxyglucose) and amino acid (α-aminoisobutyric acid) transport in striated muscle. *Fed. Proc.* 24:1051, 1965.
107. Kirschner, L. B., and Barker, J. Turnover of phosphatidic acid and sodium extrusion from mammalian erythrocytes. *J. Gen. Physiol.* 47:1061, 1964.
108. Klahr, S., and Bricker, N. S. Transport by isolated turtle bladder during anaerobiosis and exposure to KCN. *Amer. J. Physiol.* 206:1333, 1964.
109. Klahr, S., and Bricker, N. S. Energetics of anaerobic sodium transport by the fresh water turtle bladder. *J. Gen. Physiol.* 48:571, 1965.
110. Klein, R. L., and Breland, A. P. Active cation transport and ATP hydrolysis in acanthamoeba sp. *Comp. Biochem. Physiol.* 17:39, 1966.
111. Koefoed-Johnsen, V., and Ussing, H. H. The nature of the frog skin potential. *Acta Physiol. Scand.* 42:298, 1958.
112. Krezenow, F. M., and Hoffman, J. F. Characterization of new energy dependent cation transport process in red blood cells. *Ann. N.Y. Acad. Sci.* 137:566, 1966.
113. Kül, F., Auckland, K., and Refsum, H. Renal sodium transport and oxygen consumption. *Amer. J. Physiol.* 201:511, 1961.
114. Kuhn, W., and Ryffel, K. Herstellung konzentrierter Lösungen aus verdunnten durch blosse Membranwirkung. Eim Modellversuch zur Funktion der Niere. *Hoppe Seyler Z. Physiol. Chem.* 276:145, 1942.
115. Leaf, A. On mechanism of fluid exchange of tissues in vitro. *Biochem. J.* 62:241, 1956.
116. Leaf, A., and Hays, R. M. Permeability of the isolated toad bladder to solutes and its modification by vasopressin. *J. Gen. Physiol.* 45:921, 1962.
117. Leaf, A., Page, L. B., and Anderson, J. Respiration and active sodium transport of isolated toad bladder. *J. Biol. Chem.* 234:1625, 1959.
118. LeFevre, P. G. Sugar transport in the red blood cell: Structure-activity relationships in substrates and antagonists. *Pharmacol. Rev.* 13:39, 1961.
119. Lehninger, A. L. *The Mitochondrion.* New York: Benjamin, 1964.

120. Levine, M., Oxender, D. L., and Stein, W. D. The substrate-facilitated transport of the glucose carrier across the human erythrocyte membrane. *Biochim. Biophys. Acta* 109:151, 1965.
121. Levinsky, N. G., and Berliner, R. W. The role of urea in the urine concentrating mechanism. *J. Clin. Invest.* 38:741, 1959.
122. Lichtenstein, N. S., and Leaf, A. Effect of amphotericin B on the permeability of the toad bladder. *J. Clin. Invest.* 44:1328, 1965.
123. Ling, G. N. Physiology and anatomy of the cell membrane: The physical state of water in the living cell. *Fed. Proc.* 24 (Suppl. 15): S103, 1965.
124. Lundergårdh, H. Untersuchungen über die Anionenatmung. *Biochem. Z.* 290:104, 1937.
125. McLaughlin, S. G. A., and Hinke, J. A. M. Sodium and water fibers of the giant barnacle. *Canad. J. Physiol. Pharmacol.* 44:837, 1966.
126. Mawe, R. C., and Hempling, H. G. The exchange of C14 glucose across the membrane of the human erythrocyte. *J. Cell. Comp. Physiol.* 66:95, 1965.
127. Miller, D. M. The kinetics of selective biological transport: I. Determination of transport constants for sugar movements in human erythrocytes. *Biophys. J.* 5:407, 1965.
128. Miller, D. M. The kinetics of selective biological transport: II. Equations for induced uphill transport of sugars in human erythrocytes. *Biophys. J.* 5:417, 1965.
129. Minakami, S., Kakinuma, K., and Yoshikawa, H. The control of erythrocyte glycolysis by active cation transport. *Biochim. Biophys. Acta* 90:434, 1964.
130. Mitchell, P. Metabolic Flow in the Mitochondrial Multiphase System: An Appraisal of the Chemi-Osmotic Theory of Oxidative Phosphorylation. In Tager, J. M., Papa, S., Quagliariello, E., and Slater, E. C. (Eds.), *Regulation of Metabolic Processes in Mitochondria.* Amsterdam: Elsevier, 1966.
131. Mitchell, P. Translocations through natural membranes. *Advances Enzym.* 29:33, 1967.
132. Mitchell, P., and Moyle, J. Stoichiometry of proton translocation through the respiratory chain and adenosine triphosphatase systems of rat liver mitochondria. *Nature* (London) 208:147, 1965.
133. Moore, C., and Pressman, B. C. Mechanisms of action of Valinomycin on mitochondria. *Biochem. Biophys. Res. Commun.* 15:562, 1964.
134. Mullins, L. J., and Awad, M. Z. The control of the membrane potential of muscle fibers by the sodium pump. *J. Gen. Physiol.* 48:761, 1965.
135. Nagano, K., Kanayawa, T., Miyuno, N., and Nakao, M. Some acyl phosphate-like properties of P32-labeled sodium-potassium-activated adenosine triphosphates. *Biochem. Biophys. Res. Commun.* 19:759, 1965.
136. Newey, H., and Smyth, D. H. Effects of sugars on intestinal transfer of amino-acids. *Nature* (London) 202:400, 1964.
137. Norman, A. W., Bieber, L. L., Lindberg, O., and Boyer, P. D. Relationships of Ca^{++} to "protein-bound" phosphate. *J. Biol. Chem.* 240:2855, 1965.

138. Ohnishi, T., and Kawamura, H. Rôle des phosphatides dans l'adénosine triphosphatase sensitive à l'ouabaine localisée dans les membranes d'erythrocyte. *J. Biochem.* (Tokyo) 56:377, 1964.
139. Opit, L. J., and Charnock, J. S. A molecular model for a sodium pump. *Nature* (London) 208:471, 1965.
140. Page, E. Cat heart muscle in vitro: VII. The temperature dependence of steady state K exchange in presence and absence of NaCl. *J. Gen. Physiol.* 48:949, 1965.
141. Page, E., Goerke, R. J., and Storm, S. R. Cat heart muscle in vitro: IV. Inhibition of transport in quiescent muscles. *J. Gen. Physiol.* 47:531, 1964.
142. Pappius, H. M. Water transport at cell membranes. *Canad. J. Biochem.* 42:945, 1964.
143. Philpott, C. W., and Copeland, D. E. Fine structure of chloride cells from three species of *Fundulus*. *J. Cell Biol.* 18:389, 1963.
144. Podolsky, R. J., and Constantin, L. L. Regulation by calcium of the contraction and relaxation of muscle fibers. *Fed. Proc.* 23:933, 1964.
145. Ponder, E. Permeability of human red cells to cations after treatment with resorcinol, *n*-butyl alcohol, and similar lysins. *J. Gen. Physiol.* 32:53, 1948.
146. Post, R. L., Merritt, C. R., Kinsolving, C. R., and Albright, C. D. Membrane adenosine triphosphatase as a participant in the active transport of sodium and potassium in the human erythrocyte. *J. Biol. Chem.* 235:1796, 1960.
147. Post, R. L., and Sen, A. K. An enzymatic mechanism of active sodium and potassium transport. *J. Histochem. Cytochem.* 13:105, 1965.
148. Post, R. L., Sen, A. K., and Rosenthal, A. S. A phosphorylated intermediate in adenosine triphosphate-dependent sodium and potassium transport across kidney membranes. *J. Biol. Chem.* 240:1437, 1964.
149. Pressman, B. C. Metabolic function of phosphohistidine. *Biochem. Biophys. Res. Commun.* 15:556, 1964.
150. Pressman, B. C. Induced active transport of ions in mitochondria. *Proc. Nat. Acad. Sci. U.S.A.* 53:1076, 1965.
151. Quastel, J. D. Transport at Cell Membranes and Regulation of Cell Metabolism. In Kleinzeller, A., and Kotyk, A. (Eds.), *Membrane Transport and Metabolism.* New York: Academic, 1961.
152. Rasmussen, H., Chance, B., and Ogata, E. A mechanism for the reactions of calcium with mitochondria. *Proc. Nat. Acad. Sci. U.S.A.* 53:1069, 1965.
153. Rhodin, J. A. G. Structure of the Kidney. In Strauss, M. B., and Welt, L. G. (Eds.), *Diseases of the Kidney.* Boston: Little, Brown, 1963.
154. Roberts, J. S., and Schmidt-Nielsen, B. Renal ultrastructure and excretion of salt and water by three terrestrial lizards. *Amer. J. Physiol.* 211:476, 1966.
155. Robinson, J. W. L., and Felber, J. P. The absorption of dibasic aminoacids by rat intestinal slices. *Biochem. J.* 343:1, 1965.
156. Rosenberg, T., and Wilbrandt, W. Uphill transport induced by counterflow. *J. Gen. Physiol.* 41:289, 1957.

157. Rossi, C. S., Bielawski, J., Carafoli, E., and Lehninger, A. L. The relationship of the Ca^{++} efflux rate to the "super-stoichiometry" of respiration-linked Ca^{++} accumulation by mitochondria. *Biochem. Biophys. Res. Commun.* 22:206, 1966.
158. Rossi, C. S., and Lehninger, A. L. Stoichiometric relationships between accumulation of ions by mitochondria and the energy-coupling sites in the respiratory chain. *Biochem. Z.* 338:698, 1963.
159. Saunders, St. J., and Isselbacher, K. J. Inhibition of intestinal amino acid transport by hexoses. *Biochim. Biophys. Acta* 102:397, 1965.
160. Schatzmann, H. Lipoprotein nature of red cell adenosine triphosphatase. *Nature* (London) 196:677, 1962.
161. Schatzmann, H. J. The role of Na^+ and K^+ in the ouabain-inhibition of the $Na^+ + K^+$-activated membrane adenosine triphosphatase. *Biochim. Biophys. Acta* 94:89, 1965.
162. Schmidt-Nielsen, B. Urea excretion in mammals. *Physiol. Rev.* 38:139, 1958.
163. Schultz, S. G., and Solomon, A. K. Cation transport in *Escherichia coli*: I. Intracellular Na and K concentrations and net cation movement. *J. Gen. Physiol.* 45:355, 1961.
164. Sedar, A. W. Fine structure of the stimulated oxyntic cell. *Fed. Proc.* 24:1360, 1965.
165. Sjöstrand, F. S. A new ultrastructural element of the membranes in mitochondria and of some cytoplasmic membranes. *J. Ultrastruct. Res.* 9:340, 1963.
166. Sjöstrand, F. S. A new repeat structural element of mitochondrial and certain cytoplasmic membranes. *Nature* (London) 199:1262, 1963.
167. Skou, J. C. The influence of some cations on an adenosine triphosphatase from peripheral nerves. *Biochim. Biophys. Acta* 23:394, 1957.
168. Skou, J. C. Enzymatic basis for active transport of Na^+ and K^+ across cell membrane. *Physiol. Rev.* 45:596, 1965.
169. Slater, E. C., and Cleland, K. W. Effect of calcium on respiratory and phosphorylative activities of heart-muscle sarcosomes. *Biochem. J.* 55:566, 1953.
170. Snell, F. M., and Chowdhury, T. K. Contralateral effects of sodium and potassium on the electrical potential in frog skin and toad bladder. *Nature* (London) 207:45, 1965.
171. Stahl, W. L., Sattin, A., and McIlwain, H. Separation of adenosine diphosphate–adenosine triphosphate-exchange activity from the cerebral microsomal sodium-plus-potassium ion-stimulated adenosine triphosphatase. *Biochem. J.* 99:404, 1966.
172. Swanson, P. D., and Stahl, W. L. The adenosine diphosphate–adenosine triphosphate-exchange reaction of cerebral microsomes and its relation to the sodium ion-stimulated adenosine-triphosphatase reaction. *Biochem. J.* 99:396, 1966.
173. Thier, S., Fox, M., Rosenberg, L., and Segal, S. Hexose inhibition of amino acid uptake in the rat-kidney-cortex slice. *Biochim. Biophys. Acta* 93:106, 1964.
174. Tormey, J. M. Significance of the histochemical demonstration of

ATPase in epithelia noted for active transport. *Nature* (London) 210: 820, 1966.
175. Tosteson, D. C. Regulation of Cell Volume by Sodium and Potassium Transport. In Hoffman, J. F. (Ed.), *The Cellular Functions of Membrane Transport.* Englewood Cliffs, N.J.: Prentice-Hall, 1964.
176. Tosteson, D. C. Some properties of HK and LK sheep red cell membrane fragments and their relevance to active transport of K and Na. *Trans. N.Y. Acad. Sci.* 27:970, 1965.
177. Tosteson, D. C., Cook, P., and Blount, R. Separation of adenosine triphosphatase of HK and LK sheep red cell membranes by density gradient centrifugation. *J. Gen. Physiol.* 48:1125, 1965.
178. Tosteson, D. C., and Hoffman, J. F. Regulation of cell volume by active cation transport in high and low potassium sheep red cells. *J. Gen. Physiol.* 44:169, 1960.
179. Ussing, H. H. Distinction by means of tracers between active transport and diffusion. The transfer of iodide across the isolated frog skin. *Acta Physiol. Scand.* 19:43, 1949.
180. Ussing, H. H. Active and Passive Transport Across Epithelial Membranes. In Coursaget, J. (Ed.), *The Method of Isotopic Tracers Applied to the Study of Active Ion Transport.* New York: Pergamon, 1959.
181. Ussing, H. H. Relationship between osmotic reactions and active sodium transport in the frog skin epithelium. *Acta Physiol. Scand.* 63:141, 1965.
182. Ussing, H. H. Transport of electrolytes and water across epithelia. *Harvey Lect.* 59:1, 1965.
183. Ussing, H. H., Biber, T. V., and Bricker, N. S. Exposure of the isolated frog skin to high potassium concentrations at the internal surface: II. Changes in epithelial cell volume, resistance, and response to antidiuretic hormone. *J. Gen. Physiol.* 48:425, 1965.
184. Ussing, H. H., and Windhager, E. E. Nature of shunt path and active sodium transport path through frog skin epithelium. *Acta Physiol. Scand.* 61:484, 1964.
185. Van Rossum, G. D. V. The effect of oligomycin on cation transport in slices of rat liver. *Biochem. J.* 84:35P, 1962.
186. Van Rossum, G. D. V. The effect of oligomycin on net movement of sodium and potassium in mammalian cells—in vitro. *Biochim. Biophys. Acta* 82:556, 1964.
187. Vasington, F. D., and Murphy, J. V. Ca ion uptake by rat kidney mitochondria and its dependence on respiration and phosphorylation. *J. Biol. Chem.* 237:2670, 1962.
188. Vidavar, G. A. Glycine transport by hemolyzed and restored pigeon red cells. *Biochemistry* (Washington) 3:795, 1964.
189. Vidavar, G. A. Mucate inhibition of glycine entry into pigeon red cells. *Biochemistry* (Washington) 3:799, 1964.
190. Vidavar, G. A. Some tests of the hypothesis that the sodium-ion gradient furnishes the energy for glycine-active transport by pigeon red cells. *Biochemistry* (Washington) 3:803, 1964.
191. Vidavar, G. A. Inhibition of parallel flux and augmentation of counter

flux shown by transport models not involving a mobile carrier. *J. Theor. Biol.* 10:301, 1966.

192. Whittam, R. Permeability of kidney cortex to chloride. *J. Physiol.* (London) 131:542, 1956.
193. Whittam, R. The asymmetrical stimulation of a membrane adenosine triphosphatase in relation to active cation transport. *Biochem. J.* 84:110, 1962.
194. Whittam, R. The Interdependence of Metabolism and Active Transport. In Hoffman, J. F. (Ed.), *The Cellular Functions of Membrane Transport.* Englewood Cliffs, N.J.: Prentice-Hall, 1964.
195. Whittam, R., and Ager, M. E. The connection between active cation transport and metabolism in erythrocytes. *Biochem. J.* 97:214, 1965.
196. Whittam, R., Ager, M. E., and Wiley, J. S. Control of lactate production by membrane adenosine triphosphatase activity in human erythrocytes. *Nature* (London) 202:1111, 1964.
197. Whittam, R., and Blond, D. M. Respiratory control by an adenosine triphosphatase involved in active transport in brain cortex. *Biochem. J.* 92:147, 1964.
198. Whittam, R., and Willis, J. S. Ion movements and oxygen consumption in kidney cortex slices. *J. Physiol.* (London) 168:158, 1963.
199. Wirz, H. Location of Antidiuretic Action in Mammalian Kidney. In Heller, H. (Ed.), *Colston Research Society. The Neurohypophysis: Proceedings of the 8th Symposium of the Society Held in the University of Bristol, April 9th–12th, 1956.* New York: Academic, 1957.
200. Wojtczak, L., and Lehninger, A. L. Formation and disappearance of an endogenous uncoupling factor during swelling and contraction of mitochondria. *Biochim. Biophys. Acta* 51:442, 1961.
201. Wool, I. G. Relation of effects of insulin on amino acid transport and on protein synthesis. *Fed. Proc.* 24:1060, 1965.
202. Zerahn, K. Oxygen consumption and active sodium transport in the isolated and short-circuited frog skin. *Acta Physiol. Scand.* 36:300, 1956.

5

Toward a Molecular Description of Active Transport

David Schachter[1]

WHAT WOULD a complete molecular description of active transport entail? Nothing less than a description in three dimensions of the structural components of active transport mechanisms and their mode of operation. To a biologist today this goal seems a peak to be scaled far in the future, when other fundamental biological processes, such as oxidative phosphorylation or the mechanism of enzyme action, are also understood in terms of three-dimensional mechanisms. It seems wiser at present to approach the problem of active transport by asking questions capable of solution by two-dimensional, rather than three-dimensional, representations.

Let us consider here two such questions. What is the nature, or possible nature, of transport carriers? What are the energy-coupling mechanisms which permit such carriers to bring about uphill transport? To review some of the experimental evidence now available to answer these questions, I will describe two transport mechanisms, the galactoside-accumulating mechanism of *Escherichia coli* and the active transport of calcium across the small intestinal mucosa of mammals. These mechanisms were chosen because our laboratory has worked actively with both and because they may be among the first to be understood at a biochemical level.

Supported by U.S. Public Health Service grants AM-01483 and AM-04407.

[1] Career Scientist of the Health Research Council of the City of New York (I-183).

GALACTOSIDE ACCUMULATION IN *E. COLI*

The study of transport in bacteria offers the physiologist many technical advantages and will, I think, play an increasingly important role in the field. Bacteria are easily grown in bulk, they provide uniform cell populations of very high density, and mutants can be designed by a variety of techniques. All these advantages are illustrated by the transport system which accumulates galactosides in *E. coli*. This transport was discovered by Monod and collaborators at the Pasteur Institute in the course of their studies on the metabolism of lactose (a β-galactoside) in *E. coli*.

The work of the Pasteur Institute group, summarized well in the Nobel Laureate Lectures of Monod [14] and Jacob [9], has demonstrated that lactose metabolism is under the control of three structural genes and two regulator genes. The three structural genes are the *z* gene, which controls the synthesis of β-galactosidase, the *y* gene, which controls the synthesis of a cellular component essential for galactoside accumulation, and the *a* gene, which controls the synthesis of the enzyme thiogalactoside transacetylase. In many strains of *E. coli* the products of the structural genes are found only when the bacteria are grown in the presence of a suitable β-galactoside, and the compound is said to induce the formation of these products. Studies designed to explore the nature of the induction process led to the discovery of two regulatory genes, *i* and *o*. The *o*, or operator, gene is required to turn the three structural genes on or off in concert. The entire unit of operator and three structural genes is termed an *operon*. The *i* gene elaborates a product, called the repressor, which binds to the operator gene, thereby preventing transcription of the structural genes. The repressor is a cytoplasmic protein, recently isolated [7] and demonstrated to have a relatively high affinity for β-galactosides, which induce the lactose operon. Thus an inducing galactoside binds the repressor, removes it from the operator gene, and thereby initiates transcription of the three structural genes of the operon.

The result of "derepressing" the lac operon can be followed experimentally by studying the metabolic fate of orthonitrophenylgalactoside (ONPG), a compound which yields orthonitrophenol on hydrolysis by β-galactosidase within the bacterial cell. Orthonitrophenol is readily detected and quantified by its bright yellow color in alkaline solution. Many strains of uninduced *E. coli* do not hydrolyze ONPG significantly, whereas following induction there is rapid hydrolysis.

Careful studies of the kinetics of hydrolysis indicate that the maximal rate of hydrolysis (V_m) and the apparent Michaelis-Menten constant (K_m) do not correspond to the respective values for β-galactosidase isolated from the bacterial cell. Rather, the maximal velocity of hydrolysis by intact bacteria is considerably slower and corresponds to an initial, rate-limiting step, penetration of the ONPG into the bacterial cell. Genetic studies have shown that transport into the bacterial cell requires a functioning *y* gene in the lactose operon, and considerable information is available concerning the kinetics of entry, the specificity of the mechanism, the effects of competitive inhibitors, and the effect of sulfhydryl inactivating agents such as N-ethylmaleimide, to indicate that the entry is a carrier-mediated process [10, 12]. It is noteworthy that the hydrolysis of ONPG by whole bacterial cells thus provides an estimate of the carrier-mediated transport function [12].

Another aspect of the transfer mechanism became evident when Monod and his collaborators studied the metabolic fate of ^{14}C-labeled thiogalactosides, such as thiomethylgalactoside (TMG). These compounds are not hydrolyzed by β-galactosidase, but a number of them are very effective inducers of the lactose operon, and they are transferred rapidly across the cell boundary and accumulated within the cell to concentrations considerably in excess of those in the surrounding medium. The features of this active transport have been reviewed by Kepes and Cohen [10]; it appears to be a true, active transport mechanism, capable of net transfer against a chemical potential. Accumulation of ^{14}C-TMG also provides a convenient experimental method for studying galactoside transport, and two good controls are available for such studies, cells uninduced for the lactose operon and mutant strains which have lost *y* gene but not *z* gene function. The latter strains have β-galactosidase within the cell but no efficient way of taking up galactoside from the medium, and they are called cryptic.

The transport system just described provides a useful didactic illustration of various types of transport across cell boundaries. Cryptic *E. coli* cells lack the specific transport carriers, and galactosides penetrate mainly by *simple diffusion* across a membrane largely impermeable to these water-soluble compounds. When suspensions of these cells are incubated with ONPG, little hydrolysis is observed until very high external concentrations of ONPG are used.

Now consider suspensions of a strain of *E. coli* induced for the lactose operon but treated with a metabolic inhibitor such as cyanide or 2,4-dinitrophenol. Although no accumulation (i.e., internal concentration of galactoside in excess of the external concentration) can be shown under these conditions, the function of the carriers, the products of the *y* gene, is demonstrable by estimating ONPG hydrolysis. In contrast to the cryptic cells, the rate of hydrolysis is relatively rapid and follows Michaelis-Menten kinetics. Here, therefore, is an example of facilitated or carrier-mediated diffusion. Finally, if, in the suspension just studied, the cells are washed to remove the metabolic inhibitor, they will transport ^{14}C-TMG actively; removal of the metabolic inhibitor has changed facilitated diffusion to active transport. Active transport, therefore, can result from facilitated diffusion coupled to energy-yielding biochemical reactions.

Identification of the y *Gene Product*

An important contribution to the subject was made by Fox and Kennedy [5, 6], who devised a chemical method for labeling the product of the *y* gene and subsequently isolated the material as a highly purified lipoprotein. The labeling procedure of Fox and Kennedy utilizes the phenomenon of protection of the essential sulfhydryl group of the protein. It had been known that galactoside transport is inhibited by reagents which inactivate —SH groups, e.g., N-ethylmaleimide, and that the inhibition is at least partially prevented by thiodigalactoside (TDG). The latter compound is a β-galactoside which is transported and accumulated and which has a high affinity for the transport system. The procedure adopted, therefore, was to prepare an induced and an uninduced culture of *E. coli,* the latter serving as a control. Both were treated with nonradioactive N-ethylmaleimide (NEM) in the presence of TDG. In this step NEM bound sulfhydryl groups in both cultures by covalent linkage. In the induced cells, however, some of the specific sulfhydryl groups of the product of the *y* gene were protected. The cultures were then washed. The induced cells were treated with ^{14}C-NEM and the uninduced with ^{3}H-NEM. The cell suspensions were combined, fractionated, and purified, using the ratio of ^{14}C/^{3}H to follow the isolation of the desired material. Enrichment for ^{14}C was initially found in the membrane particulate fractions of the preparations, and a soluble lipoprotein was eventually isolated after treating the particulate material with detergents.

The same workers demonstrated that this specifically labeled protein, which they call the M protein, is present only in cultures capable of transporting galactoside. In uninduced or cryptic cultures the protein is not found, and in strains with multiple copies of the *y* gene the amounts of protein are proportionately increased. Thus the genetic evidence supports the hypothesis that the protein is an essential part of the transport mechanism, either the carrier itself or a portion thereof.

Physiology of the Transport Carrier

These findings pose challenging questions for the physiologist. If the M protein is a carrier for galactoside transport, how does it bring about accumulation in the bacterial cell? We approached this question experimentally in the following way [25]. Consider a conventional model of a reversible carrier system (Fig. 5-1). The carrier-galactoside complex may be considered to have a dissociation constant at the outer border of the cell membrane (K_o) and at the inner border of the membrane (K_i). If the cell is incubated with cyanide or another metabolic inhibitor so that accumulation cannot occur, there will be a rapid

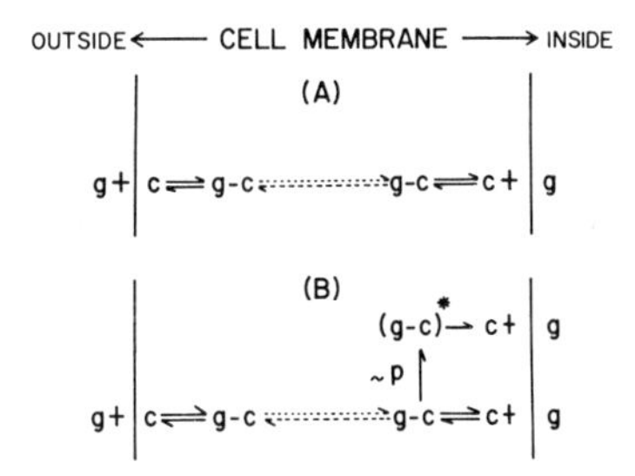

FIG. 5-1. Model of thiogalactoside transport in *Escherichia coli.* The equilibration of thiogalactoside across the cell membrane, in either direction, may result from a conventional carrier mechanism, as shown in (A), and represents the situation in the absence of energy metabolism. Free carrier (*c*) at the inner or outer aspect of the membrane combines with thiogalactoside (*g*) to form a complex (*g-c*) which can traverse the membrane in either direction. Dissociation of the complex at the opposite surface liberates free carrier, which can then either combine with and transfer a thiogalactoside molecule in the opposite direction or itself traverse the membrane to repeat the cycle. (The latter path is not shown in the diagram for simplicity.) The final result is equilibration of the inside and outside concentrations of thiogalactoside. This reversible shuttle mechanism is again shown in (B) and is referred to as path A in the text. The additional path shown in (B) is the proposed energy-coupling mechanism. High-energy phosphate intermediates (~p) convert the carrier-thiogalactoside complex to an intermediate (*g-c*)*, which dissociates at the inner aspect of the membrane. The energy-dependent influx is designated path B in the text.

equilibration of galactoside across both sides of the cell membrane owing to the shuttle action of the carrier. The dissociation constants for the carrier-galactoside complexes on the inner and outer boundaries of the membrane are equal, and the concentrations of galactoside on each side of the membrane are also equal. Now the culture is washed free of the metabolic inhibitor, and oxidative metabolism and accumulation of galactoside are allowed to proceed. How could metabolic energy be used for the active transport? At least two mechanisms seem feasible: (1) energy could decrease the affinity of the carrier for the galactoside at the inner border of the membrane, so that the dissociation constant would be increased and the fraction of total carriers bound to galactoside in the membrane decreased. The energy-coupling mechanism could therefore be said to "pull" galactoside into the cell; (2) energy could act at the outer border of the cell membrane, increasing the affinity of the carrier for galactoside, thereby reducing K_0 and increasing the fraction of carriers in the membrane bound to galactoside. This sort of mechanism could be termed a "push" mechanism.

Several investigators have attempted to distinguish between these alternative hypotheses. Koch [12] observed that metabolic inhibitors decrease influx of ONPG relatively little (although some reduction is observed) as compared to a marked increase in efflux of TMG. He suggested, therefore, that energy is required to prevent formation of the carrier-galactoside complex at the inner boundary of the membrane. The same conclusion was drawn by Winkler and Wilson [35], who studied influx and efflux of ONPG and lactose in a mutant of *E. coli* deficient in β-galactosidase. The latter authors noted a pronounced decrease in the apparent Michaelis-Menten dissociation constant for efflux on treatment with metabolic inhibitors. The results are not entirely exclusive, however, inasmuch as they used preloaded cells for the corresponding influx studies and considered that no counterflow occurred. As indicated below, counterflow is observed under specified conditions and may, in fact, become evident after addition of a metabolic inhibitor.

Recent studies in our laboratory have focused on the possibility of estimating the proportion of membrane carriers bound to galactoside, in order to define more conclusively the role of energy metabolism in accumulation. The method devised has been reported [25]; it utilizes the capacity of galactosides to protect carrier —SH from N-ethylmaleimide (NEM). Inasmuch as protection depends on formation of the carrier-galactoside complex, the carriers which survive treatment with NEM give an estimate of the proportion complexed in the membrane. The initial studies were with thiodigalactoside (TDG) alone, for other thiogalactosides seemed to be incapable of protecting the carrier. When the carrier was exposed to NEM for 20 minutes, at 23° C, distinct protection was observed in the presence of TDG, and, as tested with three different strains of *E. coli,* metabolic inhibition (cyanide or 2,4-dinitrophenol) markedly reduced the protection. Consequently it appeared that energy metabolism might be required to enhance formation of the TDG-carrier complex. Generalization to the role of energy in thiogalactoside transport, however, was difficult because other thiogalactosides did not protect the carrier under these conditions. Lately it has been possible to modify the technique so as to demonstrate protection with thiomethylgalactoside (TMG) and isopropylthiogalactoside (IPTG) as well as TDG. The modifications consist of exposing the cells to NEM at a lower temperature (10° C) for only one minute. TMG and IPTG protect much more poorly than TDG, but do so consistently under these conditions. (Contamination of the preparations of TMG, IPTG, or TDG with sulfhydryl-containing impurities has in each instance been excluded by estimating —SH according to the method of Ellman [3].) Preliminary experiments with TMG as protector indicate that cyanide inhibition increases protection. They suggest, therefore, that energy coupling for accumulation involves dissociation of the carrier-thiogalactoside complex, probably near the inner face of the cell membrane.

The model illustrated in Figure 5-1 was devised as a possible mechanism for the role of energy in thiogalactoside accumulation. It incorporates all the prior results and, more significantly, has predicted some new findings and explained some puzzling older ones. Two influx paths are predicted, one (path A) involving the reversible carrier mechanism and the other (path B) an energy-coupled essentially irreversible reaction which alters the galactoside-carrier configuration and leads to dissociation. Efflux, on the other hand, is via path A alone. To test these predictions, we first assumed that path A would show the phenomenon of counterflow, as a reversible, carrier-mediated transfer, and we hoped that path B might not. Thus counterflow would be demonstrable for efflux under all conditions and for influx only when transfer is via path A predominantly. Influx would be shifted from path B to path A in the presence of a metabolic inhibitor. Moreover, it seemed reasonable to predict from prior studies that influx would shift from path A to path B at low temperatures. Kepes and Cohen [10] pointed out that the plateau level of accumulation of ^{14}C-TMG may increase as the ambient temperature decreases, suggesting that efflux is more temperature sensitive than influx. Koch [12] noted marked reductions in exit rate constants for ^{14}C-TMG at lower temperatures. The increase in plateau level of accumulation of ^{14}C-TMG as the temperature is lowered is readily understood in terms of the proposed model, particularly if one considers the limiting conditions. If low temperature completely inhibited path A, accumulation via path B would approach very high plateau values, being limited only by the relatively inefficient process of passive leak from the cell. With increasing temperature, in contrast, path A would predominate, and if influx over path B became insignificant, no accumulation at all would be observed, only a rapid equilibration between the interior and exterior of the cell.

Experiments were designed to test for counterflow in the efflux direction and in the influx direction at temperatures varying from 5° to 35° C. To test in the efflux direction, cells were preloaded with ^{14}C-TMG, washed at 2° C, and subsequently added to medium containing either ^{12}C-TMG or none; the efflux rate constants were estimated [12] by filtering aliquots on millipore filters at appropriate intervals. To test in the influx direction, cells were either preloaded or not with ^{12}C-TMG, and subsequently ^{14}C-TMG was added and the initial influx estimated from a series of samples filtered within the first 15–60 seconds; linear uptake was observed throughout. The results of studies with *E. coli* strains ML 30 and ML 308 support the model proposed in Figure 5-1B. Counterflow was easily demonstrable in the efflux direction at all the temperatures tested. In the influx direction, however, counterflow was observed only at temperatures greater than either 15°–20° C for strain ML 308 or 10°–15° C for ML 30. The results clarify an apparent discrepancy in the literature, Koch [12] having found counterflow in the efflux direction whereas Winkler and Wilson [35] could not observe counterflow at 10° C in the influx direction.

It is instructive to consider the fluxes of TMG at 10° C in *E. coli* ML 308, conditions under which no counterflow is observed in the influx direction, yet marked counterflow is noted for efflux. One explanation of the mechanism of counterflow frequently advanced is that it results from more rapid mobility of loaded as compared to unloaded membrane carriers. If this be a principal mechanism, the foregoing results clearly exclude a single pathway for thiogalactoside transport such as is illustrated in Figure 5-1A, inasmuch as the difference in mobility would influence the flux in both directions. The model in Figure 5-1B, on the other hand, suggests additional mechanisms for the observed results. Consider the branch point in the reversible path A at which the energy-dependent path B is assumed to begin. The energy-coupled path returns thiogalactoside taken up by carrier at the inner face to the interior of the cell, in effect preventing efflux, as noted by Koch [12], and reducing the *apparent* Michaelis-Menten dissociation constant for efflux as described by Winkler and Wilson [35]. Now suppose the cell were loaded with ^{14}C-TMG and tested for counterflow in the efflux direction as described above. Delivery of carrier-^{12}C-TMG (via path A) from the exterior of the cell to the branch point would dilute the carrier-^{14}C-TMG at this point and compete with it for the energy-coupled reaction. Hence efflux of ^{14}C-TMG would be favored and counterflow observed. Repetition of the experiment, testing for counterflow in the influx direction, reveals the following. At 10° C and with no prior preloading of the cell, influx of ^{14}C-TMG is mainly along path B. If the cell is preloaded with ^{12}C-TMG, the delivery of carrier-^{12}C-TMG from the interior of the cell to the branch point dilutes the carrier-^{14}C-TMG at this point and competes for the energy-dependent influx path. Hence counterflow, which might be expected to result from increased mobility of the carrier, can be counteracted. At higher temperatures influx over path A becomes enhanced, the increase in carrier mobility with preloading may predominate, and counterflow in the influx direction is observed.

The preceding suggests that a shift in influx from path B to path A, at 10° C, might result in demonstrable counterflow (influx direction) at that temperature. We have tested this possibility by treating the cells with cyanide to inhibit path B. Influx of ^{14}C-TMG was estimated with cells either preloaded or not with ^{12}C-TMG, exactly $2\frac{1}{2}$ minutes after addition of cyanide. Whereas no counterflow was observed with uninhibited cells, a marked counterflow was observed after cyanide treatment. This is additional evidence in favor of the proposed model and suggests also that carriers involved in the energy-coupled pathway may become available for influx via the energy-independent pathway in the presence of a metabolic inhibitor.

In contrast to the product of the *y* gene, which appears to be a membrane-bound component of an active transport mechanism for galactosides, a number of soluble proteins capable of binding transported materials have recently been isolated from bacteria. These proteins are obtained by the "osmotic shock" procedure of Neu and Heppel [15], which releases enzymes and other proteins into solution. Soluble proteins which bind sulfate [17], leucine [18], and galactose [1] have thereby been identified. In the case of the sulfate-binding protein a good correlation exists between the occurrence of the binding protein and the capacity of the *Salmonella* strain to accumulate sulfate. While these proteins seem clearly related to the respective transport mechanisms involved, the precise relationship needs further investigation. They could be membrane carriers released during the process of osmotic shock, or parts of membrane transport mechanisms, or they may play some ancillary role in bacterial transport not yet defined.

CALCIUM TRANSPORT IN THE SMALL INTESTINE

The attempt to isolate a carrier for active transport led us some years ago to initiate studies on the absorption of calcium by mammalian small intestine. It had long been recognized that vitamin D is necessary for normal absorption of the mineral, and in vitamin D deficiency a generalized calcium deficiency results, leading to abnormally soft bones and the syndrome of rickets in growing children. Inasmuch as vitamin D is a sterol, it seemed intuitively reasonable that, like cholesterol, it might be found primarily in the cell membranes of the mucosal cells and function there as a carrier or prosthetic group for calcium transport. The hypothesis was attractive too because studies in whole organisms have shown that calcium absorption is at least partially adaptive, changing with the needs of the organism. Calcium requirements vary widely throughout life and are increased in the growth period and in the final stages of pregnancy when the mineral is required for fetal bones. Moreover, low dietary calcium results in increased efficiency of the absorptive mechanism, and the reverse is true with high calcium diets. The necessity for control suggested that active transport plays a role in the absorption, for a pump mechanism would provide the flexibility needed for adaptation.

At the outset we adapted and developed a number of in vitro techniques to define the nature of the intestinal calcium transport. Initially, using the everted gut sac technique [34], we demonstrated net transport of calcium from the mucosal to the serosal surface of rat duodenum in vitro against both concentration and electrical gradients [21, 23, 26]. Many of the features of active transport were also characteristic of the calcium transport. Oxygen deprivation, metabolic inhibitors, low temperatures, failure to include a metabolizable substrate in the medium—all inhibited the transfer. The transport was found to be relatively specific for calcium ion as compared to magnesium, barium, strontium, and potassium. Competition for transport was observed with a number of other substances including potassium, strontium, magnesium, actively transported hexoses, and divalent iron.

Because calcium is complexed by various anions, it is also necessary to consider whether transport of the mineral is a true cation pump or merely a metabolically dependent transfer of a calcium anion complex. Recently we have used the calcium-specific divalent electrode to examine this question [30]. During the course of a typical experiment in which net transfer of calcium occurs across everted gut sacs in vitro, the chemical potential of the calcium ion on the serosal surface, estimated directly with the specific electrode, increases relative to that at the mucosal surface. The result is the development of high activity gradients for calcium ion serosal/mucosal, with values generally in the range of 2.5–3.5. Thus the active transport of calcium involves a cation pump.

In addition to the everted gut sac technique, a number of other methods have been developed to study calcium transport in vitro. Full-thickness slices of duodenum, for example, are incubated in a medium containing radioactive calcium, and the uptake of the isotope is estimated. Uptake under oxygen minus uptake under nitrogen affords an estimation of the active accumulation [24]. Fluxes of Ca across segments of rat duodenum have also been studied in short-circuited preparations [21]. Influx of calcium—i.e., transfer from the mucosal to the serosal surface in vitro—greatly exceeded transfer in the opposite direction with no electrical potential across the intestine, and influx/efflux ratios of approximately 10 were observed.

Many of the transport phenomena observed in vitro have been confirmed by experiments in vivo. These include loop experiments in which segments of rat intestine are filled with a solution containing radioactive calcium, the loops being subsequently removed and the residual isotope estimated. Recirculation techniques, in which segments of intestine are intubated and a solution containing radioactive calcium is recirculated for a period of time, have also been used [31].

These techniques have demonstrated active transport of calcium across the intestine in vivo.

Finally, preparations of intestinal villi [19] or isolated mucosal cells [8] have been studied. One should bear in mind that uptake or efflux of radioactive calcium from isolated cells is somewhat difficult to interpret, inasmuch as these cells are normally polarized with respect to entry and exit.

Adaptive Role of the Calcium Pump

It is easily demonstrable that changes in the active transport mechanism underlie adaptive changes in calcium absorption to meet the needs of the organism [23]. Thus everted gut sacs prepared from young, growing animals transfer calcium much more readily than do similar preparations from older, nongrowing rats. Gut sacs prepared from pregnant rats in the last week of pregnancy transfer the cation much more readily than do segments from nonpregnant female controls. A low-calcium diet is a most effective means of greatly stimulating the active transport of calcium by segments of rat intestine in vitro. Whereas normal animals on a casual laboratory diet demonstrate active transport of calcium only with the proximal one-quarter of the small intestine, after a period of four–five weeks on a low-calcium diet almost all segments of the small intestine show active transport in vitro [11].

The pathways mediating this response to a low-calcium diet are not known. The possibility that endocrine pathways are involved was investigated by ablating either the pituitary, adrenal, or parathyroid glands. None of the ablation experiments completely prevented the usual response of increased active transport in vitro on a low-calcium diet. However, hormonal control of the absorptive mechanism was demonstrated. The most striking effect was a marked decrease in the calcium transport observed two–three weeks after hypophysectomy in the rat [4]. A systematic study of replacement with various anterior pituitary and end-organ hormones showed that only bovine growth hormone, and to a small extent prolactin, could restore the active transport mechanism in the hypophysectomized rats [4]. Removal of the adrenal glands resulted in a small increase in the calcium transport, and, correspondingly, large doses of cortisone or hydrocortisone decreased the transport. Total thyroparathyroidectomy resulted in a 20–30 percent decrease in calcium transport, as observed in vitro, approximately two–three weeks after the operation. Although large doses of vitamin D or dihydrotachysterol restore the transport in such animals, we have to date not been able to restore the mechanism consistently by administration of parathyroid extract. The foregoing adaptive effects of growth, pregnancy, or calcium deprivation appear to be mediated by increases in the active *cation* pump for calcium, i.e., by increases in the activity gradients for Ca^{++} serosal/mucosal as recently demonstrated with the specific calcium electrode [30].

The localization of the calcium transport mechanism in the small intestine provides a clear example of specialization of various segments of this organ. In most small mammals, including the rat, rabbit, guinea pig, and mouse, maximal transport is seen in the proximal duodenum with a decided decrease in the distal segments. In addition, there is evidence in both the chicken [20] and man [33] that the duodenum is specialized for calcium absorption too. In the golden hamster [23], however, and in the frog [22], maximal activity was observed in the distal ileum. Although the significance of segmental location is not yet understood, it is reasonable to suggest that the duodenal and ileal sites may have arisen at different stages in evolution. Recently it has been observed in our laboratory that the ileal mechanism appears to predominate in newborn rats and mice, with a shift to the adult pattern at approximately two–three weeks of age [2]. Moreover, when rats are placed on a low-calcium diet, the maximal increase in calcium transport is always observed in the duodenum, but the area of next greatest stimulation is the distal ileum. As indicated below, the duodenal and ileal mechanisms may differ functionally as well as in their anatomic localization.

Role of Vitamin D

Vitamin D plays an essential role in the calcium transport mechanism. When rats are placed on a vitamin D–deficient diet and maintained in cages shielded from the light, a marked decrease in the active transport mechanism is observed [26]. Restoration of the calcium transport follows large doses of vitamin D, with onset of the process at approximately two hours after the sterol and almost complete restoration by six hours; small, physiological doses of vitamin D require approximately 48 hours. Ultraviolet irradiation is also effective.

The lag period required for the action of vitamin D suggested that a sequence of reactions would be necessary for restoration of the mechanism. Moreover, direct addition of vitamin D in various vehicles to intestinal segments from deficient animals in vitro consistently failed to produce an effect on the transport mechanism. Thus it was conceivable that the vitamin might require transformation in some organ other than the small intestine prior to the ultimate action in the mucosa. The use of radioactive vitamin D labeled with either ^{14}C or tritium for studies on the metabolic fate of the sterol provided background information necessary to pursue its physiological role further. It could be shown, for example, that the time lag in the action of the vitamin was not necessary for accumulation of the sterol in the intestinal mucosa [22]. In addition, it was possible to devise an indirect experiment which demonstrated that vitamin D need not be metabolized in an organ outside of the intestinal mucosa prior to its action [27]. If the active metabolites of vitamin D are formed, as the data of DeLuca and his colleagues suggest [13], they can presumably be formed within the intestinal mucosa. The lag period in the action of the vitamin suggested, further, that protein biosynthesis might be involved in the restoration of the calcium transfer. The administration of actinomycin D, an inhibitor of DNA-dependent RNA synthesis, or puromycin, which inhibits protein biosynthesis at a subsequent stage, strongly inhibited the action of vitamin D on calcium transport [16, 27, 36]. These observations in turn led to the hypothesis that the function of the vitamin is to control or maintain the activity of an essential protein component of the transport mechanism itself or a precursor thereof [22].

The foregoing hypothesis appeared to be strikingly confirmed by the observations of Wasserman and Taylor [32], which opened a new chapter in the area. These investigators discovered a calcium-binding protein in the soluble fraction of mucosal homogenates prepared from chicken duodenum. The protein is dependent on vitamin D and corresponds in distribution within the chicken intestine to the calcium transport mechanism. In our laboratory a similar calcium-binding protein activity was detected in the soluble portion of rat mucosal homogenate. In the rat the calcium-binding activity corresponds to the calcium transport mechanism with regard to the following variables: distribution within the small intestine, adaptive variation with age, pregnancy, and calcium level in the diet [30].

What is the role of the calcium-binding protein? Is it a carrier for calcium transport or part of a carrier? These questions are not yet answered, and indeed final proof will be difficult. Evidence that the calcium-binding protein may not be the carrier itself has come from a study of calcium transport in the golden hamster. As indicated above, in this species transport is maximal in the distal ileum rather than in the duodenum. In contrast, the calcium-binding activity, estimated in supernatants of intestinal homogenates by a method similar to that used for the chicken and rat, is maximal in the duodenum. The ileal mechanism of the hamster is vitamin D dependent, and studies with the calcium electrode show that it is a true cation pump. Further studies are needed to exclude a particulate-bound calcium-binding protein in the ileal mucosa of the hamster, inasmuch as the assay for calcium-binding activity has thus far been applied only to the soluble supernatant fraction of mucosal homogenates. It seems prudent, however, to consider an alternative role for the calcium-binding protein. The protein may function primarily to regulate the intracellular free calcium concentration and would be increased, therefore, in cells particularly concerned with transcellular calcium transport.

WORKING MODEL OF THE CALCIUM TRANSPORT MECHANISM

Since the initial studies with everted gut sacs, a number of observations have suggested that the calcium transport involves at least two steps. When the disappearance of calcium from the mucosal medium was compared to the net appearance of calcium in the serosal medium, the two net transfers were seen to have behaved differently. For example, when the transport was studied as a function of the initial calcium concentration in the ambient mucosal and serosal medium, net uptake of calcium at the mucosal surface remained proportional to the initial calcium concentration in the range 0.4–1.0 mM [23]. In contrast, net transport to the serosal surface reached a plateau at approximately 0.4 mM and was the rate-limiting step. When the effects of various hexoses on the transport were studied, high calcium gradients serosal/mucosal were found to require a metabolizable hexose in the medium, e.g., fructose or mannose. The hexose was required primarily for transfer to the serosal surface, and the mucosal uptake showed little dependence. The results suggested that the transport might involve two distinct steps, (1) *entry* into and (2) *exit* from the mucosal cell. Inasmuch as exit was more readily rate limited and required a metabolizable hexose in the medium, it appeared to be the active, energy-requiring step. Entry was also found to involve rate limitation when calcium concentrations in excess of 1 mM were studied. Thus entry could represent facilitated (carrier-mediated) diffusion.

Observations with duodenal gut sacs prepared from vitamin D–deficient and –repleted rats, incubated with various initial concentrations of calcium in the ambient media, indicated that vitamin D is required for both entry and exit. More recent experiments provide additional evidence to support the entry-exit model. In one set of experiments everted duodenal gut sacs were incubated with radioactive calcium in the usual manner except that at the end of the incubation the sacs were drained and the mucosal tissue was scraped from the underlying intestinal coats [29]. The final concentrations of radioactive calcium in the mucosal medium, mucosal tissue layer, underlying coats, and serosal medium were estimated. A characteristic pattern was found for duodenal segments capable of active transport of calcium. Following incubation the concentration of radioactive calcium was much higher in the mucosal tissue as compared to the mucosal medium, and the isotope concentration in the underlying coats and serosal medium was in turn higher than that in the mucosal tissue. A downhill concentration difference was noted from the underlying coats to the serosal medium [29]. This pattern is stable after two hours of incubation, and under such circumstances the isotope concentration in the serosal medium reflects that in the interstitial water of the underlying coats and mucosal tissue. Inasmuch as the total concentration of isotope in the mucosal tissue (i.e., in cells plus interstitial fluid) is well below that of the underlying coats and serosal medium, it is reasonable to conclude that the concentration of radioactive calcium in the mucosal cells alone is less than that in the interstitial fluid of the mucosal tissue. Hence the transfer of calcium from the mucosal cell to the mucosal interstitial fluid apparently involves an uphill transfer. As would be expected, the uphill concentration difference between the mucosal tissue and the underlying coats and serosal medium was seen only under conditions in which net, active transport of calcium across the gut was observed. Thus in jejunal and ileal segments the concentration of radioactive calcium decreased from the mucosal tissue to the underlying coats and serosal medium. Similarly, metabolic inhibitors completely abolish the uphill concentration difference which reflects the exit step of the transport.

In a second series of experiments the effects upon calcium transport of replacing sodium chloride in the medium with isosmotic mannitol were studied. When 25 percent of the sodium chloride had been replaced with mannitol, net uptake of calcium at the mucosal surface and net transport to the serosal surface were both increased. With further replacement of sodium chloride, however, net transport to the serosal surface progressively decreased almost to zero whereas net uptake at the mucosal surface remained constant. Thus replacing sodium chloride with mannitol strikingly dissociates the two stages of the calcium transport and leads to a marked reduction in the exit step. The observation suggests that sodium chloride may be required for energy-dependent, active transfer of calcium out of mucosal cells.

In further experiments the concentrations of radioactive calcium in the tissue compartments were studied after replacement of sodium chloride by mannitol. In confirmation of the observations on net transport, the usual uphill concentration difference from mucosal tissue to underlying coats and serosal medium is reversed when sodium chloride is replaced by mannitol. The observations are in accord with the hypothesis that the exit step is the uphill transport.

We can now attempt to synthesize the experimental evidence into a working model of calcium transport (Fig. 5-2). We will assume for the present that the calcium-binding protein is a soluble, cytoplasmic carrier for calcium transport, dependent upon vitamin D for maintenance of its function. The vitamin might be necessary for the synthesis or activation of this protein, or to prevent its destruction. The first or entry stage of calcium transport, then, would be diffusion of calcium across the mucosal border of the intestinal cell and its subsequent binding to calcium-binding protein within the cytoplasm. Because of the finite number of calcium-binding protein molecules, the entry would be rate limited. The complex of calcium and calcium-binding protein would then diffuse to the serosal face of the cell where an energy-dependent process would effect extrusion of calcium against a chemical potential. One possible mechanism for the extru-

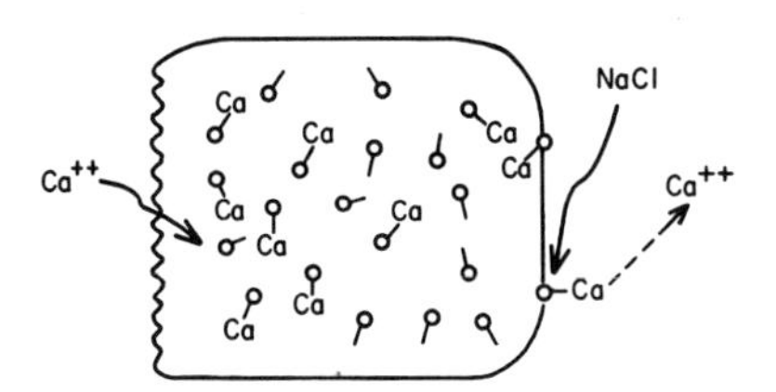

FIG. 5-2. A mucosal cell containing calcium-binding protein (o⁻) in the cytoplasm. Calcium diffuses across the luminal border and is bound to the protein to form a complex (Ca-o), which diffuses to the opposite side and enters the plasma membrane. Calcium is released when the complex is exposed to the extracellular NaCl concentration, accounting for the uphill transfer out of the cell. Vitamin D is required to maintain the carriers for calcium transport.

sion might involve incorporation of the entire complex into the membrane at this interface, followed by displacement of calcium by the extracellular sodium to which the complex is now exposed. Indeed, experiments with partially purified preparations of calcium-binding protein from rat duodenal mucosa show that sodium ion in the range of 50–145 mM does compete effectively with calcium for binding to the protein. Further studies are now under way to isolate the rat calcium-binding protein and to develop immunochemical procedures for its identification and quantification, in order to evaluate the role of the protein in the calcium transport mechanism.

SUMMARY AND CONCLUDING REMARKS

The experimental results described have provided a firmer basis for the carrier theory of active transport by identifying a number of proteins which may be transport carriers. It seems relatively safe to predict that within a few years several transport carriers will have been satisfactorily identified and isolated. In the next stage of the campaign, therefore, at least two important questions will have to be solved experimentally: first, by what mechanisms transport carriers are coupled to exergonic metabolic reactions to effect active transport; and second, by what mechanisms membrane-bound carriers become incorporated into cell membranes and function within the architecture of those organelles. At present the bacterial systems described above would appear to offer the best opportunities for further experimental investigations aimed toward a complete definition of the molecular basis for active transport.

REFERENCES

1. Anraku, Y. The reduction and restoration of galactose transport in osmotically shocked cells of *Escherichia coli*. *J. Biol. Chem.* 242:793, 1967.
2. Batt, E. R., and Schachter, D. Unpublished data, 1967.
3. Ellman, G. L. Tissue sulfhydryl groups. *Arch. Biochem.* 82:70, 1959.
4. Finkelstein, J. D., and Schachter, D. Active transport of calcium by intestine: Effects of hypophysectomy and growth hormone. *Amer. J. Physiol.* 203:873, 1962.
5. Fox, C. F., Carter, J. R., and Kennedy, E. P. Genetic control of the membrane protein component of the lactose transport system of *Escherichia coli*. *Proc. Nat. Acad. Sci. U.S.A.* 57:698, 1967.

6. Fox, C. F., and Kennedy, E. P. Specific labeling and partial purification of the M protein, a component of the β-galactoside transport system of *Escherichia coli*. *Proc. Nat. Acad. Sci. U.S.A.* 54:891, 1965.
7. Gilbert, W., and Müller-Hill, B. Isolation of the lac repressor. *Proc. Nat. Acad. Sci. U.S.A.* 56:1891, 1966.
8. Hashim, G. A., and Clark, I. Effect of vitamin D on the uptake and release of calcium by rat intestinal mucosal cells. *Fed. Proc.* 26:697, 1967.
9. Jacob, F. Genetics of the bacterial cell. *Science* 152:1470, 1966.
10. Kepes, A., and Cohen, G. N. Permeation. In Gunsalus, I. C., and Stanier, R. (Eds.), *The Bacteria*. New York: Academic, 1962. Vol. 4.
11. Kimberg, D. V., Schachter, D., and Schenker, H. Active transport of calcium by intestine: Effects of dietary calcium. *Amer. J. Physiol.* 200:1256, 1961.
12. Koch, A. L. The role of permease in transport. *Biochim. Biophys. Acta* 79:177, 1964.
13. Lund, J., and DeLuca, H. Biologically active metabolite of vitamin D_3 from bone, liver and blood serum. *J. Lipid Res.* 7:739, 1966.
14. Monod, J. From enzymatic adaptation to allosteric transitions. *Science* 154:475, 1966.
15. Neu, H. C., and Heppel, L. A. The release of enzymes from *Escherichia coli* by osmotic shock and during the formation of spheroplasts. *J. Biol. Chem.* 240:3685, 1965.
16. Norman, A. W. Vitamin D mediated synthesis of rapidly labeled RNA from intestinal mucosa. *Biochem. Biophys. Res. Commun.* 23:335, 1966.
17. Pardee, A. B. Purification and properties of a sulfate-binding protein from *Salmonella typhimurium*. *J. Biol. Chem.* 241:5886, 1966.
18. Piperno, J. R., and Oxender, D. L. Amino acid-binding protein released from *Escherichia coli* by osmotic shock. *J. Biol. Chem.* 241:5732, 1966.
19. Rasmussen, H., Waldorf, A., Dziewiatkowski, D. D., and DeLuca, H. F. Calcium exchange in isolated intestinal villi. *Biochim. Biophys. Acta* 75:250, 1963.
20. Sallis, J. D., and Holdsworth, E. S. Influence of vitamin D on calcium absorption in the chick. *Amer. J. Physiol.* 203:497, 1962.
21. Schachter, D. Vitamin D and the Active Transport of Calcium by the Small Intestine. In Wasserman, R. H. (Ed.), *The Transfer of Calcium and Strontium Across Biological Membranes*. New York: Academic, 1963.
22. Schachter, D. Vitamin D and the Intestinal Transport of Calcium and Phosphate. In Hioco, D. J. (Ed.), *L'Ostéomalacie*. Paris: Masson, 1967.
23. Schachter, D., Dowdle, E. B., and Schenker, H. Active transport of calcium by the small intestine of the rat. *Amer. J. Physiol.* 198:263, 1960.
24. Schachter, D., Dowdle, E. B., and Schenker, H. Accumulation of Ca^{45} by slices of the small intestine. *Amer. J. Physiol.* 198:275, 1960.
25. Schachter, D., Johnson, N., and Kirkpatrick, M. A. Energy coupling for galactoside accumulation in *Escherichia coli*. *Biochem. Biophys. Res. Commun.* 25:603, 1966.
26. Schachter, D., Kimberg, D. V., and Schenker, H. Active transport of calcium by intestine: Action and bioassay of vitamin D. *Amer. J. Physiol.* 200:1263, 1961.

27. Schachter, D., and Kowarski, S. Radioactive vitamin D: Preparation, metabolism and mechanism of action. *Bull. N.Y. Acad. Med.* 41:241, 1965.
28. Schachter, D., Kowarski, S., and Finkelstein, J. D. Vitamin D_3: Direct action on the small intestine of the rat. *Science* 143:143, 1964.
29. Schachter, D., Kowarski, S., Finkelstein, J. D., and Ma, R.-I. W. Tissue concentration differences during active transport of calcium by intestine. *Amer. J. Physiol.* 211:1131, 1966.
30. Schachter, D., Kowarski, S., and Reid, P. Molecular basis for vitamin D action in the small intestine. *J. Clin. Invest.* 46:1113, 1967.
31. Wasserman, R. H., Kallfelz, F. A., and Comar, C. L. Active transport of calcium by rat duodenum *in vivo. Science* 133:883, 1961.
32. Wasserman, R. H., and Taylor, A. N. Vitamin D_3-induced calcium-binding protein in chick intestinal mucosa. *Science* 152:791, 1966.
33. Wensel, R. H., Rich, C., and Volwiler, W. Studies of calcium absorption and endogenous secretion in normal man. *Gastroenterology* 46:768, 1964.
34. Wilson, T. H., and Wiseman, G. The use of sacs of everted small intestine for the study of the transference of substances from the mucosal to the serosal surface. *J. Physiol* (London) 123:116, 1954.
35. Winkler, H. H., and Wilson, T. H. The role of energy coupling in the transport of β-galactosides by *Escherichia coli. J. Biol. Chem.* 241:2200, 1966.
36. Zull, J. E., Czarnowska-Misztal, E., and DeLuca, H. F. Actinomycin D inhibition of vitamin D action. *Science* 149:182, 1965.

6
Intracellular Transport

I. Robert Fenichel and Samuel B. Horowitz

OUR KNOWLEDGE OF THE ROLE of intracellular barriers in the material transactions of the cell is in an early stage. One reason is that the preeminent role played by the plasma membrane in cellular material transactions has directed attention away from intracellular contributions. Furthermore, the presence of a surface membrane barrier greatly increases the difficulty of performing and interpreting experiments intended to elucidate intracellular transport processes. However, with the application of new techniques it has become increasingly evident that internal material transactions are not trivial. Not only may these on occasion dominate the cortical membrane in determining transactions with the environment, but also intracellular transport processes reveal themselves to be of great complexity, specificity, and importance to the functioning of the cell.

The body of observations that reflect on intracellular transport is not, at present, sufficient to support a coherent model of general applicability. Rather than have this chapter become a compendium of numerous weakly related phenomena in an attempt to include all intracellular processes which subserve a transport function, we have chosen to discuss work that reflects on what may be expected to be the most general features of such a model. We have had to exclude from consideration, therefore, processes like cytoplasmic streaming and saltatory movements [76], as well as transport in, for example, mitochondria and chloroplasts. We will be concerned primarily with three

This work was supported in part by grants from the U.S. Public Health Service and the National Science Foundation.

areas: diffusion in the cytoplasmic matrix, intracellular binding, and the transport functions of intracellular membranes.

DIFFUSION

The diffusional mixing of intracellular soluble molecules in the volumes accessible to them occurs at a finite rate. When some type of diffusional mixing forms part of a cellular process, the possibility exists that the rate of the latter is limited by the rate of mixing—i.e., that the process is diffusion controlled. Whether diffusional control occurs depends upon the diffusing molecules, the cell, and the nature of the process itself. More specifically, the determinants are the diffusion coefficient, D; the initial and final distribution of diffusing mole-

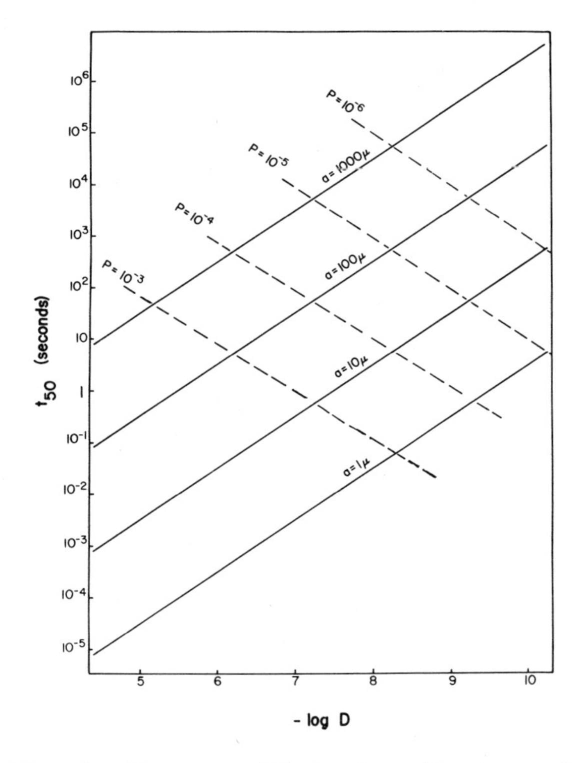

FIG. 6-1. Time for 50 percent diffusional equilibration, and membrane permeation half-time, in spherical cells with indicated *a, D,* and *P*. See text.

cules and the diffusional path length; and the nondiffusional rate constants involved in the process.

Figure 6-1 shows the conditions for diffusional control when mixing occurs inward from the cell surface. Several examples from biological systems can be discussed with the aid of this diagram, which is constructed as follows. We start with a spherical cell of radius a initially free of the solute in question and exposed to the solute in the extracellular solution. For any value of a, and with an intracellular diffusion coefficient D, the time for 50 percent diffusional equilibration of the intracellular space, t_{50}, is given by the solid lines. The times are not strictly half-times because diffusion is a multi-exponential process. For other conditions, such as the production of a material at an intracellular source, or its consumption at an intracellular sink, similar lines relating the diffusion from or to the source or sink can be drawn; the time constants involved would be similar to those in Figure 6-1.

The time constants range over nine orders of magnitude, from about 0.1 msec to 10^5 sec. This range seems extreme, yet it corresponds to a conventional range of diffusion coefficients which might be encountered in a biological milieu, and to a familiar range of cell types, including, for example, bacteria at the lower limit of size and amphibian oocytes at the upper. The reason for the great range of time constants is that they depend on the square of the diffusional path length: $t_{50} \sim a^2/D$. For this reason the influence of diffusion on any cellular rate process will be most sensitively dependent on the diffusional path length involved.

To assess the importance of the diffusional time constants for a biological process one needs, in addition, the appropriate nondiffusional constants. For the particular example of transmembrane material flow, the required constants are determined by the membrane permeability. These are given in Figure 6-1 by the broken lines, which include a conventional range of permeability coefficients, P, in centimeters per second. The ordinate corresponding to the intersection of a solid and a broken line gives the half-time of membrane permeation, $t_{1/2}$ (in this case a true half-time), for a spherical cell with the designated values of a and P. The abscissa gives the value of D for which t_{50} is the same. From the two sets of lines it is then possible to determine, for any cell size, the ranges of values of P and D for which intracellular diffusion is or is not important in determining transmembrane fluxes.

Detailed analysis shows that for $t_{1/2} \approx t_{50}$, diffusion is almost entirely rate limiting, whereas for $t_{1/2} \geqq 10\ t_{50}$, diffusional mixing is negligibly rapid compared to membrane permeation.

It is apparent from Figure 6-1 that for diffusion coefficients of a magnitude found in aqueous solution, even allowing for some slowing due to the presence of intracellular dissolved macromolecules [89], in the vast majority of cells one would expect diffusional mixing to be unimportant in determining the rate of permeation of external solutes. There are notable exceptions, and we will be concerned with some of them.

Exceptions can arise in several ways. Important among these are two which are not, strictly speaking, purely diffusional: intracellular binding and intracellular structural transport barriers. These will be discussed in later sections. In addition, the following will favor a diffusional effect on permeation: large cell size; high membrane permeability; small cytoplasmic diffusion coefficient; and surface redundancy, such as occurs in microvillous absorptive epithelium. It is instructive to examine these points in the context of recent work on the permeability of amphibian oocytes to small molecules.

Diffusion of Water in Oocytes

The amphibian oocyte [49, 50, 92, 93] is a large cell; the radius of a fully grown *Rana* oocyte is about 800μ. The surface is covered with microvilli. During the active growth of the oocyte, the surface area may be increased by as much as 40 times by the presence of microvilli. These regress, however, as growth continues, finally disappearing completely when the mature egg is ovulated. The principal inclusions are the large nucleus, or germinal vesicle, about 400μ in diameter, and the yolk platelets, which are chiefly protein and are so numerous that the water content of the oocyte is only about 50 percent. Many other vesicular inclusions are present, but there appears to be no cisternal endoplasmic reticulum.

Prescott and Zeuthen [75] determined the tracer and osmotic permeability of oocytes to water. They found the tracer permeability, P, to be 1.28×10^{-4} cm/sec and the osmotic permeability, P_f, to be 89.1×10^{-4} cm/sec. Dick [15] has calculated a pore radius of 29.9 Å from these data, one of the largest values found in biological systems. From Figure 6-1, using the tracer P and an intracellular diffusion coefficient D of 2.4×10^{-5} cm²/sec for water, it would appear that the permeability $t_{1/2}$ is well over 10 times larger than the diffusional t_{50}, so that the rate of diffusion would seem unimportant. On the other hand, the rate of diffusion would seem to be of considerable significance in osmotic swelling.

Two recent papers [58, 62] have shown, however, that in fact the tracer flux of water is determined not by the membrane permeability but by intracellular diffusion. To understand how this comes about, we need to examine the flux in more detail than is permitted by Figure 6-1.

The exact expression [12] for the flux in the combined diffusion-permeation system of Figure 6-1 is

$$\frac{M(t)}{M(\infty)} = 1 - \sum_{n=1}^{\infty} \frac{6L^2 \exp(-\beta_n^2 Dt/a^2)}{\beta_n^2[\beta_n^2 + L(L-1)]} \tag{1}$$

where $M(t)/M(\infty)$ is the fractional uptake at time t. L is given by

$$L = \frac{aP}{D} \tag{2}$$

and β_n is the root of

$$\beta_n \cot \beta_n + L = 1 \tag{3}$$

The shape of the curve of $M(t)/M(\infty)$ as a function of time, is determined by L: for large L, the flux corresponds to a bulk diffusional curve; for small L, to a first-order exponential permeation curve. The range of L in which the shape of the flux curve changes most is shown in Figure 6-2, in which Equation 1 is plotted semilogarithmically against Dt/a^2 for $L = 1, 2$, and 5. For $L = 1$, the curve is almost first order; for $L = 5$, very close to a bulk diffusion curve. Even so, the faster process in each case continues to affect the rate constant k, defined as the slope of the linear portion of the curves in Figure 6-2. For $L = 1, k = 2.40$, whereas if diffusion were infinitely fast the permeability would give a value for k of 3.0; for $L = 5, k = 5.83$, as opposed to 9.87 for an infinitely permeable membrane.

Both Løvtrup [62] and Ling et al. [58] have shown that tracer water flux in the frog oocyte has the form of a bulk diffusion, which means that L is at least as great as 5. The exact value can be determined in principle by a careful study of flux at very early times (i.e., the first 5 to 10 percent of the flux) because the earliest flux is determined by P alone. In practice this determination is extremely difficult, and the exact value of L is unknown. Using the data of Løvtrup and Ling et al. for L in the range of 5 to ∞, D can be calculated to be between 0.25 and 0.75 times the self-diffusion coefficient of water in pure water [90]. This is reasonable in view of the presence of 50 percent solids in the oocyte: Wang et al. [89], for example, found a 40 percent reduc-

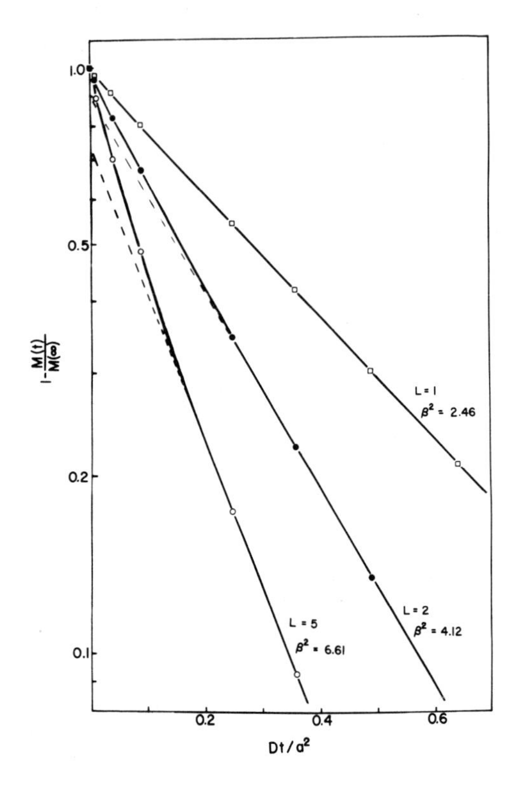

FIG. 6-2. Fractional uptake or loss in spherical cells with resistive surface. $L = aP/D$.

tion in the self-diffusion coefficient of water in 24.5 percent ovalbumin solution. Correspondingly, the minimum value of P is $3.8 - 11.4 \times 10^{-4}$ cm/sec.

It is readily seen now why the calculation using Prescott and Zeuthen's data at face value is misleading. They evaluated P directly from an observed k, assuming negligible diffusional resistance, even though their curves were not first order. Then using a value of D equal to that in water, one would find $L = 0.43$, which would confirm the assumption of negligible diffusive resistance. However, taking $L \geqq 5$, we obtain $D = 0.17$ to 0.25 times the value in pure water, and $P \geqq 2.9 \times 10^{-4}$ cm/sec, values in reasonable agreement with those from the later studies.

Osmotic and Tracer Fluxes

That cytoplasmic diffusion provides most or all of the resistance to tracer water flow in the oocyte seems secure, but we are left with considerable uncertainty as to the membrane permeability. In particular, the relationship of osmotic to tracer flow is beclouded. On the one hand, the ratio of the two permeability coefficients, P_f/P, cannot be equal to 70, as determined by Prescott and Zeuthen, because P is some undetermined value at least 2.5 times larger than originally believed and, as Dainty [13] has pointed out, the measured P_f will be less affected than P by slow cytoplasmic diffusion. On the other hand, if the tracer flux is diffusion limited, it is not easy to see how the effect of an even larger permeability coefficient can be detected.

The problem is this: As water moves osmotically across the membrane, it dilutes the osmotically active solutes just inside the membrane, so that the transmembrane osmotic gradient, and therefore the flux, will decrease unless intracellular solutes can diffuse to the membrane to restore the gradient more rapidly than the water enters. This diffusional mixing represents a mutual diffusion of water and intracellular solutes. Dick [14] has suggested that it must be quite slow in general, because the mutual diffusion is of water and macromolecular solutes, which can be expected to be at least 100 times slower than the tracer diffusion of water. However, since the major portion of the cytoplasmic osmotic activity is contributed by small ions such as K^+, it should be the mutual diffusion of these with water that is important for the restoration of the transmembrane gradient. This mutual diffusion should differ little from the tracer diffusion,[1] and one would expect intracellular mixing to be characterized by a large diffusion coefficient. Since macromolecular solutes will diffuse at a slower rate, they will lag behind during swelling, and macromolecule-deficient regions just within the membrane may form [cf. 78].

However, even assuming a mutual diffusion coefficient equal to the tracer diffusion coefficient of water, the rate of osmotic swelling may equal or exceed the calculated rate at which swelling can proceed by mutual diffusion, at all but the very earliest times. Evidence for this

[1] Electroneutrality would require that both a small cation and a small anion be available to interdiffuse with water. If they are not available in equal numbers or are of very dissimilar mobility, a diffusional potential will arise and will complicate the diffusion.

phenomenon exists for other oocytes besides the frog oocyte [75], for muscle fibers [37, 78, 97; see also below], and for toad skin (in this case transcellular diffusion) [51, 62]; in addition, in squid axon [34, 85, 88] a noticeable effect of diffusion on osmotic swelling can be expected. This problem clearly needs reevaluation.

Diffusion of Solutes in Oocytes

The question remains as to why the tracer flow of water in the oocyte is primarily impeded by intracellular diffusion rather than by the membrane. The diffusion coefficient is not strikingly smaller than in pure water; and while the large size and surface redundancy of the oocyte would appear to be reason enough, these require also a sufficiently large value of P. We would like to know which of two possibilities holds: (1) the membrane has a permeability to solutes which is not markedly different from that of other membranes, so that most solutes would be membrane limited rather than bulk limited because of smaller values of P than that for water; or (2) the membrane has an unusually high permeability to all solutes, so that one may find other cases of diffusion-limited flux. The first receives support from the observation of a high electrical resistance in the membrane [48], while the second would seem likely if the calculated pore radius of 29.9 Å is close to the correct value.

We have carried out a study [43] of the permeability of frog oocytes to glycerol, using both conventional tracer methods and autoradiography, adapting techniques for the intracellular localization of soluble materials to give results which are quantitative and of good resolution. Part of this work was directed toward the question of nuclear-cytoplasmic relationships in transport and will be examined in a later section. We will consider here the flux through the membrane and in the cytoplasm.

Oocytes were loaded with glycerol-^{3}H in 2.3 mM solution in Ringer's; the cells were washed with unlabeled solution and taken for analysis at various times of washout. For analysis, alternate cells were either extracted and the extract counted in a liquid scintillation counter, or frozen, sectioned, and autoradiographs prepared by contact of the frozen sections with frozen emulsion. Grain density profiles were determined on developed autoradiographs with the sections still superimposed.

Figure 6-3 shows the time course of tracer flux as determined by extraction and scintillation counting. All but 3.4 percent of the flux is

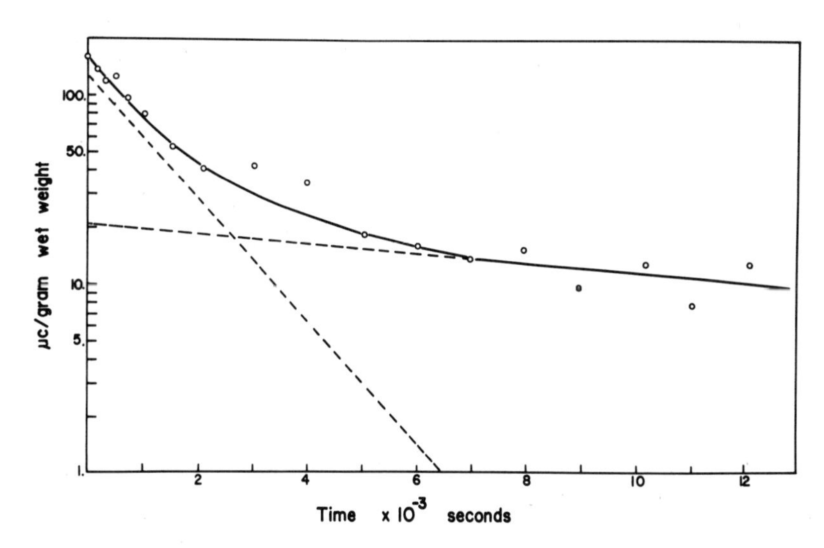

FIG. 6-3. Efflux of glycerol-^{3}H from frog oocyte at 13.6° C. Dotted lines are the two kinetic components.

accounted for by two kinetic components, constituting 77.9 percent and 18.7 percent of the total tracer, with rate constants 7.7×10^{-4} sec^{-1} and 6.0×10^{-5} sec^{-1}. The slower component may be a bound fraction; more will be said of this in the next section. The first-order kinetics of the faster fraction suggest that it is membrane limited, and this presumption is confirmed by the grain density profiles in Figure 6-4. At all times during efflux, the concentration of glycerol is uniform through the cytoplasm and falls abruptly at the cell surface. No suggestion of a diffusional profile in the cytoplasm is seen. We conclude that the flux is membrane limited and calculate the permeability to be 2.2×10^{-5} cm/sec. Assuming L in Equation 2 to be no larger than 1.0, the minimum value of D is $1.7 = 10^{-6}$ cm^2/sec, which is 0.24 times the value in dilute aqueous solution at this temperature. The similarity of this reduction in D to the reduction in the diffusion coefficient of tracer water permits reasonable confidence that diffusion in the oocyte cytoplasm is like that in water except for a relatively weakly sieving obstruction due to the solids present. The membrane appears to sieve more effectively, so that for the larger glycerol, in contrast to water, the permeability is reduced to the point at which the flux is primarily membrane limited rather than diffusion limited.

This interpretation seemed reasonable, but since the transport of

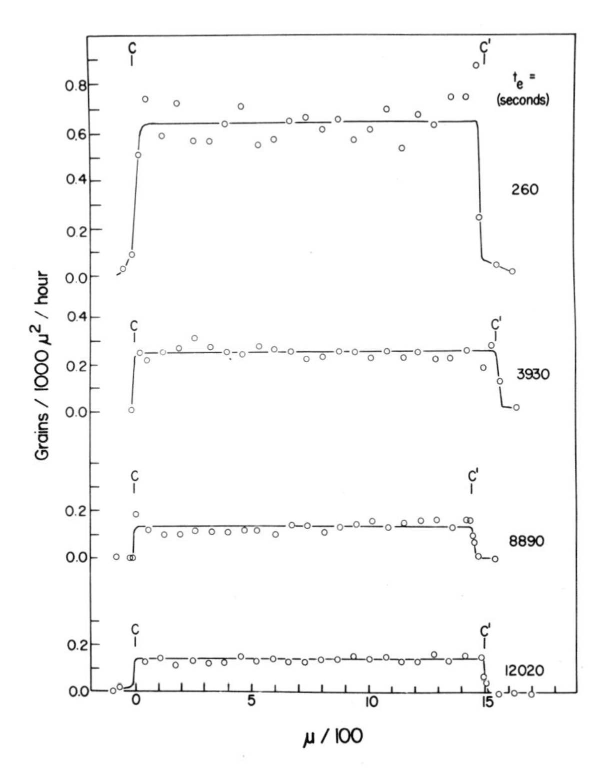

FIG. 6-4. Autoradiographic grain density profiles through oocytes during efflux of glycerol-^{3}H. C and C' are the cell edges; values of t_e on the right are durations of washout. (After Horowitz and Fenichel [43].)

glycerol may be unusual [84], we thought it desirable to seek additional support for the conclusions using another solute. Studying the tracer flux of urea in the oocyte, we have found that, as for water, the kinetics of the rapidly moving nonbound fraction are multi-exponential and correspond more closely to a bulk diffusion than to a membrane-limited process. Assuming the diffusion coefficient to be between 0.25 and 0.5 that in dilute aqueous solution, as in the case for water and glycerol, we obtained the following parameters: L = 1.9 − 6.2, P = $1.2 - 1.9 \times 10^{-4}$ cm/sec, and D = $2.5 - 5.0 \times 10^{-6}$ cm^2/sec. These fit well with the values for water and glycerol in terms of a size-dependent sieving by the membrane.

These values of P may be used to calculate an effective pore radius by the method of Renkin [77]. We obtain a value of about 4.2 Å, which is considerably smaller than that calculated from the ratio of P_f to P, even taking as a revised value of P the probable minimum value determined above. The difference in pore radius given by the two methods is similar to that seen in toad bladder and toad skin, which has been attributed to the presence of a double series membrane [16, 54]. Evidence for a like discrepancy in muscle is discussed below.

We can conclude that diffusion-limited flux in the oocyte arises primarily from the large size and redundant surface rather than from an unusually permeable membrane or from intracellular diffusion coefficients significantly smaller than in free solution. We can infer that in smaller cells diffusion will be of little importance except in the presence of membranes of substantially greater permeability or with considerably smaller values of D; and we may ask what evidence exists relevant to these properties in cases other than osmotic water flow.

Diffusion in Other Cells

The evidence may be of two types: indirect, deriving from the overall flux kinetics measured on the entire cell, and direct, the permeability or diffusion being assessed directly and independently. Kinetic analysis must be adjudged to be indirect because of the multicompartmentation of cells (even more so in tissues). While strict first-order kinetics strongly suggest an uncomplicated membrane permeation process and permit calculation of a permeability P, it is uncommon for clear-cut first-order kinetics to be seen. A theoretically uncomplicated value of P may also be calculated from the flux at very early times, but more often than not this is exceedingly difficult to determine experimentally, and even more difficult to assign unambiguously to the transmembrane flux itself rather than to some adventitious component. The estimation of diffusion coefficients from multi-exponential kinetics is, for obvious reasons, still more conducive to error.

The usual practice is to choose a major component of the flux and to determine a value of P from the time constant of this component. The common risks entailed are those of mistakenly assessing a bound component, an exponential term in a slow diffusional process, or a term in a possibly complicated expression arising from multiple intracellular membranous barriers.

Several direct techniques can be applied to the problem, but they are at present somewhat limited and have not been used extensively.

For ions, local electrical conductance measurements can be made using microelectrodes; the presumptive location of important membranous barriers, and bulk ionic mobilities in regions of known composition, can be determined. Studies of a large number of cells have, for the most part, indicated that intracellular conductivity is similar to that in a dilute salt solution [48, 59, 60]. Exceptions are studies of the dipteran salivary gland nuclear envelope [60] (see below) and the observations of Snell and Chowdhury [83], using ultrafine microelectrodes, that the potential change across the surface of frog skin and toad bladder is graded rather than abrupt.

Intracellular conductivity similar to that of extracellular fluids implies the presence within cells of mobile current carriers which, since their concentration is unlikely to be substantially greater than the concentration of extracellular fluids, can be expected to be as free to move as they are in solution.

In the squid axon, a detailed study by Hodgkin and Keynes [38] has shown that exchanged $^{42}K^+$ has both a diffusion coefficient and a mobility in a voltage gradient which are similar to those in free solution. In addition, evidence was presented that this $^{42}K^+$ was able to equilibrate with at least 80 percent of the intracellular K^+. The resistivity of the axoplasm was estimated as 46 Ω-cm, and the transference number of K^+ as 0.6–0.7. A similar value for the mobility of ^{42}K in frog muscle was estimated by Harris [28].

Intracellular conductance measurements are limited to the determination of ionic diffusivities; they also reveal only the behavior of the dominant ion—generally K^+. Thus, studies of this type have given no indication of the presence of extensive Na^+ binding, for which several recent reports have provided support (see below).

Direct observation of the diffusion of injected fluorescent materials has been carried out by Kanno and Loewenstein [59] in a study concerned with the permeability of junctional membranes in salivary glands. They found qualitatively that diffusion of fluorescent molecules of a wide range of sizes was similar to that expected in dilute aqueous solutions. This is a useful technique which could well be applied to other cell types. The principle is equally applicable, with a loss of sensitivity, to observation of the diffusion of dyes. Hodgkin and Keynes [39], employing this technique, found that methylene blue and eosin diffuse in the squid giant axon at 0.2–0.5 times their rates in free solution.

Nuclear magnetic resonance (NMR) is potentially capable of directly determining self-diffusion coefficients of intracellular materials containing certain nuclides. In practice, however, a clear-cut interpretation of the data has seldom been possible because of the complexity of interfering phenomena [4]. Cellular water has been studied by NMR on several occasions [6, 10, 69, 70] and has been inferred to be somewhat restricted in its freedom of movement, but an unambiguous quantitation is not available. The technique can, of course, detect both slow diffusion and site binding and may not differentiate them; but from the point of view of cellular transport such differentiation is important. This point, together with recent work on intracellular Na^+ binding as detected by NMR, will be discussed below.

Autoradiography, as described above, has been most frequently applied to the study of intracellular transport of macromolecules. For most soluble molecules and ions there is as yet a resolution problem which limits its use to large cells. Abelson and Duryee [1] obtained a value of 2.6×10^{-7} cm²/sec for the diffusion coefficient of Na^+ in the frog oocyte; our present estimate for glycerol is $D = 1.7 \times 10^{-6}$ cm²/sec, as given above.

Present evidence, albeit limited, suggests therefore that in most cells intracellular diffusion of small molecules can have only a minimal effect on the rate of transmembrane flux. However, when we examine the role of intracellular mixing from other points of view, its importance becomes increasingly apparent.

Diffusion and Fast Intracellular Processes

The first of these harks back to a point made early in the discussion: that the role of diffusion in any cellular process depends upon the rate constants of the other features of the process. The classic example of a process in which the cell is faced with an extreme diffusional problem is that of excitation-contraction coupling in muscle. It has long been known that the time required for a contraction-initiating substance to diffuse from the surface of a muscle fiber through the fiber is much longer than the time actually taken to develop full tension [33]. The upper limit of size of a fiber having a time constant of 10 msec for tension development, with a diffusion coefficient of 10^{-6} cm²/sec [29, 40], is, from Figure 6-1, 5μ. In order to effectively utilize thicker fibers, it is necessary to provide for release of initiator no farther from the contractile apparatus at any point than this distance, and to synchronize the release throughout the depth of the fiber. The adaptation of muscle structure to these purposes—the 1μ diameter myofibrils, the sarcoplasmic reticulum, and the transverse tubular system—is too well known to bear repeating here [cf. 30]. Although the kinetics of adenosine triphosphate (ATP) production and utilization in muscle are not fully known, it may also be of diffusional significance that in very rapid fibers, such as insect flight muscle, and in cardiac muscle mitochondria are densely packed in apposition to myofibrils [20].

Excitation-contraction coupling is an example of a transaction of a cell with its environment in which adaptation is dependent upon intracellular diffusion. A related, but less direct, example is that of the conduction velocity of nerve axons [17, 36]. The conduction velocity is proportional to the space constant λ of the axon, i.e., to the distance in which a potential change falls to 1/e of its value at a point of depolarization. For an axon embedded in a large volume of conducting medium, λ is given by $\lambda = (r_m/r_c)^{1/2}$, where r_m is the membrane resistance and r_c the axoplasm resistance. Since $r_m \alpha a^{-1}$ and $r_c \alpha a^{-2}$, where a is the axon radius, $\lambda \alpha a^{1/2}$. Thus, for a membrane of given properties the conduction velocity is limited by the core conductance, and increased velocity requires increased axon radius. Above a certain size, however, further increases are more effectively obtained by myelination.

BINDING

A major way in which intracellular mixing may affect transcellular fluxes is through binding; the question of binding is of course extremely important in considering purely intracellular material distribution as well. Unfortunately, binding in cells has rarely been handled satisfactorily, so that at present the concept is ill defined and somewhat suspect. At various times, and by various workers, it has been maintained on the one hand that there is no important binding in cells and on the other hand that almost everything in the cell is bound.

For any definition of binding—operational or conceptual—that may be offered, it is no difficult feat to provide examples of biological phenomena which either (1) do not fit the definition but have been referred to as binding by experienced biologists; or (2) fit the definition but have been considered not to represent binding, by the same or other biologists; or (3) fit or do not fit the definition but have been thought to be ambiguous as to whether they do or do not represent binding.

Two examples will suffice. When an extract of muscle—"relaxing factor"—was found to antagonize the contraction of model systems, the suggestion was made that it acted by binding Ca^{++}. When evidence was presented that the factor bound Ca^{++} by actively transporting it, usage of the term *binding* was not discontinued, even though few would apply that term to active transport into a whole cell. Subsequently, evidence indicated that Ca^{++} is "bound" by the sarcoplasmic reticulum both by active transport and by "binding." At this point rational semantic analysis fails, although the meaning is perfectly clear.

Second, no one would hesitate to speak of "membrane-bound hormone" in the case of, say, a nonpolar steroid hormone found predominantly in association with a cell-free membrane preparation rather than with an aqueous supernatant. In the case of a perhaps equally nonpolar anesthetic agent similarly distributed, however, one probably would speak, not of binding, but of differential solubility in the two phases.

Nevertheless, the situation is not so ominous as it might seem, or as it has been in the past. With increasingly sophisticated techniques of measuring and analyzing binding, and with more and more knowledge of subcellular structure, it should be possible to retain the convenient *binding* as a shorthand—subject to conventions and taboos—for a large group of phenomena which may share only the idea of a restriction, without the danger of serious error through either unwarranted inferences or impeded elucidation of the actual mechanisms.

At least the following have been or could be, under some circumstances, described as, or attributed to, binding:

1. Specific site binding of a solute to a macromolecule or particle.
2. Differential solubility between a disperse and a continuous phase.
3. Active transport into dispersed cell isolates.
4. Kinetic compartmentalization.
5. One or more first-order components of a multi-exponential flux curve.
6. Any asymmetrical transmembrane distribution of unknown mechanism.
7. An activity coefficient of a solute smaller than in an aqueous solution.
8. Evidence of an increase in some molecular relaxation time.
9. In the case of aqueous systems, almost any deviation from the bulk properties of pure water.

The list can probably be extended indefinitely.

In the following discussion of the effect of "binding" on transport it will be necessary to move between different operational and conceptual usages of the term. On the one hand, the exact mechanism involved in the "binding" is usually unknown. On the other hand, it frequently makes no difference, with respect to overall transport, what the mechanism is so long as it shows certain phenomenological characteristics. If these are made explicit, inferential errors can be avoided.

We will proceed successively in three different ways: (1) examine transport data and infer the existence and role of binding from the data; (2) postulate the phenomenological characteristics of binding in which we are interested and deduce theoretically their effects on transport kinetics; (3) examine data on binding obtained independently of transport and infer the probable consequences which would be seen in a transport study. It is necessary to proceed in this way because few studies of the role of binding in transport have concerned themselves with both the examination of transport per se and an independent measure of binding.

Binding in the Oocyte

Consider, as a relatively uncomplicated example, the efflux of glycerol from the oocyte, as shown in Figure 6-3. We have previously suggested that the slower flux component represents "bound" glycerol, and it is worthwhile, before moving to more complex problems, to describe the assumptions underlying this inference from a pure transport study:

1. That the slow fraction is glycerol, and not a metabolic product of glycerol.
2. That the existence of a slow fraction is not an artifact of the experimental technique.
3. That neither fraction is extracellular in origin.
4. That the slow fraction does not represent the flux from a large intracellular inclusion bounded by a membrane which is less permeable than the cellular membrane.

If these assumptions are satisfied (as, in fact, they have been in this case), we may entertain the following hypothesis: that glycerol in the oocyte exists in two states, "free" and "bound"; that the flux of "free" glycerol into and out of the cell is determined by the membrane permeability and the diffusion coefficient in the cytoplasmic matrix while the "bound" glycerol is not free to diffuse—except possibly over distances small compared to the cell diameter—unless it becomes "free," which it does with a rate constant small relative to that for the flux of "free" glycerol from the cell. Although this hypothesis is appropriate for the specific case considered, we would not regard it as defining, in general, the phenomenological properties of a "bound" fraction as related to transport, because we would wish to include broadening in these directions: (1) "bound" material might be able to diffuse throughout the cell, and out of it, without becoming "free"—i.e., without mixing at a significant rate with "free" material; or (2) while "bound" material is not able to diffuse as such over significant distances, the rate constant with which it becomes "free," or exchanges with "free" material, may have any value, including effectively instantaneous mixing. The effect of this broadening of admissible properties, however, is to make it possible that binding, while affecting transport and detectable by other techniques, would not be directly detectable in transport studies. This would be the case in (1) if no "free" material existed, and in (2) if mixing of "free" and "bound" were effectively instantaneous.

Glycerol transport in Figure 6-3, then, is relatively uncomplicated because a "free" component is identifiable—the fast component—having the characteristics of rapid intracellular diffusion and an intracellular space nearly equal to the water space and because the "bound" component mixes with the "free" slowly compared to the flux rate of the latter. Note, however, that the nature, and many of the properties, of the binding may not be inferred from the transport date. Nor has autoradiography, at 10μ resolution, yet provided any further clue.

The value of having made explicit these considerations becomes clear when we consider a more complex picture. Figure 6-5 shows the tracer efflux of urea-^{14}C from the frog oocyte, obtained in essentially the same way that glycerol was in Figure 6-1. Interpretation here is not obvious. There are at least three possibilities: (1) There is no bound fraction; the limiting first-order component is intracellular, and faster material is extraneous. (2) There is no bound fraction; the curve is a bulk diffusion, the single component of intracellular urea having a very small diffusion coefficient. (3) The limiting first-order component is bound urea, which constitutes 50 percent of the intracellular urea; free urea leaves the cell more rapidly (in the first 2000 sec), and the intracellular space available to free urea is 67 percent of the water space.

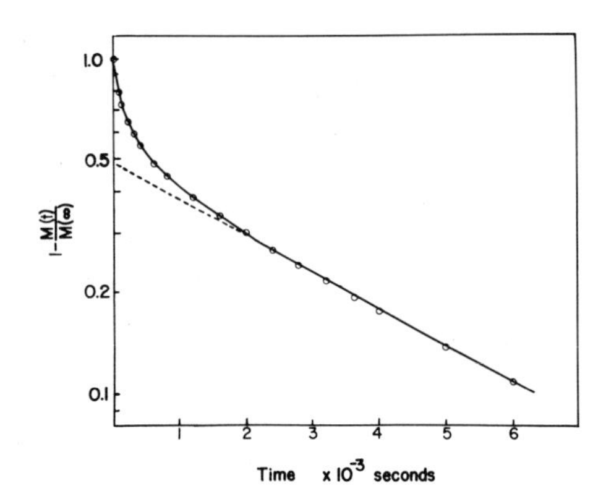

FIG. 6-5. Efflux of urea-^{14}C from frog oocyte at 13.6° C.

It is likely that, generally, (1) would be accepted; this would give a permeability P of 9.9×10^{-6} cm/sec, smaller than that of glycerol. If (2) were accepted, a diffusion coefficient D of 1.6×10^{-7} cm^2/sec would be calculated, again smaller than that of glycerol. In fact, (3) is correct, as became clear when it was found that the size of the final component is highly variable, like that of water [58], while the size of the next faster component—the "free" urea—was constant. The values of P and D calculated from the "free" urea were given earlier and shown to fit in well with those for glycerol and water, but we are left with a large and variable "bound" fraction of unknown nature.

Binding in Muscle

An even more complex problem is posed by the flux of urea and glycerol from muscle, as shown in Figure 6-6. Here we have a multicellular tissue with extracellular space which will itself contribute a flux component. If we assume, as in (1) above, that the final first-order component is "free" intracellular material, we are left with the problem of accounting for 65 percent, in the case of glycerol, and 45 percent, in the case of urea, of extraneous material. Both are greatly in excess of the extracellular space [5, 57a]. In addition, there is a problem of magnitudes, as follows. Zadunaisky et al. [97], by determining the reflection coefficients of several nonelectrolytes, estimated the pore radius of *Leptodactylus* muscle to be about 4 Å. The tracer permeabilities calculated from the final components in Figure 6-6, 2.1×10^{-7} cm/sec for urea and 0.89×10^{-7} cm/sec for glycerol, suggest the same radius in *Rana* muscle, using Renkin's equation [77], as does the observation [5] that frog muscle is almost, but not quite, impermeable to sucrose (r ≈ 4.4 Å). Then the tracer permeability to water, for a 4-Å pore radius, would be about 10^{-6} cm/sec. The osmotic permeability of muscle to water is about 10^{-2} cm/sec [15], so that the ratio P_f/P is 10^4, which strains credibility in the "double series membrane" that would need to be invoked to explain these properties.

If it is assumed that these kinetics represent uniform bulk diffusion of urea and glycerol in the fiber, new problems arise. The diffusion coefficients calculated, 2.5×10^{-10} cm²/sec for urea and 1.0×10^{-10} cm²/sec for glycerol at 5° C, are much smaller than in water [41] and must be explained. The authors have considered this possibility in detail elsewhere [25, 42]. There is as yet no direct corroborative evidence for diffusion coefficients this small. Besides, while urea conforms closely to bulk diffusion kinetics, glycerol does not; and further studies using the smaller frog toe muscle, although giving the same rate constants as the sartorius, suggest that urea also deviates from bulk diffusional kinetics.

Finally, if it is assumed that the slowest first-order component for

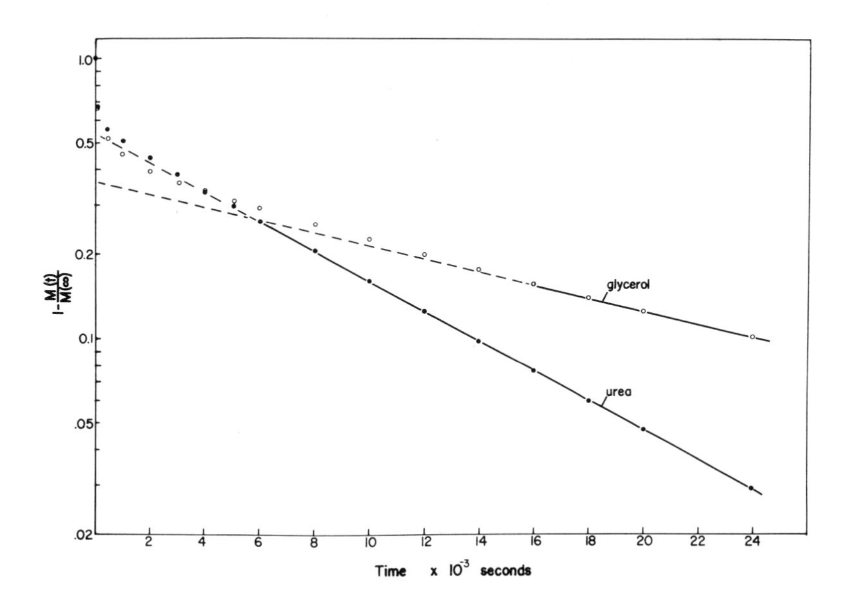

FIG. 6-6. Efflux of urea-^{14}C and glycerol-^{14}C from frog sartorius muscle at 5° C. For experimental details see reference 25.

both urea and glycerol is bound, and that the "free" material is contained in faster components, the filtration-permeation difficulty is somewhat alleviated. It becomes necessary, however, to explain both (1) the sizable apparent exclusion of "free" glycerol and urea from the total fiber water, since the total urea space is 1.0 and the total glycerol space is 0.9, and (2) the astonishingly large binding capacity of the muscle fiber for glycerol and urea. We have found that the kinetics for both components remain unchanged over the concentration range 1 to 100 mM; the largest value corresponds, in a model of site binding to muscle proteins, to 1 bound glycerol or urea per 40 amino acid residues. Since no saturation is yet seen at this value, the capacity must be substantially greater. A possibility which would account for both (1) and (2) is that a substantial part of the fiber water is sequestered in some way so as to entail slow access but does not exclude solutes. Such a distinction might, for example, exist between myofibrillar and sarcoplasmic water, or parts of these; however, there is as yet no direct evidence for it.

The point of these examples is to show how the possibility of binding may confuse the analysis of transport, so that neither the binding nor the transport parameters per se, although in principle distinct, can be unambiguously identified. Beyond this, it is possible for the existence of bound material to directly modify the measurable transport parameters. The following analysis will show how.

The Influence of Binding on Measured Transport Parameters

Consider a spherical cell of radius a immersed in a solution of a diffusing solute at concentration C_o. Assume that the solute can exist in the cell as freely diffusible material, at concentration C, and as bound nondiffusing material, at concentration S. Since the binding is reversible, in any volume over which C and S are uniform the free-bound exchange is described by

$$\frac{\partial S}{\partial t} = \xi C - \eta S \tag{4}$$

Diffusion in the cell is only of free material, as described by

$$\frac{\partial C}{\partial t} = D\left(\frac{\partial^2 C}{\partial r^2} + \frac{2}{r}\frac{\partial C}{\partial r}\right) - \frac{\partial S}{\partial t} \tag{5}$$

since part of the change of C in any volume is due to conversion into S.

Finally, let the membrane permeability be P; then the surface permeation condition is given by

$$-D\frac{\partial C}{\partial r} = P(C - KC_o) \qquad [r = a] \tag{6}$$

where we have allowed for an asymmetry in the distribution of free material between the cell and the solution; i.e., $K = C/C_o$ for $t \rightarrow \infty$.

The complete solution of this system is complicated [12]. At the extreme of very slow exchange between bound and free solute, i.e., $\eta \ll D/a^2$ or $\eta \ll P/a$, depending on whether diffusion or permeation is limiting, the processes separate, as in Figure 6-3. If exchange is very rapid compared with diffusion, i.e., $\eta \gg D/a^2$, we may set $\xi C - \eta S = 0$ or $S = (\xi/\eta)C = \rho C$ for all values of r and t, so that Equation 5 becomes

$$\frac{\partial C}{\partial t} = D\left(\frac{\partial^2 C}{\partial r^2} + \frac{2}{r}\frac{\partial C}{\partial r}\right) - \rho\frac{\partial C}{\partial t}$$

or

$$\frac{\partial C}{\partial t} = \frac{D}{\rho + 1}\left(\frac{\partial^2 C}{\partial r^2} + \frac{2}{r}\frac{\partial C}{\partial r}\right) \tag{7}$$

so that the diffusion of free solute is slowed by a factor of $\rho + 1$. Setting $Q = S + C = (\rho + 1)\,C$, and substituting in Equation 7, we find that the apparent diffusion of the total solute is equally slowed. This "chromatographic effect" is similarly applicable to the iontophoretic mobility u through the Nernst equation $D = (RT/ZF)u$ [cf. 28, 29, 38, 40].

If the flux is entirely membrane limited (i.e., $L = \frac{aP}{D} \ll 1$), Equation 6 becomes

$$\text{flux} = \frac{VdQ}{dt} = AP\ (C - KC_o)$$

$$= AP\,\frac{Q}{\rho + 1} - KC_o \tag{8}$$

Integrating this, we obtain

$$\ln\frac{Q - Q_\infty}{Q_o - Q_\infty} = -\frac{A}{V}\frac{P}{\rho + 1}t = -\frac{3}{a}\frac{P}{\rho + 1}t \tag{9}$$

where Q_∞ is the total final concentration and Q_o is the total initial concentration in the cell.

From this it is seen that the measurable permeability is reduced by a factor of $\rho + 1$ by the presence of the bound fraction. Also, only a single first-order term would occur in the flux, despite the existence of the bound fraction.

It is worth noting, too, that if the asymmetry characterized by K is imposed by the cytoplasm rather than the membrane, the measured P will be affected by the value of K. Thus if P_o is the value determined for a membrane with $K = 1$, a value of P_o/K will be found for the same membrane separating two phases having a distribution coefficient K. With binding as well, of course, the measured value will be $P_o/K(\rho + 1)$.

Direct Measurement of Binding

This effect of binding will be most pronounced when the amount of bound material is largest, and the question of binding should naturally arise most often in reference to materials which show large asymmetrical distributions across the cell membrane. The question of active transport *versus* intracellular binding is not yet laid to rest,[2] and the remainder of this section will be directed toward a few points regarding binding which relate to the problem. The essence of the question is: If a large asymmetry of distribution of a freely diffusible solute is maintained between two ostensibly dilute aqueous phases, the intracellular and extracellular, to what extent is this a nonequilibrium state maintained by the provision of energy at the membrane which results in an asymmetrical permeability, and to what extent is it an equilibrium state reflecting a change of the activity coefficient of intracellular solute as a result of the presence of other materials? The most direct approach to the question is, on the one hand, through the study of isolated, intact membranes and, on the other, through the direct measurements of the properties of intracellular solutes. Nothing will be said here of the first, which is treated elsewhere in this volume.

With regard to the second, it would be desirable to obtain a direct measure of the activity coefficient of the solute, by a method not requiring the transfer of the solute across the cell membrane or lead-

[2] It is not implied that these alternatives are mutually exclusive; evidence for the simultaneous active transport and binding of Ca in the sarcoplasmic reticulum and Na in the nucleus is discussed below.

ing to significant changes in the cytoplasm. Failing this, some measure of a nonthermodynamic property of the solute in situ may be attempted and the direction of change of the activity coefficient inferred with the aid of assumptions. Generally, the nature of these assumptions is to the effect that an observed restriction in some type of *mobility* implies a lowered activity coefficient. Such an assumption is rarely based on tight theory, but, by the same token, the inverse assumption, that a lowered activity coefficient implies a restricted mobility, is usually equally intuitive.[3]

The common small solutes which show large asymmetries of distribution between the cell and the extracellular medium are sugars, amino acids, and ions. We are not aware of any direct studies of intracellular sugars. In the case of amino acids, a recent study [87] has utilized several isomers of the fluorescent amino acid aminonaphthylalanine, which is accumulated by sarcoma-37 mouse ascites tumor cells. It was found that the fluorescence of the amino acid inside the cell showed the same excitation and emission spectra, quantum yield, temperature coefficient, and polarization (to within a small viscosity correction) as in aqueous solution, whereas three fluorescent molecules known to bind to macromolecules showed changes in several of these properties. It was concluded that no significant binding was involved in the accumulation, which in all respects studied, including competition with natural amino acids, showed the usual characteristics of active transport.

Among ions, we will restrict our attention to the monovalent cations Na^+ and K^+. With the development of cation-sensitive glasses [18] it became possible to directly estimate the activity coefficients of intracellular Na^+ and K^+ using microelectrodes. This has been done on a number of occasions in both axon and muscle. Without exception, the activity coefficient of K^+ has been found to be high: 0.61 in axon [35], 0.77 in frog muscle [55], and 1.1 (gross) in barnacle muscle [64]. The interpretation of these values is subject to two considerations: (1) Are other cations, to which the electrode is sensitive but which are not allowed for, present in the cell? (2) Are the electrode-determined ac-

[3] It is true that the driving force for translational diffusion is the gradient of chemical potential tather than of concentration, so that the activity coefficient γ is directly involved in D: we may write $D = BRT\left(1 + C\frac{\partial \ln \gamma}{\partial C}\right)$ where B is the translational mobility defined by $J = -BC\frac{\partial \mu}{\partial x}$ and may be a function of the concentration C. However, we have no knowledge of the expected variation of γ with C in the relevant systems.

tivity and the drying-plus-flame-photometry-determined concentration directly relatable? The second point covers several possible types of inhomogeneity of distribution of the ion being measured relative to the cellular water, one of which has been suggested by McLaughlin and Hinke [64]: that a large part of the water of barnacle muscle is nonsolvent for K^+, so that the apparent activity coefficient is abnormally high. An alternative suggestion derives from the first point above. The total intracellular $Na^+ + K^+$ in barnacle muscle is about 200 mM, which is much smaller than the extracellular total of about 500 mM. If this deficit of osmotically active solutes is made up by a high concentration of an organic cation to which the K^+ electrode is sensitive, the high activity coefficient would be accounted for.

Despite these problems, the burden of present direct evidence—including both the above and the conductometric and diffusional studies discussed earlier—is that K^+ in cells is of normal activity coefficient and mobility. There remain, however, important pieces of indirect evidence in less thoroughly studied cells which suggest extensive potassium binding, and they merit further study. An example is provided by the unfertilized sea urchin egg, in which only 15 percent of the K^+ is readily exchangeable with $^{42}K^+$; after fertilization, the exchangeable K^+ becomes 85 percent [86].

Direct studies of intracellular Na^+ reveal a quite different picture. By the use of Na^+-sensitive microelectrodes it has been found that the activity coefficient of Na^+ in muscle [55, 64] is considerably smaller than in the corresponding free aqueous solution; a lesser reduction is seen in squid axon [35]. While it is true that the low concentration of Na^+ in cells, and the difficulty of correcting exactly for extracellular sodium, lends some uncertainty to these estimates (which in any case imply nothing about the mechanism involved), the idea that Na^+ is held by strong secondary forces receives support from recent studies of nuclear magnetic resonance.

Without examination of the detailed theory, the observation—made in muscle, kidney, brain, and concentrated (15 percent) actomyosin by Cope [11] and in frog skin by Rotunno et al. [79]—is that the peak signal height of ^{23}Na in these systems is only 30–50 percent of what would be seen in an NaCl solution of equal concentration, in which the height is known [47] to be proportional to the concentration. Since the determinants of peak height in these complex systems are as yet far from known, one would hesitate to conclude from these data alone that Na^+ is restricted in mobility, let alone of reduced activity coefficient. Taken together with the Na^+-electrode data, and in some cases with indirect evidence derived from tracer exchange studies [79], the case for sodium binding becomes impressive. There is considerable evidence, some of which is discussed later, for multiple exchange fractions of Na^+ in a variety of cells, and it is to be expected that the work will be complemented by direct studies in the future.

Besides the obvious implications of this binding for Na^+ transport, some of which have been described earlier, two points should be mentioned. In the first place, since Na^+, but apparently not K^+, is bound, in the face of a much higher intracellular K^+ concentration, the binding process must have a very high selectivity. Second, Na^+ binding does not directly imply anything about the mechanism of sodium exclusion; but to the extent that electrometric and NMR evidence provide mutual validation, they strongly suggest that exclusion is not an equilibrium phenomenon attendant upon a very high intracellular Na^+ activity coefficient. In his NMR study Cope [11] has apparently not considered this point because he concludes otherwise.

STRUCTURAL TRANSPORT BARRIERS

The Endoplasmic Reticulum

Within the cytoplasmic matrix is found an extensive, complex system of inclusions referred to as the endoplasmic reticulum (ER). This is the most important and widespread intracellular structural component having the characteristics one might expect of a transport barrier. In its broadest sense, the system includes the membrane-enclosed perinuclear space, the granular and agranular endoplasmic reticulum, and the annulate lamellae, as well as bodies such as the Golgi apparatus and other organelles of specialized cells. Structurally, the ER is recognized by a limiting membrane that lies between the cytoplasmic matrix without and the endoplasmic matrix within. The unit form of the ER may be sheetlike cisternae, subspherical vesicles, or tubules, and these units may be aggregated into rather extensive organelles of considerable complexity and order [20, 74].

A priori reasoning about the transport properties of the ER, while speculative in any case, must be further tempered by the facts that (1) the ER membrane is not structurally identical with the cortical membrane [82], nor is it generally continuous with it, and (2) virtually nothing is known about the composition and properties of the endoplasmic matrix vis-à-vis the cytoplasmic matrix. Furthermore, if one wishes to argue teleologically, a case can be made for considering the presence of the ER qua transport barrier as logistically undesirable in many synthetic systems, inhibiting the flow of precursor to, and product away from, synthetic sites.

Nevertheless, the involvement of the ER in at least some transport functions of great importance has been securely demonstrated. Consider, for example, the intensively studied pancreatic acinar cell [8, 46, 72, 91]. The base of this truncate pyramidal cell is apposed to the vascular system, while the apex forms the lining of a secretory duct. The lamellar, ribosome-encrusted granular ER is broadly concentric with the nucleus but oriented in such a way as to provide canalicular paths from the vascular to the duct end of the cell. Amino acids entering the cell at the vascular pole are incorporated into enzyme protein by the ribosomes; the protein crosses the reticular membrane and there, sequestered from the cytoplasm, moves to the Golgi complex for additional processing into zymogen granules. With respect to the movement of free amino acids, the granular ER may be viewed as a transport sink which encourages a unidirectional flow of substrate across the cell membrane. With respect to the newly synthesized enzyme, the cisternae provide both a milieu in which they may be concentrated into intracisternal granules of prozymogen and a low-resistance, appropriately directed path in which the granules may move to the site of additional processing. A similar functional role for the granular ER has been described in other cells in which proteins are synthesized and "packaged" for export or storage [98].

Analogous to the transport function of the granular ER with respect to amino acids is the role of the reticulum in intestinal mucosal epithelium in the resorption and transcellular passage of triglycerides [7]. The ER cisternae in these cells have both granular and agranular regions. Monoglycerides and free fatty acids, produced by hydrolysis of triglycerides in the lumen of the intestine, enter the cell through the microvilli, diffuse to the reticulum, and there are reconstituted and sequestered (the order of the latter events is unknown) in the agranular region as triglycerides. The granular portion appears to synthesize protein, which enters the cisternae and coats droplets of the newly synthesized lipid to form the chylomicron. The route of movement of chylomicron to the intercellular space is uncertain but appears to involve the temporary opening of the reticulum to the extracellular space and subsequent extrusion of the chylomicron.

It is not known whether resynthesis of triglycerides takes place within the cisterna itself or in the reticular membrane. In the case of protein synthesis, we may be confident that ribosomes are the synthetic site. It seems probable, therefore, that protein synthesized on the ribosomes passes through the reticular membrane. Furthermore, the traffic in protein molecules is asymmetrical, since the granular ER concentrates protein. This process is seen not only in cases in which granules are formed, such as zymogen, but also in cases in which protein appears to remain in solution [32, 98].

The question arises: How do asymmetrical distributions of material between the ER matrix and the cytoplasmic matrix occur: It should be pointed out, first of all, that since, in the case of proteins and triglycerides, a material source—the synthetic process—is involved, which may be located in a spacially heterogeneous environment (for proteins, the ribosome-ER complex), asymmetrical movement and the consequent accumulation within the ER need not imply an active transport process even if the activity coefficient of the transported molecule is the same in both cytoplasmic and endoplasmic matrices. An analogy will make this clear. If we introduce a small amount of $H_2{}^{18}O$ without convection at an interface between ordinary ice and liquid water, we will find, after a short time, an asymmetrical distribution of tracer between the two phases because of the difference in diffusion coefficient in the two directions, even though after a sufficiently long period the specific labeling will be the same (isotope effects excluded). At no time is the transport process moving away from equilibrium.

The same might be true of the ER, and such a kinetic asymmetry would be favored by the presence of a material sink, as is provided by the secretory process per se. Membrane-dependent active transport, or asymmetry imposed by the bulk properties of the endoplasmic matrix vis-à-vis the cytoplasm, might also be involved. Clear examples of each type of asymmetry are found within cells. The fluid in vacuoles of plant cells is often markedly different in ion concentration from that of the external medium (Table 6-1), differences between the vacuole and the cytoplasm also exist. This condition holds even though the fluids are essentially simple ionic solutions to which special binding or solubility properties cannot be reasonably attributed. In these cases an intermediate phase or membrane must exist which determines the asymmetry. Membrane independent asymmetries are not uncommonly seen in cells in the case of macromolecules. An example is pro-

TABLE 6-1. *Sodium, Potassium, and Chloride Concentrations in the Vacuolar Sap of* Nitella translucens *and in the Surrounding Medium* [65, 66]

Compartment	Na*	K*	Cl*
Vacuolar sap	60	76	170
Medium	1.0	0.1	1.3

* μEq/ml.

vided by the hyaline cap and hyaline and granular ectoplasm of the amoeba. Here, granule segregation and water enrichment occur in cytoplasmic aqueous phases, without interspersed structure [2]. Similarly, the crystallization of protoplasmic fibrils from the cytoplasmic matrix involves the exclusion of nonfibrous material [96].

The above examples make it clear that the ER serves a transport function associated with the synthesis of molecules, although the mechanism of this transport may be very complex. Evidence as to the transport of indifferent molecules is still meager.

Palade and Siekevitz [71] found that granular ER (microsomes) which had been isolated by homogenization and centrifugal fractionation manifested osmotic behavior. When fixed in the presence of 0.88 M sucrose, the reticulum maintained the flattened appearance characteristic of intracellular cisternae; when fixed in more dilute media, swelling was evidenced and the reticulum appeared vesicular. Although semipermeability is thus indicated, there is no direct evidence to assign this property to selective filtration by the membrane or to selective hydration of the bulk endoplasmic matrix.

The Sarcoplasmic Reticulum

A most important transport function of the ER, not associated with synthesis, is seen in the case of the sarcoplasmic reticulum. As mentioned earlier (p. 191) the sarcoplasmic reticulum (SR), an intracellular membranous system of muscle homologous with the endoplasmic reticulum, is important in muscle contraction by virtue of its role in the transport of calcium. Contraction is initiated when the free Ca of the myofibril rises to about 10^{-6} M and is maximal at about 10^{-5} M. Relaxation occurs when concentration drops to about 10^{-7} M. The characteristics of the uptake of Ca by vesicular SR fragments in vitro and the quantity and distribution of SR in the muscle have been shown to be consistent with the functions of controlling myofibril calcium within these limits [30, 73]. It is believed that in the cyclic process of contraction and relaxation Ca is alternately released and sequestered by the SR. Release is somehow triggered by ionic movements associated with depolarization.

The mechanism whereby Ca is accumulated in the SR reflects on the transport function of intracellular membrane systems in general. Both membrane-imposed active transport and site binding by the SR have been suggested as playing a role. The role of ATP, whose presence in microsomal preparations is essential to Ca accumulation, is different in the two models. In membrane models, ATP hydrolysis provides an energy source for an active transport system [31, 94]; in the bulk model, ATP, or its hydrolysis, influences the ionic selectivity of the SR, which functions as a Ca-binding anionic exchanger [9]. There is evidence that both binding and active transport occur.

Carvalho and Leo [9] showed (Fig. 6-7) that, at constant pH, Ca uptake by SR involves a stoichiometric exchange with Mg and K. The affinity of the system for Ca relative to Mg is increased dramatically by 1.0 mM ATP. Furthermore, the fact that less than 10 percent of the

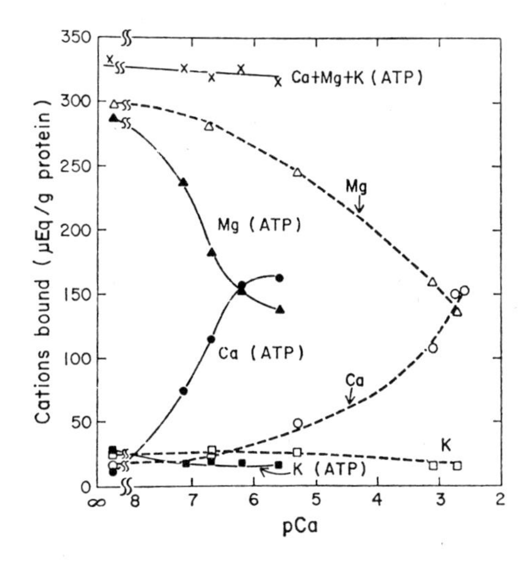

FIG. 6-7. Binding of Ca, Mg, and K by microsomal SR of rabbit skeletal muscle at 23° C as a function of free Ca in the presence (*solid lines*) and in the absence (*broken lines*) of 1.0 mM ATP. (From Carvalho and Leo [9], with permission of the *J. Gen. Physiol.*)

Ca of the microsomes is lost during repeated washings with 0.25 M sucrose indicates that the Ca counteranion is not freely diffusible.

Experiments with calcium-precipitating anions, on the other hand, provide evidence that free Ca is present in the SR at activities substantially greater than in solution [30]. The maximum concentration of soluble Ca in the presence of 5 mM oxalate, Mg, and ATP is about 5×10^{-4} M. If 5 mM oxalate is added to a sarcoplasmic reticulum system containing 10^{-4} M Ca and the other substrates, no precipitation of calcium oxalate occurs in the medium; however, crystals form promptly in the vesicles, implying the presence of Ca in excess of 5×10^{-4} M. (The possibility of active transport of oxalate has been considered and excluded.) Furthermore, precipitation of calcium oxalate continues until the free Ca of the medium is about 10^{-8} M, when the activity of Ca in the vesicle is 500 to 5000 times that of the medium. Energy-requiring active transport must maintain this asymmetry.

Additional evidence supports both binding and active transport as mechanisms involved in Ca accumulation in SR [9, 30, 94]. It is possible that, physiologically, binding is an energy-sparing device for the storage of large amounts of Ca. The binding site has little ability to distinguish Ca from Mg in the absence of ATP (Ca binds slightly in excess of Mg from solutions in which their activities are equal); this could also be the case in the presence of ATP. The ATP-dependent membrane-pump system may serve to create a high ratio of free Ca to Mg in the limited quantity of free solution within the vesicle. In this scheme equilibrium would exist between the binding sites and the vesicle solution, but not with the external medium. An analogous mechanism involving the plasma membrane sodium pump and intranuclear binding may determine nuclear sodium concentrations, as will be discussed below.

The Nucleus

The most intensively studied intracellular organelle, from the point of view of transport, is the interphase nucleus. The movement of materials into and out of the nucleus occurs not across a simple membrane but across a complex, laminar envelope. The envelope is formed in telophase from isolated elements of the endoplasmic reticulum [67]. Its most characteristic structures are the nuclear pores, which have a diameter of about 500 Å; the details of pore fine structure are contested [19, 26, 95]. It is at the pores that the cytoplasm and nuclear matrix are probably most closely apposed, but their continuity, both structural and with respect to transport, is, notwithstanding diligent investigation, an open question. The nuclear envelope itself consists of two 75-Å membranes and the contained cisternal matrix, which is continuous with the matrix of the endoplasmic reticulum.

There are several possible routes by which material may move into or out of the nucleus. Material may pass into the cisterna from the nucleus, and may distribute through the cell within the confines of the reticulum, or may cross the membrane on the cytoplasmic surface of the envelope and diffuse through the cytoplasmic matrix; material may pass directly between nucleus and cytoplasm through the pore structure. Similarly, of course, materials moving into the nucleus may tend to prefer one or the other route. Nothing is known about the relative contributions of intracisternal, transcisternal, or transpore routes to the transport of any soluble material in any cell. On the other hand, the weight of evidence seems to support the idea that some particulate matter is excluded from the cisterna and crosses between nucleoplasm and cytoplasm at the pores, either by diffusion or by some "active" process.

The microinjection of electron-opaque particles into cytoplasm, and subsequent analysis of distribution by electron microscopy, has been a useful technique in the study of particle movements. Ferritin particles of 95-Å diameter enter [21], and gold particles stabilized with a coat of polyvinylpyrrolidone (PVP) enter and are concentrated [24] in the nucleus of the amoeba *Chaos chaos*. Gold particle studies in *Amoeba proteus* [23] indicate that particles up to, but not above, 145 Å in diameter enter the nucleus; the rate of entry is influenced by particle size. In the nucleoplasm of *A. proteus*, although concentrations of gold particles do not rise above those of the cytoplasm, there is marked accumulation of gold particles near and within the nuclear pores. The precise distribution of gold particles in the vicinity of the pore, on the isolated nuclear envelope of *R. pipiens* oocyte, is influenced by the chemical properties of the polymer coat [22], which would seem to imply a specific chemical interaction between the particle and the nuclear envelope. Binding may also account for the accumulation of PVP coated gold in the nucleus of *Chaos*.

As one might expect from the ability of particulate matter to distribute between nucleus and cytoplasm, small solutes can enter and leave the nucleus. We have studied the distribution and transport of glycerol-^{3}H in the frog oocyte nucleus by extractive techniques and by quantitative, diffusible-solute autoradiography [43]. At equilibrium,

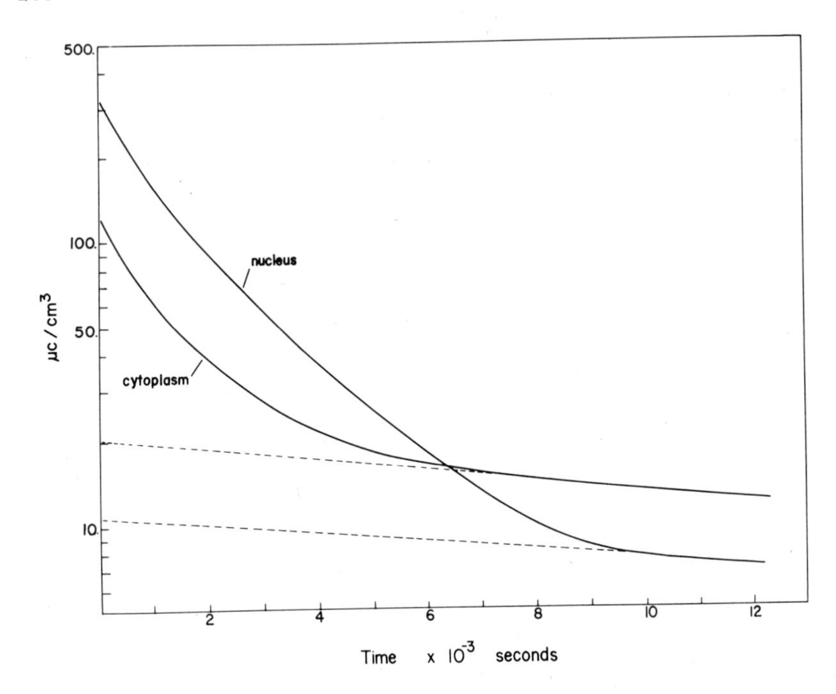

FIG. 6-8. Mean glycerol-^{3}H activity, as determined by autoradiography, in nucleus and cytoplasm of intact frog oocytes as a function of efflux time at 13.6° C. For additional details see text p. 184. (After Horowitz and Fenichel [43].)

the concentration of glycerol in the water of the nucleus was similar to that of the cytoplasm and the extracellular medium. Kinetic analysis disclosed that two fractions were present in the nucleus; they appeared to be identical in rate with those in the cytoplasm, discussed on page 184, but, as seen in Figure 6-8, differed in relative concentration. The slower fraction represents only 4.0 percent of the nuclear content as compared to about 18 percent of the cytoplasmic content. The implication of these studies was that, while the density of glycerol-binding sites was smaller in the nucleus than in cytoplasm, all of the transport parameters were otherwise similar. Hence the inference that the oocyte nuclear membrane is, as compared to the plasma membrane, a negligible barrier to glycerol transport. In contrast, recent autoradiographic evidence is provided by Stirling and Kinter [84a] that in the intestinal columnar epithelium the nuclear mem-

TABLE 6-2. *Sodium and Potassium Concentrations in Cytoplasm and Nucleus of Frog Ovarian Egg* [68]

Compartment	Na*	K*
Cytoplasm	88 ± 7	106 ± 8
Nucleus	281 ± 28	258 ± 26
Ringer's	120	2.5

* μEq/ml H_2O.

brane is less permeable to galactose than the plasma membrane.

Many similarities are found in the nuclear transport of sodium, as studied by Naora et al. [68], and of glycerol. These have been considered in detail elsewhere [43]. Unlike glycerol, Na is markedly accumulated in the nuclear water of the intact frog oocyte. The ratio of the Na concentrations in nuclear and cytoplasmic water is 3.2:1 (Table 6-2). However, as in the case of glycerol, analysis of the kinetics of ^{22}Na flux (Fig. 6-9) discloses that more than one fraction is present

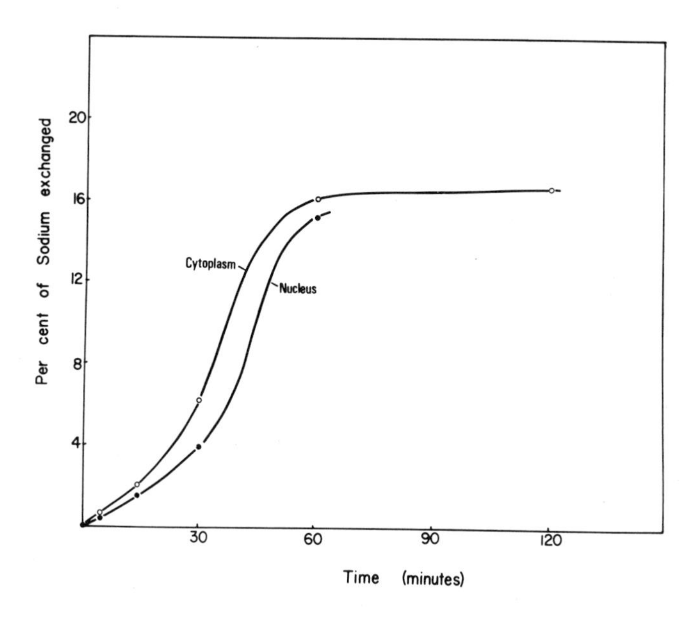

FIG. 6-9. Time course of sodium exchange in the cytoplasm and the nucleus of intact frog oocytes at 16° C. (After Naora et al. [68] with permission.)

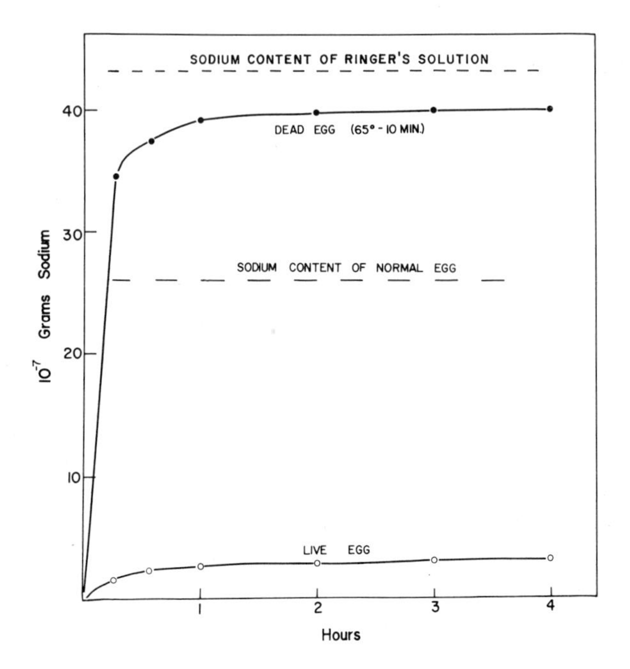

FIG. 6-10. Time course of sodium exchange in live and dead frog oocytes at 20° C. (After Abelson and Duryee [1] with permission.)

in both nucleus and cytoplasm.[4] Fast fractions account for about 15 percent of both nuclear and cytoplasmic Na while the remainder is very slowly exchanging.

Analogous behavior was reported by Abelson and Duryee [1] for the whole oocyte (Fig. 6-10). In their study the rapidly exchanging fraction constituted about 12 percent of the total Na.

The influxes of ^{22}Na into nucleus and cytoplasm in Figure 6-9 are similar. Naora et al. [68] have reported that a delay can be detected in the exchange of nuclear ^{22}Na vis-à-vis cytoplasmic. This delay is small and we cannot at present say whether it is due to diffusional delay in the cytoplasm or to the nuclear membrane. In either case, the implica-

[4] It should be noted that the sigmoid form of the influx curves in Figure 6-9 is unusual. It implies that ^{22}Na influx is slower at very early times than at intermediate times. This change in rate is irreconcilable with any uncomplicated transport mechanism. The form of the curves may reflect transient osmotic changes, as these do not seem to have been adequately controlled.

tion is that with respect to Na, as with glycerol, the oocyte nuclear envelope is a less substantial transport barrier than the plasma membrane. This conclusion is consistent with the absence of a significant electrical resistance across the nuclear envelope of immature oocytes [48]. These observations dictate caution in attributing the nuclear-cytoplasmic Na asymmetry to the operation of a nuclear membrane pump, though the existence of a pump cannot be ruled out.

Little additional information is available on the kinetics of solute transport between nucleus and cytoplasm.[5] In dipteran salivary gland nuclei an appreciable membrane resistance exists [60]. Uptake of ^{22}Na was studied in *Chironomus* by Lezzi and Kroeger [56], whose data show that the rate of Na exchange in the nucleus is increased by the trauma of cell isolation, suggesting the possibility that nuclear exchange rates are significantly determined by the cortical membrane. The authors believed that exchange is complete in vivo when the activity of ^{22}Na in the nucleus is only 15 percent of that in the hemolymph. That this interpretation must be viewed with caution is clear from the demonstration by Naora et al. that ^{22}Na exchange in frog oocyte nuclei appears to reach a plateau when only 15 percent of the total Na has exchanged.

Table 6-2 shows that cytoplasmic Na in the frog oocyte is somewhat less concentrated than in Ringer's solution, whereas in the nucleus it is more concentrated. However, if, as seems reasonable, the rapidly exchanging Na fraction (Fig. 6-9) is Na in free solution, we can conclude that there is a marked exclusion of free Na from both nuclear and cytoplasmic water, the concentrations of rapidly exchanging Na being only 13μEq/ml and 42μEq/ml in cytoplasm and nucleus, respectively. Similar exclusion of free sodium occurs in the *Chironomus* salivary gland nucleus, as shown by Lezzi and Kroeger.

Sodium exclusion commonly occurs in cells and is usually accounted for by an energy-dependent sodium pump, although other mecha-

[5] The results of osmotic studies on nuclei are so complex as to prohibit definitive interpretation. Sources of difficulty appear to be: (1) stretching of the nuclear membrane with swelling, leading to changes of permeability as well as loss of nuclear material; (2) specific interactions of the permeant solute with nuclear constituents, modifying the osmotic response of the nucleus; and (3) dependence of the mechanical state of the nuclear membrane on the specific ionic milieu. However, investigators in the area [27, 44, 61, 63] would, we believe, generally subscribe to the conclusions that (1) small polar and ionic solutes (i.e., not larger than sucrose) freely enter the nucleus of all cells studied and (2) substantially larger molecules such as serum albumin are excluded or enter so slowly as to support osmotic shrinkage for appreciable periods of time.

nisms have been suggested [3, 11, 57]. A sodium pump may be reasonably assumed to be operative in the oocyte cortex but not, as pointed out above, in the nuclear envelope. If there is no Na pump associated with the nuclear membrane, an explanation must be provided for the 29μEq/ml of rapidly exchanging nuclear Na that is present in excess of that in the cytoplasm. It can be suggested that this fraction serves to maintain nuclear electroneutrality in the presence of high concentrations of nuclear anions.

Still unaccounted for is the 85 percent of very slowly exchanging nuclear and cytoplasmic Na observed by Naora et al. and by Abelson and Duryee. The fact that it is present in the nucleus rules out the possibility that the fraction can be attributed to Na sequestered within one or more of the dispersed membrane-bounded inclusions, as these are restricted to the cytoplasm. The only explanation that appears consistent with all the evidence is that this large Na fraction is tightly bound in some manner to a nondiffusible component of the cell. Recent studies with sodium-sensitive intracellular electrodes and with nuclear magnetic resonance, reviewed above, support the existence of such a mechanism.

A convincing test of the ability of the nuclear envelope to "pump" Na in the absence of the intact cell membrane is, in principle, provided by isolated nuclear preparations. In fact, interpretation of experiments with isolated nuclei is difficult because the loss of substrates and enzymes, as well as physical damage to the nuclear envelope, is essentially uncontrolled. For these reasons we have largely excluded from this discussion consideration of the few permeability studies on

TABLE 6-3. *Sodium and Potassium Concentrations in Calf Thymus Cells* [45]

Compartment	Na*	K*
Cytoplasm†	12.0–31.0	138.0–118.0
Nuclei (isolated in nonaqueous media)	40.5	112.4
Blood serum	140.0	5.0
Nuclei (isolated in Ca-sucrose sol.)	14.9	41.5
Nuclei (extracted with TCA)	6.0	0

* μEq/gm wet wt.

† The first figure in each column assumes 10% and the second figure 5% extracellular space.

isolated nuclei. Nevertheless, interesting results on the monovalent cations have been forthcoming. Itoh and Schwartz [45] studied Na and K distribution in isolated thymus nuclei; Table 6-3 gives some of their data. The Na concentration in nuclei isolated in nonaqueous media may be taken as representative of in vivo values. Unlike the oocyte, thymus has a nuclear Na concentration below that of the extracellular medium; it is, however, higher than in the cytoplasm. In nuclei isolated in sodium-free Ca^{++}-sucrose medium, Na is about one-third of its original value, but repeated washings have little additional effect. In fact, appreciable Na is maintained by the nucleus even after trichloroacetic acid (TCA) treatment. Bound Na fractions appear to be present.

It was suggested above that the free Na concentration of the nucleus is equal to that of the cytoplasm, and that differences between cytoplasm and nucleus are determined by bound fractions. Itoh and Schwartz have provided isotherm data which permit a test of this hypothesis. In Figure 6-11 the points are experimental and the dashed line is fitted. The solid line is given by

$$C_{Na}{}^{n} = 14.9\mu\mathrm{Eq/gm} + 0.83\ C_{Na}{}^{o} \qquad (10)$$

where $C_{Na}{}^{n}$ and $C_{Na}{}^{o}$ are the Na concentrations in the nucleus and medium respectively. The value 14.9μEq/gm wet weight is the Na concentration that cannot be extracted by repeated washings with Na-free Ca^{++}-sucrose solution, and 0.83 is the water content of the nucleus. It appears that at $C_{Na}{}^{o} < 200\mu$Eq/ml a constant bound fraction and a sodium distribution coefficient of unity between nuclear and extranuclear water accounts for the observed results. Either a specific in vivo nuclear Na exclusion mechanism existed and has failed in vitro or it did not exist. The latter is the parsimonious explanation in that the sum of cytoplasmic and unextractable nuclear sodium (Table 6-3) is about sufficient to account for the total in vivo nuclear sodium.

A striking deviation from simple expectation represented by Equation 10 is seen at values of $C_{Na}{}^{o} > 200\mu$Eq/ml. $C_{Na}{}^{n}$ rises sharply so that at $C_{Na}{}^{o} = 300\mu$Eq/ml, $C_{Na}{}^{n}$ is more than 1.7 times expectation. This excess sodium binding is, in the words of the authors, "associated with gelation of the nuclear constituents."

An almost identical picture is seen in the potassium isotherm. The equation for K, analogous to Equation 10, is

$$C_{K}{}^{n} = 41.5\mu\mathrm{Eq/gm} + 0.83\ C_{K}{}^{o} \qquad (11)$$

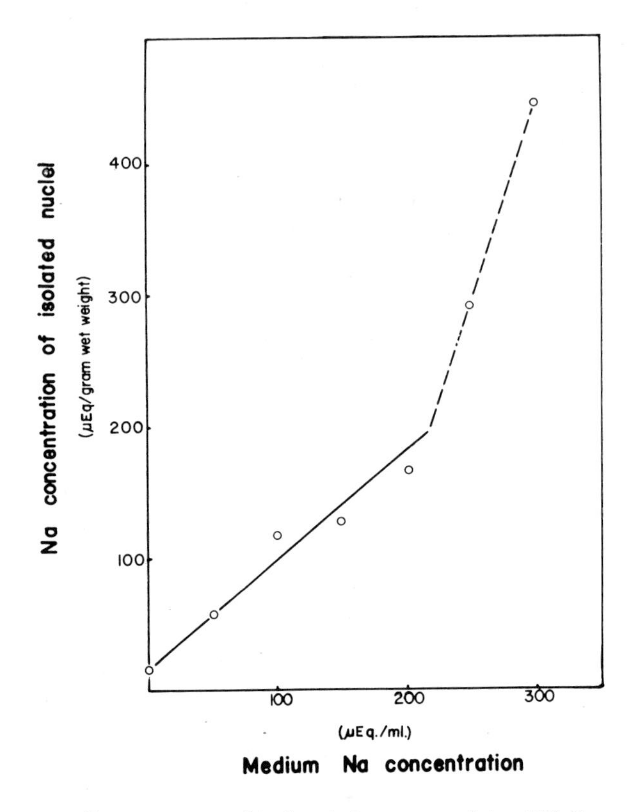

FIG. 6-11. Sodium content of isolated thymus nuclei at 37° C as a function of Na concentration in the suspending medium. See text for additional details. (After Itoh and Schwartz [45] with permission.)

The rise in slope above $C_K{}^o = 200\mu Eq/ml$ is less sharp than in the case of Na, so that at $C_K{}^o = 300\mu Eq/ml$, $C_K{}^n$ is 1.3 times expectation. Assuming the transition that occurs at $C^o = 200\mu Eq/ml$ is the same for K and Na (the possibility that the chloride concentration is the determinant of state cannot be ruled out), the difference in slope between Na and K implies that the additional binding sites, which become available at the time of nuclear "gelation," preferentially bind Na with respect to K.

Recent studies by Siebert and his associates on monovalent cations in isolated rat liver nuclei provide data that round out some aspects

TABLE 6-4. *Monovalent Cation Concentration Changes in Nuclei Isolated from Rat Liver After Perfusion with Different Isotonic Solutions* [81]. Figures are the change in μEq/gm dry weight from the normal cation concentration of the cell

Cell Fraction	Sucrose	Perfused with Isotonic		
		KCl	NaCl	LiCl
Nonfractionated liver				
Na	−105	− 99	+ 68	−112
K	−147	+154	− 67	− 77
Li	—	—	—	+217
Nuclei				
Na	−129	−111	+250	−120
K	−195	+780	+ 66	+175
Li	—	—	—	+440

of the picture. The ratio of Na in nuclear and cytoplasmic water is 13:1, while that of K is only 1.6:1 [52]. This is in qualitative agreement with data from other cells but quantitatively represents the greatest nuclear-cytoplasmic Na asymmetry reported, as well as the greatest relative Na/K selectivity. The total Na of the liver nucleus exchanges rapidly: Following intravenous injection of sodium into rats, the specific activity of nuclear Na is, within 10 minutes (the time of the first point), higher by about 5 percent than the specific activity of serum Na [53].

A phenomenon that was puzzling to Siebert and Humphrey [80] deserves comment. The rat liver was perfused with different isotonic solutions, and changes in whole tissue and nuclear concentrations were determined; the results are shown in Table 6-4. Sucrose perfusion caused an expected decrease in total and nuclear cations. Perfusion with each of the cations brought about a relative increase of that cation in the total tissue, with a more or less complementary decrease of the other cations. In the nucleus the situation was different. Perfusion in each case brought about a large increase in total nuclear cation, with K increasing even when NaCl and LiCl were the perfusants. This finding may be analogous to the results of Itoh and Schwartz in thymus nucleus, where the increase was probably attributable to newly available binding sites.

REFERENCES

1. Abelson, P. H., and Duryee, W. R. Radioactive sodium permeability and exchange in frog eggs. *Biol. Bull.* 96:205, 1949.
2. Allen, R. D., and Francis, D. W. Cytoplasmic Contraction and the Distribution of Water in the Amoeba. In Fogg, G. E. (Ed.), *The State and Movement of Water in Living Organisms.* (*Sympos. Soc. Exp. Biol.,* Vol. 19.) New York: Academic, 1965.
3. Baird, S. L., Jr., Karreman, G., Mueller, H., and Szent-Györgyi, A. Ionic semipermeability as a bulk property. *Proc. Nat. Acad. Sci. U.S.A.* 43:705, 1957.
4. Balazs, E. A., Bothner-By, A. A., and Gergely, J. Proton magnetic resonance studies of water in the presence of various macromolecular substances. *J. Molec. Biol.* 1:147, 1959.
5. Bozler, E. Determination of extracellular space in amphibian muscle. *J. Gen. Physiol.* 50:1459, 1967.
6. Bratton, C. B., Hopkins, A. L., and Weinberg, J. W. Nuclear magnetic resonance studies of living muscle. *Science* 147:667, 1965.
7. Cardell, R. R., Jr., Badenhausen, S., and Porter, K. R. Intestinal triglyceride absorption in the rat. *J. Cell Biol.* 34:123, 1967.
8. Caro, L. G., and Palade, G. E. Protein synthesis, storage, and discharge in the pancreatic exocrine cell. *J. Cell Biol.* 20:473, 1964.
9. Carvalho, A. P., and Leo, B. Effects of ATP on the interaction of Ca^{++}, Mg^{++}, and K^{+} with fragmented sarcoplasmic reticulum isolated from rabbit skeletal muscle. *J. Gen. Physiol.* 50:1327, 1967.
10. Cerbon, J. Nuclear magnetic resonance of water in microorganisms. *Biochim. Biophys. Acta* 88:444, 1964.
11. Cope, F. W. NMR evidence for complexing of Na^{+} in muscle, kidney, and brain, and by actomyosin. The relation of cellular complexing of Na^{+} to water structure and to transport kinetics. *J. Gen. Physiol.* 50:1353, 1967.
12. Crank, J. *The Mathematics of Diffusion.* London: Oxford University Press, 1956.
13. Dainty, J. Water Relations of Plant Cells. In Preston, R. D. (Ed.), *Advances in Botanical Research.* New York: Academic, 1963.
14. Dick, D. A. T. The permeability coefficient of water in the cell membrane and the diffusion coefficient in the cell interior. *J. Theor. Biol.* 7:504, 1964.
15. Dick, D. A. T. *Cell Water.* Washington: Butterworth, 1966.
16. Durbin, R. P. Osmotic flow of water across permeable cellulose membranes. *J. Gen. Physiol.* 44:315, 1960.
17. Eccles, J. C. *The Neurophysiological Basis of Mind.* London: Oxford University Press, 1953.
18. Eisenman, G., Rudin, D. O., and Casby, J. U. Glass electrode for measuring sodium ion. *Science* 126:831, 1957.
19. Fawcett, D. W. On the occurrence of a fibrous lamina on the inner

aspect of the nuclear envelope in certain cells of vertebrates. *Amer. J. Anat.* 119:129, 1966.
20. Fawcett, D. W. *The Cell.* Philadelphia: Saunders, 1966.
21. Feldherr, C. M. The intracellular distribution of ferritin following microinjection. *J. Cell Biol.* 12:159, 1962.
22. Feldherr, C. M. Binding within the nuclear annuli and its possible effect on nucleocytoplasmic exchanges. *J. Cell Biol.* 20:188, 1964.
23. Feldherr, C. M. The effect of the electronopaque pore material on exchanges through the nuclear annuli. *J. Cell Biol.* 25:43, 1965.
24. Feldherr, C. M., and Marshall, J. M. The use of colloidal gold for studies of intracellular exchanges in the ameba *Chaos chaos. J. Cell Biol.* 12:640, 1962.
25. Fenichel, I. R., and Horowitz, S. B. The transport of nonelectrolytes in muscle as a diffusional process in cytoplasm. *Acta Physiol. Scand.* 60 (Suppl. 221):1, 1963.
26. Gall, J. G. Octagonal nuclear pores. *J. Cell Biol.* 32:391, 1967.
27. Harding, C. V., and Feldherr, C. Semipermeability of the nuclear membrane in the intact cell. *J. Gen. Physiol.* 42:1155, 1959.
28. Harris, E. J. Ionophoresis along frog muscle. *J. Physiol.* (London) 124:248, 1954.
29. Harris, E. J. The output of ^{45}Ca from frog muscle. *Biochim. Biophys. Acta* 23:80, 1957.
30. Hasselbach, W. Relaxing Factor and the Relaxation of Muscle. In Butler, J. A. V., and Huxley, H. E. (Eds.), *Progress in Biophysics and Molecular Biology.* New York: Macmillan, 1964. Vol. 14.
31. Hasselbach, W., and Makinose, M. Über den Mechanismus des Calciumtransportes durch die Membranen des sarkoplasmatischen Reticulums. *Biochem. Z.* 339:94, 1963.
32. Hendler, R. W., Dalton, A. J., and Glenner, G. G. A cytological study of the albumin-secreting cells of the hen oviduct. *J. Biophys. Biochem. Cytol.* 3:325, 1957.
33. Hill, A. V. On the time required for diffusion and its relation to processes in muscle. *Proc. Roy. Soc.* [*Biol.*] 135:446, 1948.
34. Hill, D. K. The volume change resulting from stimulation of a giant nerve fibre. *J. Physiol.* (London) 111:304, 1950.
35. Hinke, J. A. M. The measurement of sodium and potassium activities in the squid axon by means of cation-selective glass micro-electrodes. *J. Physiol.* (London) 156:314, 1961.
36. Hodgkin, A. L. A note on conduction velocity. *J. Physiol.* (London) 125:221, 1954.
37. Hodgkin, A. L., and Horowicz, P. The influence of potassium and chloride ions on the membrane potential of single muscle fibres. *J. Physiol.* (London) 148:127, 1959.
38. Hodgkin, A. L., and Keynes, R. D. The mobility and diffusion coefficient of potassium in giant axons from *Sepia. J. Physiol.* (London) 119:513, 1953.
39. Hodgkin, A. L., and Keynes, R. D. Experiments on the injection of

substances into squid giant axons by means of microsyringe. *J. Physiol.* (London) 131:592, 1956.
40. Hodgkin, A. L., and Keynes, R. D. Movements of labelled calcium in squid giant axons. *J. Physiol.* (London) 138:253, 1957.
41. Horowitz, S. B., and Fenichel, I. R. Solute diffusional specificity in hydrogen-bonding systems. *J. Phys. Chem.* 68:3378, 1964.
42. Horowitz, S. B., and Fenichel, I. R. Diffusion and the transport of organic nonelectrolytes in cells. *Ann. N.Y. Acad. Sci.* 125:572, 1965.
43. Horowitz, S. B., and Fenichel, I. R. Analysis of glycerol-^{3}H transport in the frog oocyte by extractive and autoradiographic techniques. *J. Gen. Physiol.* 51:703, 1968.
44. Hunter, A. S., and Hunter, F. R. Studies of volume changes in the isolated amphibian germinal vesicle. *Exp. Cell Res.* 22:609, 1961.
45. Itoh, S., and Schwartz, I. L. Sodium and potassium distribution in isolated thymus nuclei. *Amer. J. Physiol.* 188:490, 1957.
46. Jamieson, J. D., and Palade, G. E. Intracellular transport of secretory proteins in the pancreatic exocrine cell. *J. Cell Biol.* 34:577, 1967.
47. Jardetsky, O., and Wertz, J. E. Weak complexes of the sodium ion in aqueous solution studied by nuclear spin resonance. *J. Amer. Chem. Soc.* 82:318, 1960.
48. Kanno, Y., and Loewenstein, W. R. A study of the nucleus and cell membranes of oocytes with an intracellular electrode. *Exp. Cell Res.* 31:149, 1963.
49. Kemp, N. E. Electron microscopy of growing oocytes of *Rana pipiens. J. Biophys. Biochem. Cytol.* 2:281, 1956.
50. Kemp, N. E., and Istock, N. L. Cortical changes in growing oocytes and in fertilized or pricked eggs of *Rana pipiens. J. Cell Biol.* 34:111, 1967.
51. Koefoed-Johnsen, V., and Ussing, H. H. The contributions of diffusion and flow to the passage of D_2O through living membranes. *Acta Physiol. Scand.* 28:60, 1953.
52. Langendorf, H., Siebert, G., Hannover, R., Lorenzi, I., and Beyer, R. Kationenverteilung in Zellkern und Cytoplasma der Rattenleber. *Biochem. Z.* 335:273, 1961.
53. Langendorf, H., Siebert, G., and Nitz-Litzow, D. Participation of rat liver nuclei in movements of sodium. *Nature* (London) 204:888, 1964.
54. Leaf, A., and Hays, R. M. Permeability of the isolated toad bladder to solutes and its modification by vasopressin. *J. Gen. Physiol.* 45:921, 1962.
55. Lev, A. A. Determination of activity and activity coefficients of potassium and sodium ions in frog muscle fibres. *Nature* (London) 201:1132, 1964.
56. Lezzi, M., and Kroeger, H. Aufnahme von ^{22}Na in die Zellkerne de Speicheldrusen von *Chironomus thummi. Z. Naturforsch.* [B] 21:274, 1966.
57. Ling, G. N. *A Physical Theory of the Living State: The Association-Induction Hypothesis.* New York: Blaisdell, 1962.
57a. Ling, G. N., and Kromash, M. H. The extracellular space of voluntary muscle tissues. *J. Gen. Physiol.* 50:677, 1967.
58. Ling, G. N., Ochsenfeld, M. M., and Karreman, G. Is the cell membrane a universal rate-limiting barrier to the movement of water between

the living cell and its surrounding medium? *J. Gen. Physiol.* 50:1807, 1967.
59. Loewenstein, W. R. Permeability of membrane junctions. *Ann. N.Y. Acad. Sci.* 137:441, 1966.
60. Loewenstein, W. R., Kanno, Y., and Ito, S. Permeability of nuclear membranes. *Ann. N.Y. Acad. Sci.* 137:708, 1966.
61. Löfgren, B. The lyophilic properties of the red blood cell nucleus. *Acta Physiol. Scand.* 6:266, 1943.
62. Løvtrup, S. On the rate of water exchange across the surface of animal cells. *J. Theor. Biol.* 5:341, 1963.
63. MacGregor, H. C. The behavior of isolated nuclei. *Exp. Cell Res.* 26:520, 1962.
64. McLaughlin, S. G. A., and Hinke, J. A. M. Sodium and water binding in single striated muscle fibers of the giant barnacle. *Canad. J. Physiol. Pharmacol.* 44:837, 1966.
65. MacRobbie, E. A. C. Ionic relations of *Nitella translucens*. *J. Gen. Physiol.* 45:861, 1962.
66. MacRobbie, E. A. C. Factors affecting the fluxes of potassium and chloride ions in *Nitella translucens*. *J. Gen. Physiol.* 47:859, 1964.
67. Moses, M. J. The Nucleus and Chromosomes: A Cytological Perspective. In Bourne, G. H. (Ed.), *Cytology and Cell Physiology* (3rd ed.). New York: Academic, 1964.
68. Naora, H., Naora, H., Izawa, M., Allfrey, V. G., and Mirsky, A. E. Some observations on differences in composition between the nucleus and cytoplasm of the frog oocyte. *Proc. Nat. Acad. Sci. U.S.A.* 48:853, 1962.
69. Odeblad, E. Some investigations with nuclear magnetic resonance on water associated with the vaginal cells. *Acta Obstet. Gynec. Scand.* 39:528, 1960.
70. Odeblad, E., and Ingelman-Sunberg, A. Proton magnetic resonance studies on the structure of water in the myometrium. *Acta Obstet. Gynec. Scand.* 44:117, 1965.
71. Palade, G. E., and Siekevitz, P. Liver microsomes. *J. Biophys. Biochem. Cytol.* 2:171, 1956.
72. Palade, G. E., Siekevitz, P., and Caro, L. G. Structure, Chemistry, and Function of the Pancreatic Exocrine Cell. In de Reuck, A. V. S., and Cameron, M. P. (Eds.), *Ciba Foundation Symposium on the Exocrine Pancreas.* Boston: Little, Brown, 1962.
73. Peachey, L. D. The sarcoplasmic reticulum and transverse tubules of the frog's sartorius. *J. Cell Biol.* 25:209, 1965.
74. Porter, K. R. The Ground Substance; Observations from Electron Microscopy. In Brachet, J., and Mirsky, A. E. (Eds.), *The Cell.* New York: Academic, 1961. Vol. II.
75. Prescott, D. M., and Zeuthen, E. Comparison of water diffusion and water filtration across cell surfaces. *Acta Physiol. Scand.* 28:77, 1953.
76. Rebhun, L. I. Structural aspects of saltatory particle movement. *J. Gen. Physiol.* 50:223, 1967.
77. Renkin, E. M. Filtration, diffusion, and molecular sieving through porous cellulose membranes. *J. Gen. Physiol.* 38:225, 1955.

78. Reuben, J. P., Girardier, L., and Grundfest, H. Water transfer and cell structure in isolated crayfish muscle fibers. *J. Gen. Physiol.* 47:1141, 1964.
79. Rotunno, C. A., Kowalewski, V., and Cereijido, M. Nuclear spin resonance evidence for complexing of sodium in frog skin. *Biochim. Biophys. Acta* 135:170, 1967.
80. Siebert, G., and Humphrey, G. B. Enzymology of the Nucleus. In Nord, F. F. (Ed.), *Advances in Enzymology*. New York: Interscience, 1965. Vol. XXVII.
81. Siebert, G., Langendorf, H., Hannover, R., Nitz-Litzow, D., Pressman, B. C., and Moore, C. Untersuchungen zur Rolle des Natrium-Stoffwechels im Zellkern der Rattenleber. *Z. Physiol. Chem.* 343:101, 1965.
82. Sjöstrand, F. S. A comparison of plasma membrane, cytomembranes, and mitochondrial membrane elements with respect to ultrastructural features. *J. Ultrastruct. Res.* 9:561, 1963.
83. Snell, F. M., and Chowdhury, T. K. Intracellular Potentials in Frog Skin and Toad Bladder. In Warren, K. B. (Ed.), *Intracellular Transport*. New York: Academic, 1966.
84. Stein, W. D. Spontaneous and enzyme-induced dimer formation and its role in membrane permeability: II. The mechanism of movement of glycerol across the human erythrocyte membrane. *Biochim. Biophys. Acta* 59:47, 1962.
84a. Stirling, C. E., and Kinter, W. B. High-resolution radioautography of galactose-^{3}H accumulation in rings of hamster intestine. *J. Cell Biol.* 35:585, 1967.
85. Tasaki, I., Teorell, T., and Spyropoulos, C. S. Movement of radioactive tracers across squid axon membrane. *Amer. J. Physiol.* 200:11, 1961.
86. Tyler, A., and Monroy, A. Changes in rate of transfer of potassium across the membrane upon fertilization of eggs of *Arbacia punctulata*. *J. Exp. Zool.* 142:675, 1959.
87. Udenfriend, S., Zaltzman-Nirenberg, P., and Guroff, G. A study of cellular transport with the fluorescent amino acid, aminonaphthylalanine. *Arch. Biochem.* 116:261, 1966.
88. Villegas, R., and Villegas, G. M. Characterization of the membranes in the giant nerve fiber of the squid. *J. Gen. Physiol.* 43 (Suppl. 1):73, 1960.
89. Wang, J. H., Anfinsen, C. B., and Polestra, F. M. The self-diffusion coefficients of water and ovalbumin in aqueous ovalbumin solutions at 10°. *J. Amer. Chem. Soc.* 76:4763, 1954.
90. Wang, J. H., Robinson, C. V., and Edelman, I. S. Self-diffusion and structure of liquid water: III. Measurement of the self-diffusion of liquid water with H^{2}, H^{3}, and O^{18} as tracers. *J. Amer. Chem. Soc.* 75:466, 1953.
91. Warshawsky, H., LeBlond, C. P., and Droz, B. Synthesis and migration of proteins in the cells of the exocrine pancreas as revealed by specific activity determinations from radioautographs. *J. Cell Biol.* 16:1, 1963.
92. Wartenberg, H. Elektronenmikroskopische und histochemische Studien über die Oogenese der Amphibieneizelle. *Z. Zellforsch.* 58:427, 1962.
93. Wartenberg, H., and Schmidt, W. Elektronenmikroskopische Untersuchungen der strukturellen Veränderungen im Rindenbereich des

Amphibieneies im Ovar und nach der Befruchtung. *Z. Zellforsch.* 54:118, 1961.

94. Weber, A., Herz, R., and Reiss, I. Study of the kinetics of calcium transport by isolated fragmented sarcoplasmic reticulum. *Biochem. Z.* 345:329, 1966.
95. Wischnitzer, S. The ultrastructure of the nucleus of the developing amphibian egg. In Abercrombie, M., and Brachet, J., *Advances Morph.* 6:173, 1967.
96. Wohlfarth-Bottermann, K. E. Differentiations of the Ground Cytoplasm and Their Significance for the Generation of the Motive Force of Ameboid Movement. In Allen, R. D., and Kamiya, N. (Eds.), *Primitive Motile Systems in Cell Biology.* New York: Academic, 1964.
97. Zadunaisky, J. A., Parisi, M. N., and Montoreano, R. Effect of antidiuretic hormone on permeability of single muscle fibers. *Nature* (London) 200:365, 1963.
98. Zeigel, R. F., and Dalton, A. J. Speculations based on the morphology of the Golgi systems in several types of protein-secreting cells. *J. Cell Biol.* 15:45, 1962.

7

Model Membrane Systems

John P. Reeves

DURING THE LATTER PART of the last century it was shown that the rate of penetration of many substances into cells could be correlated with their lipoid solubility. Thus, hydrocarbons, alcohols, and ketones penetrated quite rapidly while organic and inorganic ions or highly polar nonionized substances penetrated much more slowly. These results led Overton [71] to propose that each cell is covered with a thin layer of oil through which substances must diffuse before they can enter the cell. This led to the development of "paucimolecular" theory by Davson and Danielli [22], in which the overall structure of the membrane was regarded as a very thin, perhaps bimolecular, layer of lipid molecules oriented so that their polar head groups were directed outward toward the aqueous solution on either side of the membrane. Layers of protein were thought to be adsorbed on both surfaces of the lipoid layer in order to increase the stability and reduce the surface tension of the lipid; the protein was not considered to be playing an important role in determining the permeability properties of the membrane. The physical evidence that led to this formulation has been summarized by Davson and Danielli [22] and more recently by Davson [21]. The model has been widely accepted and today is regarded by many as the most firmly established of all the proposed membrane structures. Robertson [79] has examined a large variety of membrane systems by x-ray diffraction and electron microscopy and has suggested that all natural membranes share a

This work was supported in part by Grant GB-5474 from the National Science Foundation.

common structure which is quite similar to that proposed by Davson and Danielli for the plasma membrane.

Recently, however, this model has come under criticism. It appears that the membrane protein is much more important in maintaining the structure and determining the permeability characteristics of natural membranes than the Davson-Danielli model would indicate. For example, the model cannot easily account for the apparent subunit structure of certain membranes [31, 76], the probable existence of hydrophobic bonds between membrane proteins and lipids [14, 104], the effect of lipid solvents on membrane structure [18, 26], or the high rates of penetration of certain inorganic ions through many cell membranes (see below). Thus there is a need to specify the nature of the membrane lipid-protein interactions much more clearly than in the Davson-Danielli model. Nevertheless, the possibility that the lipids are arranged in the bilayer configuration throughout much of the membrane is not necessarily excluded. If the importance of the lipid bilayer has been overemphasized in this model, it is only because the nature of lipid-lipid interactions has been fairly well defined through studies of lipid monolayers and liquid crystalline systems, whereas very little is known about lipid-protein interactions. Therefore, while the Davson-Danielli model can certainly be questioned, it may be useful to consider it a working hypothesis and to examine the implications of the model with regard to its permeability properties.

Phospholipids are a major component of all natural membranes so far analyzed [100] and appear to be an important factor in maintaining their functional and structural integrity [97]. Furthermore, the phospholipid head groups,[1] being ionic in nature, provide pos-

[1] Different phospholipids are generally distinguished by the nature of the polar head groups. The phospholipids that will be considered here are phosphatidic acid and the phosphatidylesters of choline, serine, ethanolamine, and inositol. These phospholipids, when isolated from natural sources in a pure state, will be referred to as phosphatidylcholine, phosphatidylserine, etc. The fatty acid composition of pure synthetic phospholipids will also be specified, as in dioleoyl-phosphatidylcholine. Sphingomyelin will also be mentioned occasionally and should be taken to signify a preparation isolated from natural sources with negligible impurities. The phospholipids used in early experiments were usually a mixture of several different kinds of phospholipids; they will be designated as *lecithin* and *cephalin* throughout this chapter. Lecithin is a preparation containing mostly phosphatidylcholine and minor amounts of sphingomyelin. Cephalin contains mostly phosphatidylethanolamine, phosphatidylserine, and phosphatidylinositol in proportions which vary with the preparative technique used [3]. Phosphatidylethanolamine, phosphatidylserine, phosphatidylinositol, and phosphatidic acid will also be referred to collectively as the *acidic phospholipids* because they all possess a net negative charge within the pH range of 2.0–11.0. The *neutral phospholipids*, phosphatidylcholine and sphingomyelin, have no net charge in the above pH range.

sible sites for the ion exchange phenomena which may play a role in the electrical activity of cell membranes. It is not surprising, then, that much attention has been focused on artificial membranes containing phospholipids in the hope that these comparatively simple models might exhibit properties analogous to those observed in natural membranes. Moreover, the lipid bilayer which forms the core of the Davson-Danielli model is the phase structure preferred by phospholipids in water over a wide range of concentrations [57, 77], and an examination of the permeability properties of such structurally organized membranes might provide some indication of the ability of the Davson-Danielli model to describe the permeability properties of natural membranes.

ION EXCHANGE PROPERTIES OF PHOSPHOLIPIDS

Acidic phospholipids bind divalent cations, particularly calcium, much more strongly than monovalent cations, as can be demonstrated in a number of ways. Calcium can be transferred from an aqueous solution to an organic solvent containing acidic phospholipids and will not be displaced by monovalent ions until their concentration in the aqueous phase exceeds that of Ca^{++} by two orders of magnitude [25, 27, 61, 112]. Calcium is about 100 times more effective than monovalent cations in coagulating phospholipid suspensions [1, 72]. Suspensions of acidic phospholipids dialyzed exhaustively against various electrolyte solutions preferentially retain divalent cations [17, 46]. The pH of an aqueous suspension of acidic phospholipids is lowered when electrolytes are added to the medium; calcium salts will produce a given change in pH at a concentration approximately 100 times less than that required for salts of monovalent cations [1]. The addition of Ca^{++} to the aqueous phase under a monolayer of acidic phospholipids produces noticeable changes in the surface area and the surface potential at all pressures while monovalent cations are without effect [80, 87].

Although it has been relatively easy to show that lipids preferentially bind Ca^{++}, attempts to demonstrate selectivity between different monovalent cations, especially Na^{+} and K^{+}, have led to conflicting results [1, 17, 27, 61]. Reasons for the disagreements are not clear, but all the data taken together indicate that phospholipids in solution or suspension show no *large* differences in affinity for Na^{+} and K^{+}.

The selectivity of phospholipids in vivo, however, may be quite different. Kirschner [50] extracted acidic phospholipids from erythrocytes and found a higher content of sodium in the phosphatidylserine, phosphatidylethanolamine, and sphingolipid fractions than potassium although there was 5 to 20 times more K^+ than Na^+ in the intracellular fluid. Care was taken to wash away all the extracellular fluid with choline Ringer's solution before extraction. Katzman and Wilson [47] extracted frozen brain tissue at such temperatures that lipid-associated cations could not exchange and found 1.8 times more sodium than potassium in the extract, even though the potassium content of brain tissue is higher than that of sodium.

Phospholipid membranes were first prepared from a phospholipid-collodion solution [30, 105]. The properties of these early membranes, however, were not sufficiently well defined to yield any information on the ability of lipids to regulate ion fluxes through membranes. A more detailed series of experiments was undertaken by Tobias, Agin, and Pawlowski [98], who prepared membranes by repeatedly dipping a millipore filter into a benzene solution of cholesterol and cephalin and drying the membrane in air. The membranes showed no selectivity between Na^+ and K^+. In the presence of $CaCl_2$, however, the membrane resistance increased gradually to a value several times greater than its resistance in NaCl or KCl [60, 98]. The resistance returned to a low value again when K^+ was electrically driven into the membrane (rectification) or when the entire membrane was soaked in a Ca^{++}-free medium. Direct analysis of the membrane showed that the resistance changes could be correlated with its Ca^{++} content and that the percent of the total cation concentration as Ca^{++} could be described by the Donnan relation [60]. Phosphatidylserine membranes resembled cephalin membranes in their behavior while phosphatidylcholine and phosphatidylethanolamine membranes did not; apparently the negatively charged phosphatidylserine provided ion exchange sites within the cephalin membranes.

The increase in membrane resistance in the presence of Ca^{++} is the result of two factors. First, the mobility of Ca^{++} ions within the membrane, as determined by the limiting values for the membrane conductance at low electrolyte concentrations, is about 100 times less than the mobility of K^+ or Na^+ ions [69]. Second, the presence of Ca^{++} leads to a decrease in water content of the membrane [53, 60, 98], probably because of a reduction of the swelling pressure in the interior. This factor may be partially responsible for the reduced membrane permeability to water [53] and galactose [5] in the presence of

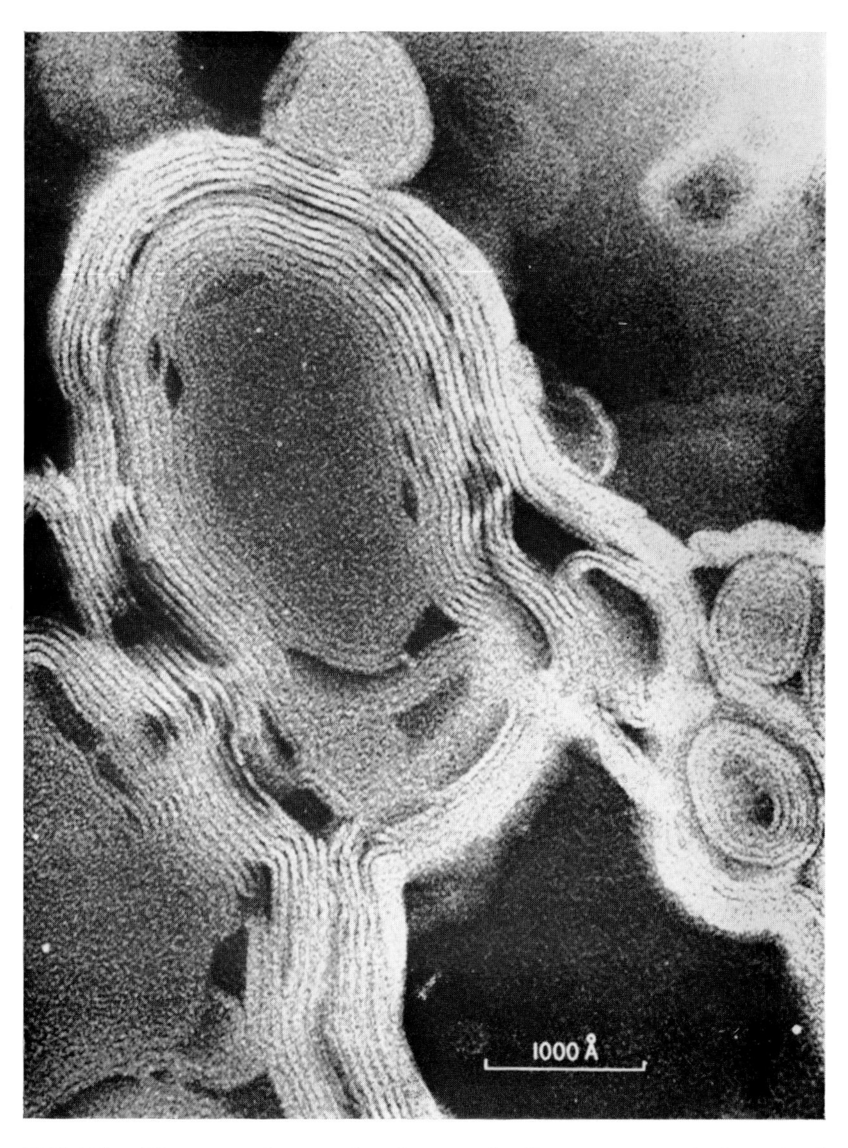

FIG. 7-1. Electron micrograph of a myelin fragment negatively stained with potassium phosphotungstate. Note the multilamellar structure of the fragment. (From Bangham and Horne [6a] with permission of Academic Press, Inc., New York.)

Ca^{++}. The permeability to electrolytes is similarly reduced and, furthermore, remains low even after 94 percent of the Ca^{++} has been removed from the membrane [60].

Rosano and Schulman [81, 82] constructed a liquid ion exchange membrane consisting of cephalin dissolved in a butanol-chloroform mixture, which was floated on top of two stirred aqueous electrolyte solutions separated by a partition. Stirring of the butanol phase insured rapid transfer through its bulk so that the rate of transport was limited by the ion-exchange and desorption processes at the butanol-water interfaces. Diffusion of ions through the oil phase required the presence of cephalin in the organic phase and exchangeable cations in both aqueous phases. Potassium exchanged with a particular cation more quickly than sodium. The rate of exchange increased with the electrolyte concentration in both aqueous phases until saturation of the cephalin in the oil phase occurred. Transport of sodium against its concentration gradient occurred when one aqueous chamber contained sodium as the only exchangeable cation and the second chamber contained a higher concentration of sodium in the presence of additional exchangeable cations (e.g., K^{+}).

PHOSPHOLIPID MYELIN FIGURES

Phospholipids swell in an aqueous solution of electrolyte or nonelectrolyte to form tubular myelin figures (Fig. 7-1). These can be fragmented by gentle agitation of the swollen suspension to yield multilamellar spherical structures which float freely in the medium. The interior lamellae are separated by a layer of the aqueous solution. The solute appears to be trapped within the spherules and diffuses very slowly into the external medium. Bangham, Standish, and Watkins [9] have suggested that each lamella makes a closed surface since the formation of energetically unfavorable water-hydrocarbon interfaces is avoided by this configuration. Thus, the spherules are pictured as a series of concentric vesicles, which implies that trapped solute must diffuse through the vesicle walls in order to find its way to the external phase. Therefore, the initial rates of solute leakage should reflect diffusion across the outermost lamella and should provide a direct measure of the permeability of the lipid membrane. The large external surface area generated by a concentrated suspension of phospholipid enables one to determine leakage rates easily, even for very slowly penetrating substances. The spherules thus provide an excellent system for studying the permeability of lipid membranes.

Bangham et al. [9] prepared phosphatidylcholine spherules in the presence of radioactive cations or anions and observed the rate of leakage of the trapped ions by dialysis of the suspension after the external labeled ions had been removed by dialysis or column chromatography. Potassium leaked out of the structures very slowly, but the rate was much increased when a long-chain anion, such as dicetyl phosphoric acid, was added to the phospholipid solution used to prepare the suspension. Leakage was almost completely blocked when long-chain cations were incorporated. The increase in the cation diffusion rate induced by the long-chain anions was not entirely unexpected because cations are attracted electrostatically to the surface of a negatively charged membrane. However, the increase is much greater than would be predicted on the basis of a Boltzmann distribution of cations within the double layer [9]. Considering values recently published for the external surface area of phospholipid suspensions [6], it appears that the increase in surface area resulting from the high degree of swelling of the charged spherules cannot account for the increased cation flux either. Bangham et al. [9] suggested that the incorporated anions decreased the stability of the bilayer configuration of the lipids and led to transient micelle formation within the membranes. The surfaces of the micelles were thought to be covered with negative charges so that cation diffusion could proceed without restriction through the interstices between micelles. On the other hand, diffusion would be blocked when the micelles were covered with positive charges derived from long-chain cations.

In contrast to the relatively slow leakage rates of cations, anions diffused across the membranes so quickly that the rates could not be determined accurately. Instead, the amount of labeled anion trapped in the spherules after passage through a Sephadex column in less than five minutes was taken as an indication of the leakage rate. In this way it was determined that Cl^- and I^- exchanged very rapidly with external anions, and F^-, NO_3^-, $SO_4^=$, and $HPO_4^=$ exchanged somewhat more slowly. The incorporation of long-chain anions or cations into the membranes had no significant effect on the rate of anion leakage.

The leakage rates of Cl^- and K^+ from spherules prepared from a variety of lipids and combinations of lipids are shown in Figure 7-2 [74]. The leakage rates of K^+ show a marked increase when the zeta

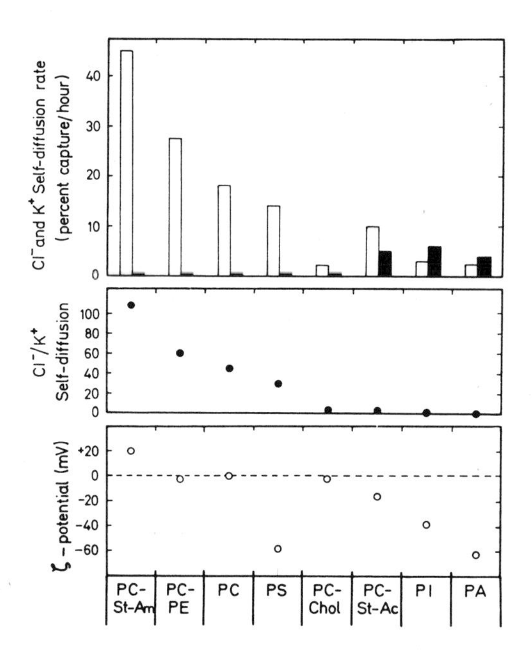

FIG. 7-2. Leakage rates of Cl^- (open bars) and K^+ (solid bars) from spherules prepared from various phospholipids. The solid circles indicate the ratio of the Cl^- leakage rate to the K^+ leakage rates. The open circles show the zeta potential of the spherules in millivolts. The leakage rates are expressed as the percentage of total captured ion diffusing from the spherules during one hour. PC, phosphatidylcholine; PE, phosphatidylethanolamine; PS, phosphatidylserine; PI, phosphatidylinositol; PA, phosphatidic acid; St-Am, stearylamine; St-Ac, stearic acid; Chol, cholesterol. (From Papahadjopoulos and Watkins [74], *Biochim. Biophys. Acta* 135:645, 1967, Fig. 2).

potential of the particles becomes negative. The leakage rates of Cl^- are generally much higher than those of K^+ except in the case of phosphatidylinositol and phosphatidic acid spherules. The Cl^- rates and the Cl^-/K^+ rate ratio tend to decrease as the zeta potential of the particles decreases. However, the Cl^-/K^+ rate ratio remains high in some spherules even though the zeta potential is near zero (Fig. 7-2—PC; PC + PE) or negative (PS), indicating that the zeta potential is not the only factor which determines the anion-cation selectivity of the membranes. The diffusion of Cl^- is markedly depressed when cholesterol is incorporated into phosphatidylcholine spherules although the diffusion of K^+ is not affected greatly.

The rate of Na^+ efflux from phosphatidylserine spherules is slightly higher than that for K^+, a selectivity which is not observed with other phospholipids [73, 74]. When the concentration of Ca^{++} exceeds 1.0 mM, this selectivity is reversed and K^+ diffuses out of the spherules at a faster rate than Na^+. In addition, the rate of diffusion for both cations is markedly increased at these Ca^{++} concentrations, an effect also noted in phosphatidic acid and phosphatidylinositol spherules [74].

The effect of Ca^{++} is not thoroughly understood. Surface potential measurements of phosphatidylserine monolayers show that at a surface concentration of 1 Ca^{++} ion for every phosphatidylserine molecule (about 1.0 mM Ca^{++} in the bulk phase) a transition occurs in the phospholipid-calcium complex which yields a product with a much higher affinity for Ca^{++} [7]. It is not known whether the transition is directly involved in the increase in cation permeability noticed at these Ca^{++} concentrations. Palmer and Schmitt [72] have shown that Ca^{++} causes a drastic reduction in the thickness of the water layers in phosphatidylserine suspensions. This appears to be the result of the efficient screening of the net negative changes of the phospholipid head groups, which decreases the electrical repulsion between adjacent lamellae [48]. Papahadjopoulos and Watkins [74] have suggested that the increase in apparent K^+ permeability may be partially the result of the extrusion of water and electrolyte brought about by the collapse of the particles in this manner. The implication, of course, is that the membranes have become permeable to Ca^{++}, because it must penetrate the lamellae to cause the collapse of the water layers in the interior of the particle [74].

Papahadjopoulos and Watkins [74] have estimated the permeability coefficient of phosphatidylcholine and phosphatidylinositol membranes using the initial leakage rates of $^{42}K^+$ from spherules prepared in 0.145 M KCl and an estimate of the external surface area of the spherules. The values obtained were 0.5×10^{-15} moles/cm^2·sec for phosphatidylcholine and 2×10^{-15} moles/cm^2·sec for phosphatidylinositol. These correspond to 3.4×10^{-12} cm/sec and 1.3×10^{-11} cm/sec, respectively, assuming that the driving force for diffusion is a K^+ concentration gradient of 0.145 M. Although the rapid leakage rates of Cl^- did not permit an accurate measure of the initial rates, it appeared that the Cl^- permeability might not be more than about 100 times that of K^+, or approximately 10^{-9}–10^{-10} cm/sec (Fig. 7-2). These low values are an indication that ion permeation does, in fact, take place by diffusion across the lipid membranes rather than through structural faults in the particle. The coefficients are compared with the permeabilities of a number of cell membranes to K^+, Na^+, and Cl^- in Table 7-1. It can be seen that the permeabilities of the lipid membranes are much lower than those quoted for the biological membranes.

A study of the temperature dependence of K^+ leakage from neutral and negative phospholipid suspensions between 37° and 50° C yields a value of 15–18 kcal/mole for the activation energy of the diffusion process [9, 74]. The activation energy becomes somewhat lower at temperatures below 37° C for phosphatidylcholine, phosphatidylcholine-stearic acid, and phosphatidylserine suspensions, suggesting the occurrence of a phase transition within the lipids [73]. In phosphatidylserine suspensions, the activation energy for Na^+ diffusion is 12–14 kcal/mole, or somewhat lower than that for K^+ [73, 74]. These values are much higher than the activation energy for counterion diffusion within ion exchange membranes (6–10 kcal/mole [40]) or for electrolyte diffusion in aqueous solutions (3–6 kcal/mole [40]), but they are similar to the reported values of 12–18 kcal/mole for the activation energy of Na^+ and K^+ flux in erythrocytes [110]. In contrast to the high values for cations, the activation energy for Cl^- diffusion is only 4–8 kcal/mole, the negatively charged spherules generally having somewhat higher values than the neutral or positive ones [74]. The low activation energy for Cl^- considered with its high diffusion rate provides an indication that anions may interact with the lipid membranes in a different manner than cations do.

Phospholipid spherules undergo volume changes in response to an osmotic gradient, indicating that the lipid lamellae are permeable to

TABLE 7-1. *Membrane Permeability to Na, K^+, and Cl^-:* Permeability Coefficient in cm/sec

Membrane	K^+	Na^+	Cl^-	Ref.
Phosphatidylcholine	3.4×10^{-12}	—	10^{-9}–10^{-12}	74
Phosphatidylinositol	2.7×10^{-11}	—	10^{-9}–10^{-12}	74
Human erythrocyte	2.4×10^{-10}	—	2×10^{-4}	110
Cat erythrocyte	3.4×10^{-9}	—	—	110
Squid axon	5.6×10^{-7}	1.5×10^{-8}	1.0×10^{-8}	13
Frog sartorius*	1.6×10^{-7}	1.4×10^{-7}	9.5×10^{-8}	99

* Calculated from the data summarized by Troshin [99, p. 243].

water. Rendi [78] prepared suspensions of mixed phospholipids in distilled water and measured the changes in volume of the spherules by several different methods when the suspensions were placed in sucrose solutions of varying concentration. The volume of the particles was found to be a linear function of the reciprocal of the sucrose concentration, and Rendi concluded that the particles behaved as perfect osmometers. Similar results were obtained by Bangham et al. [6], who also investigated the volume changes of spherules prepared in dilute KCl solutions when they were placed in salt solutions of higher or lower osmolarity. In more concentrated solutions the particles behaved as osmometers, but at very low salt concentrations the equilibrium volumes deviated from those predicted by the van't Hoff relation, probably because some of the spherules burst as swelling became excessive.

Using turbidity measurements, Bangham et al. [6] measured the initial volume changes of phosphatidylcholine–phosphatidic acid particles during osmotic shrinking or swelling and obtained a water permeability coefficient of 8×10^{-5} cm/sec. This value is considerably smaller than the osmotic water permeability of phospholipid bilayer membranes ($1.7–10.4 \times 10^{-3}$ cm/sec [33, 44]) or most biological membranes ($0.3–7 \times 10^{-3}$ cm/sec [24]). The values obtained by the procedure of Bangham et al. probably underestimate the true permeability because the standard curve used to relate turbidity to particle volume was obtained using suspensions in osmotic equilibrium. If the initial volume changes involve swelling or shrinking of only the outermost compartment, as assumed by Bangham et al., the corresponding turbidity changes would not be as great as turbidity changes associated with swelling and shrinking of the entire particle. This is because the light-scattering properties of the spherules are related not only to their volume but also to their form-refractive index, i.e., the degree of separation of the membranes within the particles.

Rendi [78] and Bangham et al. [6] have also used turbidity changes to evaluate the rate of penetration of several nonelectrolytes in the interior of phospholipid suspensions. Ethylene glycol, methylurea, ethylurea, and water were all found to penetrate at approximately the same rate; urea, propionamide, and glycerol penetrated more slowly, and malonamide and erythritol still more slowly. A similar order of permeabilities has been observed in erythrocytes and mitochondria [6].

The permeabilities of the spherules to cations, anions, and nonelectrolytes can be increased greatly by exposure to detergents [108], *n*-alkyl alcohols [8], steroids [10, 108, 109], polyene antibiotics [83, 106], and certain proteins [10, 107], all of which increase the permeability of natural membranes.

BIMOLECULAR LIPID MEMBRANES

Formation and Physical Properties of Bilayer Membranes

Mueller, Rudin, and their associates [65, 66, 67] have described a method for the production of single bimolecular lipid leaflets in the aperture of an inert barrier separating two aqueous solutions. Their procedure consists of transferring a small amount of a chloroform-methanol solution containing phospholipids to the aperture using a small paintbrush. A film forms across the hole which gradually thins as the solvent diffuses into the aqueous phase and flows into a thick annular film at the edges of the support. A few small portions of the film become invisible and appear black as the film thins to the point where the hydrocarbon chains of the oppositely oriented lipid molecules come into chance contact. This contact is strongly reinforced by London-van der Waals interactions between the hydrocarbon chains and the membrane continues to thin in a "zipper-like" fashion [96], giving rise to a sudden spread of the black areas across the film. The entire process takes place within a few minutes. The black film has been shown to be about 70 Å thick, the dimension of a bimolecular leaflet of phospholipids.

The lipid solution used by Mueller, Rudin, et al. consisted of α-tocopherol plus a complex mixture of lipids obtained from a chloroform-methanol extract of brain [67]. Hanai, Haydon, and Taylor [34] prepared black membranes using a solution of phosphatidylcholine in a hydrocarbon solvent, and Huang et al. [45] later described membranes prepared from phosphatidylcholine and *n*-tetradecane dissolved in chloroform-methanol. Sphigomyelin [63] and phosphatidylinositol [55] have also been used instead of phosphatidylcholine. The original composition of a variety of solutions used to prepare black membranes is given in Table 7-2. It should be emphasized that these compositions are not necessarily the same as the composition of the black membrane itself.

The phospholipids used in these studies were obtained from natural sources and contained a variety of fatty acid residues. Oxidation

TABLE 7-2. *Composition of Solutions Used to Prepare Black Membranes*

Lipid	% (w/v)	Hydrocarbon Additive	% (v/v)	Solvent	Ref.
Brain lipids: cholesterol	2:2	α-Tocopherol	17	Chloroform:methanol 2.2:2.8	67
Phosphatidylcholine	2	*n*-Tetradecane	18	Chloroform:methanol 3:2	45
Phosphatidylcholine	1–2	—		*n*-Decane	34, 52
Sphingomyelin	2.5	α-Tocopherol	50	Chloroform:methanol 3:2	63
Phosphatidylinositol	1	—		*n*-Decane	55

of the phosphatidylcholine used in the studies of Huang et al. [45] resulted in a failure to form membranes. Synthetic dipalmitoyl-phosphatidylcholine and other saturated phosphatidylcholine and phosphatidylethanolamines also failed to form membranes, but an unsaturated lipid, dioleoyl-phosphatidylcholine, yielded membranes which were stable for hours [45, 52]. Tien, however, using an infusion pump technique, was able to form membranes from the fully saturated substances glycerol distearate and sorbitan tristearate [94, 96], and also from oxidized cholesterol [94, 95], which shows that unsaturation of the hydrocarbon chains is not an absolute requirement for membrane formation. Formation of black membranes from egg phosphatidylcholine did not occur at temperatures below 20° C [61]. This suggests that bilayer membranes will form only when the phospholipid hydrocarbon chains possess a certain degree of fluidity, a general condition required for the formation of phospholipid liquid crystalline phases [16]. The fluid character of the membrane itself [68] also indicates that the hydrocarbon chains are in a liquid state.

The α-tocopherol used in Mueller's procedure was present in relatively high proportions (17% v/v) [67] and was required to prevent solidification of the film during the thinning process [68]. It could be replaced by silicon fluid, tetradecane, mineral oil, or butyl stearate [68]. Tetradecane was required in the system of Huang et al. although it could be replaced by methyl oleate or cholesterol plus decane [45]. Hanai, Haydon, and Taylor [34] successfully prepared membranes using a large variety of hydrocarbon solvents, including the homologous series from *n*-decane through *n*-hexadecane and also the bulky 2:2:4-trimethylpentane.

A survey of the literature shows that no stable membrane has been prepared without the inclusion of a hydrocarbon component either as an additive or as the solvent itself. Tien [96] has suggested that perhaps the hydrocarbon serves as a "space filler" which enhances the van der Waals interactions in the membrane interior. Data on the actual chemical composition of the membranes are not available, owing to the difficulties inherent in isolating and analyzing such small amounts of material. The low surface tension of the membrane [89], however, suggests that the concentration of the hydrocarbon stabilizer within the black areas is not high, certainly less than 50 percent. Leslie and Chapman [54] prepared phosphatidylcholine–β-carotene membranes in the beam of a spectrophotometer and estimated the concentration of the β-carotene within the black areas to be 1–2 percent.

Membranes prepared from complex mixtures of lipids appear to be more stable than membranes of simpler composition. Pure phosphatidylcholine membranes have lifetimes of a few hours in the absence of vibrations, agitation of the aqueous phase, or a hydrostatic head. Hanai et al. [37] reported that addition of cholesterol to the phospholipid solution resulted in a more stable membrane. Mueller et al. [68], using a very complex mixture of lipids obtained from a chloroform-methanol extract of brain, prepared membranes which were stable for more than 24 hours and did not break even when the membrane was impaled with a thin glass pipet or subjected to rather violent mechanical agitation.

Phosphatidylcholine membranes are stable in KCl, NaCl, or $CaCl_2$ solutions, even at concentrations as high as 4.0 M [34]. In D_2O the membranes are slightly less stable than in H_2O [44]. Saturation of the surrounding aqueous phase with air results in oxidation of the lipids and a drastic shortening of membrane lifetime [45]. The presence of proteins in the surrounding solutions seems to have little effect on the stability of the membranes although in one case the membrane resistance is lowered by a factor of 10^3 [68]. Certain detergents will destroy the membrane completely, others merely lower the resistance, and still others are without any effect at all [68, 84]. Cholesterol-containing membranes are destroyed by exposure to polyene antibiotics, which are known to interact specifically with cell membranes containing cholesterol [49].

The membranes break down when a potential difference of more than about 100–200 mv is imposed across the membrane [45, 52]. The resulting field strength within the membrane is more than 10^5 v/cm,

TABLE 7-3. *Membrane Thickness*

Thickness (Å)	Method of Determination	Ref.
74 ± 15	Optical	4
72 ± 10	Optical	59, 90
69	Optical	92, 93
45–47	Capacitance	34
56	Capacitance	52
60–90	Electron microscopy	68
37.5–90	Electron microscopy	41

a value which corresponds to the dielectric breakdown of hydrocarbons [52].

The lipid membranes have been examined by electron microscopy and are found to be 40–90 Å thick (Table 7-3). In an early study [68], a membrane prepared from brain lipids was fixed and stained with OsO_4 and appeared as a single black line, varying in thickness with several portions as thin as 60 to 90 Å. A different fixing and staining technique utilizing $La(NO_3)_3$ and $KMnO_4$ yielded membranes which showed the trilaminar appearance characteristic of cell membranes [41]. The membrane varied in thickness between 37.5 and 90 Å. Thicker segments of membrane occurring primarily at the edges of the film were also seen and appeared multilayered with a regular repeat distance of 38.6 Å. The film was also shadowed and viewed face on, but no evidence of substructure was seen.

An average value for the membrane thickness can also be obtained by an optical method, in which the membrane is not subjected to the dehydration and embedding processes used in electron microscopy. A geometrical analysis of the path differences between light reflected from each of the membrane interfaces shows that the membrane thickness may be calculated from the intensity of the reflected light and the membrane refractive index [92, 93]. The refractive index has been obtained from a measurement of Brewster's angle, the angle of incidence at which the reflected light intensity is at a minimum, and found to be much higher than the refractive index of any of the membrane components, probably a consequence of the highly organized state of the membrane [92]. The values obtained by this procedure for the membrane thickness (Table 7-3) agree well with each other and with the value expected for a bimolecular leaflet of fully extended phosphatidylcholine molecules [102].

The thickness of the membrane can also be calculated from capacity measurements by assuming that the membrane acts as a parallel plate condenser. Taylor and Haydon [88] measured the capacitance of films prepared from glycerol or sorbitan esters of a number of carboxylic acids with chain lengths varying from 12 to 22 carbon atoms and calculated the thicknesses expected for a bimolecular leaflet of the appropriate lipid. Thicknesses calculated from capacity measurements of phospholipid membranes are given in Table 7-3. Because the polar head groups do not contribute substantially to the measured membrane capacitance, these thicknesses represent only the hydrocarbon part of the membrane [35, 38], and another 10–20 Å should be added to account for the polar regions on each side [34, 35, 52].

Electrical Properties of Bilayer Membranes

For the measurement of electrical properties, the membranes are formed across a small aperture in a support which completely separates two aqueous compartments. Suitable electrodes are placed in each compartment, and the electrical properties of the system are measured using standard techniques. The values obtained by a number of workers for the specific resistance and capacitance of the membranes are given in Table 7-4.

The values for the specific resistance of the membranes are in general very high, much higher than those reported for cellular membranes. The values themselves are not very reproducible and will vary between membranes prepared from identical lipid solutions and even during the lifetime of a single membrane [34]. Hanai et al. [37, 38] and MacDonald [58] found that the resistance increased linearly with the bilayer area for the most poorly conducting membranes while the more highly conducting membranes did not show this relationship. These authors concluded that the highest specific resistances

TABLE 7-4. *Electrical Properties of Black Membranes*

Membrane Lipids	Resistance (ohm·cm²)	Capacitance (μf/cm²)	Ref.
Brain lipids	10^7–10^8	0.7–1.3	65, 66, 67, 68
Brain lipids	10^8	—	84
Sheep erythrocyte lipids	10^8	0.3–0.4	2
Phosphatidylcholine	0.2–4×10^6	0.69 ± 0.05	34, 59
Phosphatidylcholine	10^7–10^9	0.38–0.4	34, 35, 36, 38
Phosphatidylcholine	10^6–10^8	0.33 ± 0.02	52
Phosphatidylcholine	10^8	0.33	58
Phosphatidylcholine	—	0.37 ± 0.01	4
Phosphatidylcholine	10^8	—	103
Phosphatidylinositol	5–50×10^7	—	55
Sphingomyelin	10^6	—	63
Glycerol or sorbitan esters of C_{14}-C_{22} carboxylic acids	10^6	0.32–0.57*	88
Biological membranes	10^3–10^5	0.5–1.3	59

* Increases with decreasing chain length of carboxylic acid.

obtained under a given set of conditions are most likely to represent the true membrane resistance whereas the lower values are probably the result of leaks in the border region of the membrane.

Dividing the specific resistance (10^8–10^9 ohm·cm^2) by the thickness of the hydrocarbon portion of the membrane (5×10^{-7} cm) yields a resistivity of 10^{14}–10^{15} ohm·cm for the membrane material, a value which corresponds to the resistivity of bulk hydrocarbon [52, 88, 103]. Extremely purified hydrocarbons have resistivities of the order of 10^{19} ohm·cm, but under practical conditions the resistivity is reduced to 10^{14}–10^{15} ohm·cm because of the dissociation of dissolved water [28, 103]. The similarity between the resistivities of the membrane material and bulk hydrocarbon indicates that the membranes are without pores, since even a minute fraction of membrane area occupied by aqueous pores would result in a much lower resistance. For example, if aqueous pores filled with 0.1 M NaCl occupied only 10^{-8} of the total membrane area, the specific resistance would be reduced to about 3×10^3 ohm·cm^2 at 36° C.

The specific resistance of the membrane appears to be independent of the nature of its phospholipid or hydrocarbon components (Table 7-4). Hanai et al. [34] used the homologous series from *n*-decane to *n*-hexadecane, and also 2:2:4-trimethyl pentane [37] as solvents with no significant effect on the measured resistance. Addition of chloroform or benzene to the solution likewise had no effect [34]. On the other hand, in some cases the membrane resistance is highly dependent on the composition of the aqueous phase. Addition of a certain protein to the external solution can lower the specific resistance by a large amount [65], while others that adsorb on the membrane surface have no effect at all [36, 38]. Dissolved oxygen [45], 2,4-dinitrophenol [12], and certain antibiotics [49] all lower membrane resistance in addition to reducing the stability of the membrane. Hanai et al. [37] reported that the membrane conductance appeared to increase with the external NaCl concentration, approaching a limiting value at high concentrations. Cations such as Na^+, K^+, Mg^{++}, Ca^{++}, or H^+, and anions such as Cl^-, $SO_4^=$, CH_3COO^-, and IO_3^- have little effect on the membrane resistance [34, 52]. However, in a solution of NaI or KI, the resistance of phosphatidylcholine membranes drops by a factor of 1000, perhaps because of a high permeability to I^- [52].

In contrast to the resistance, measurements of the membrane capacitance are reproducible and yield values of somewhat less than 0.4 μf/cm^2 (Table 7-4). The membrane capacitance is independent of frequency over a range of 0.001 cps to 5×10^6 cps [38, 59]. Although one would predict a dispersion due to the polar head groups at the membrane surface, Hanai, Haydon, and Taylor [35] have shown that the capacitance of this region is large and, being in series with the lower capacitance of the hydrocarbon region, contributes practically nothing to the membrane capacitance. The measured capacitance increases with an increase in the applied electrical field, but this may be a result of an increase in black membrane area in the border regions [4] rather than a decrease in thickness under the influence of the field, as some have supposed [52].

Most of the reported values for the membrane capacitance are significantly smaller than the capacitance of biological membranes (0.5–1.3μf/cm^2). Hanai, Haydon, and Taylor [36] reported that the membrane capacitance was not affected by the adsorption of a layer of protein onto the surface of the bilayer. However, the inclusion of cholesterol in the membrane increased the capacitance to 0.56μf/cm^2. The authors suggested that the remaining difference may be due to the existence of aqueous pores in natural membranes, which introduce an additive capacitance of about 0.16μf for every 1 percent increase in total membrane pore area [36]. It should be pointed out, however, that values for bilayer capacitance which fall within the range of natural membranes have been reported (Table 7-4), even for membranes of simple composition [59]. The reason for the difference between these values and the lower ones obtained by other investigators is not clear.

Excitability

Under normal conditions bilayer membranes are highly impermeable to both anions and cations, as the resistance data indicate. Mueller, Rudin, and co-workers [65, 66, 68] have found that exposing brain lipid membranes to a proteinaceous substance obtained from the growth media of *Aerobacter cloacae* lowers the membrane resistance by a factor of 10^3, makes the membrane cation selective, and confers "excitability" upon the membrane. An excitable membrane under the appropriate conditions displays resistance and potential changes which are quite similar to those occurring in nerve during the production of an action potential.

The nature of the resistive changes can be seen most clearly by inspection of the membrane current-voltage curve [62] (Fig. 7-3). The membrane resistance is constant until a certain "threshold" potential

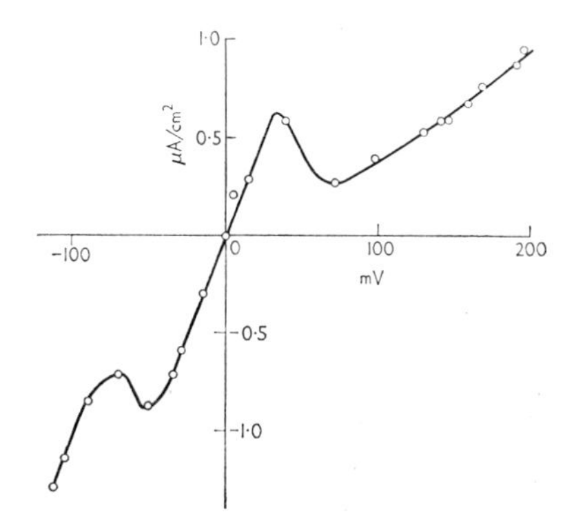

FIG. 7-3. Current-voltage curve of excitable phospholipid membrane. (From Mueller and Rudin [62] with permission of Academic Press, Inc., New York.)

is reached, when the membrane undergoes a transition to a second, stable resistive state. Transitions will occur at both positive and negative potentials, but the magnitude of the threshold potentials need not be the same. In the transition region the current decreases with increasing voltage (negative resistance), and neither the potential nor the resistance can be held constant anywhere between the limiting values given by the two resistive states. Similar behavior has been recorded for the squid axon, frog skin, and the alga *Valonia* [62, 65].

In the presence of a KCl gradient, membranes activated by the *excitability-inducing material* (EIM) develop a large resting potential which is positive on the side containing the dilute solution [63], indicating that the membrane has been made cation selective. When a constant current is passed through the membrane under these conditions, the resistance decreases and the potential tends to return to its resting level (delayed rectification). No selectivity between Na^+ and K^+ is found in EIM-activated membranes.

Mueller [62] has suggested that the EIM creates a finite number of ionic channels within the membrane which can exist in either one of two configurations with different conductivities. The free energy difference between these two configurations is thought to be a function of the transmembrane potential so that the number of channels in each configuration can be related to the applied potential and the magnitude of the free energy difference at zero potential. Assuming that the transition between channel configurations is a first-order reaction, Mueller was able to calculate the number of channels in the high resistive state, and the membrane resistance as a function of time and applied voltage. By choosing appropriate values for the activation energies of the forward and backward transitions, and for the resistivities of the two channel configurations, a calculated current-voltage curve was obtained that agreed very well with the observed curve. Mueller [62] has suggested that similar transitions occur during excitation in neuron membranes accompanied by a second, slower transition which tends to return the system to its initial state.

This sequence of transitions can be produced in sphingomyelin bilayers by the addition of protamine to the aqueous phase in precise proportion to the EIM [63]. In the presence of a salt gradient containing divalent anions, such membranes exhibit fully developed action potentials in response to suprathreshold stimulation. In some cases a rhythmic series of action potentials is seen which can continue for hours. Protamine cannot be replaced by other basic proteins such as histones or cholinesterase and will not act in the absence of EIM or sphingomyelin. Divalent anions are necessary to produce action potentials; monovalent anions generate bistable flip-flop kinetics. Mueller and Rudin [63] have suggested that the protamine acts by transforming a fraction of the ionic channels in the activated membrane so that they selectively conduct anions while leaving the resistive transitions unaltered. Thus, the reversal of potential and the decay back to the resting level appear to be brought about as the applied suprathreshold voltage alters the resistive states of the two types of channels so that the membrane becomes permeable first to anions and then to cations. The sequence of events is similar to that occurring during excitation in nerves, with an increase in anion permeability substituting for the increase in Na^+ permeability.

Permeability of Bilayer Membranes

The permeability of phospholipid bilayers to nonelectrolytes depends largely on the lipid solubility of the nonelectrolyte. Bean and Sepherd [11] have found that the introduction of polar groups onto an organic indole or benzene nucleus markedly reduces the ability of the substance to penetrate the membrane. For example, the permeability coefficient for indole is $0.5–2.5 \times 10^{-4}$ cm/sec and for 5-OH indole it is 2.0×10^{-4} cm/sec; but for indol-3-acetic acid the value drops to 5.9×10^{-7} cm/sec. Carboxylic acid derivatives will penetrate only when the carboxyl group is not ionized. In contrast, ionization seems to have little effect on the rate of penetration of most amines, even though inorganic cations hardly pass through the membrane at all.

Vreeman [103] has measured the rate of penetration of labeled urea, glycerol, and erythritol and obtained permeability coefficients of 4.2×10^{-6}, 4.5×10^{-6}, and 0.75×10^{-6} cm/sec, respectively. The permeability coefficient for mannitol is less than 3×10^{-8} cm/sec [103], and for glucose it is about 5×10^{-8} cm/sec [111]. It is difficult to compare these values with the permeability coefficients of natural membranes since such a wide range of values has been observed in different natural systems. Vreeman [103] has summarized the observed permeabilities of several organisms for urea, glycerol, and erythritol. The permeability coefficients of bilayer membranes for these substances tend to be larger—by a factor of 100 in some cases—than those observed in plant cells but fall within the range of values observed in bacteria and erythrocytes. The permeability to glucose is similar to its permeability in ox erythrocytes but is much smaller than the rate of glucose penetration in human erythrocytes, which occurs by means of a carrier mechanism [110, 111].

There has been only one report of an attempt to measure the ionic permeability of membranes by direct chemical means. Vreeman [103] was unable to detect any penetration of ^{22}Na through phosphatidylcholine membranes indicating that the permeability was less than 10^{-9} cm/sec.

An alternative measure of the permeability to ions is provided by the membrane resistance. Hodgkin [42] has derived an equation relating the conductance to the unidirectional ion flux across the membrane at equilibrium, obtaining

$$g_M = \frac{\mathbf{F}^2}{RT} \sum_{i=0}^{n} z_i^2 \, \mathbf{J}_i$$

where g_M is the membrane conductance, $\mathbf{J}_i$ is the unidirectional flux of the *i*th ion at equilibrium in moles/cm^2·sec, z_i is the valence of the *i*th ion, and $\mathbf{F}$ is the Faraday. Substituting 10^{-8}–10^{-9} mho/cm^2 in the above equation, one obtains an ion flux of 3×10^{-15} to 3×10^{-16} moles/cm^2·sec, assuming that the entire current is carried by a single univalent ion. If the ion flux were driven by a concentration gradient of 0.15 M across the membrane, this value would correspond to a permeability coefficient of 2.0–20×10^{-12} cm/sec, in good agreement with the univalent cation permeability of phospholipid suspensions (Table 7-1).

In all probability, the membrane conductance is due primarily to the migration of H^+ and OH^- ions from dissociated water in the membrane so that the values calculated above represent only an upper limit for the membrane permeability to other ions. However, the calculation does serve to illustrate the analytical problems inherent in measuring the ionic permeability of these membranes. For example, even if the flux were as high as 3×10^{-15} moles/cm²·sec, this value combined with the small area of the membranes (10^{-2} cm²) would yield an observed flux of only 10^{-14} moles/hour, which is difficult to measure with the standard isotopic techniques. Ionic permeabilities can be measured much more easily in the case of suspended phospholipid particles because their total membrane area may be 10^4–10^5 times larger than that provided by the bilayer membranes.

Attempts to detect any significant selectivity between different ions have generally been unsuccessful. Läuger et al. [52, 55] observed that I^- may penetrate phosphatidylcholine and phosphatidylinositol membranes very rapidly, but this is certainly not the case for Cl^- or other anions. Andreoli [2] has reported that the transport number of Cl^- in bilayer membranes prepared from sheep erythrocyte lipids may be as low as 0.15 relative to K^+.

Mueller and Rudin [64] reported that the membrane resistance can be lowered to less than 10^3 ohm·cm² by the addition of small amounts (10^{-5}–10^{-7} gm/cc) of the antibiotics valinomycin, monactin, dinactin, gramicidins A, B, and C, and enniatins A and B. The membranes became highly cation selective and, in addition, showed a permeability for K^+ 400 times that for Na^+. The antibiotics are cyclic structures composed of various carboxylic and amino acids in which the polar carbonyl groups face the interior of the ring while the hydrocarbon groups are directed outward. Mueller and Rudin [64] have proposed that the external surface of the ring interacts with the membrane lipids by van der Waals forces while the interior of the ring or aggregates of rings provide a polar pore. A single cation can be fitted in the interior of such a ring, and the selectivity of the membrane for different cations may be related to the ability of the inner carbonyl groups to coordinate with the cations in either the primary or the secondary hydration layer. This interpretation provides a mechanism by which high degrees of ionic selectivity can be generated without the presence of fixed ionic charges and may afford some insight into the structure of pores in natural membranes.

Del Castillo et al. [23], in an interesting paper, showed that antigen-antibody interactions occurring at the surface of brain lipid membranes can lower the impedance by a factor of 10^3. Antigen or antibody added alone, or both together, or on opposite sides of the membrane had no effect on the impedance. The impedance was reduced only when the proteins were added one at a time, on the same side of the membrane, the order of addition being unimportant. Similar changes were noticed when enzymes and their substrates were added to one side of the membrane. In this case the impedance returned to its normal high value as the substrate at the surface of the membrane was exhausted, but the changes could be regenerated by adding more substrate. The authors proposed that this phenomenon may serve as a model for synaptic transmission. Recently, Howard and Burton [43] have reported that similar impedance changes occur when acetylcholine is hydrolyzed by bovine erythrocyte acetylcholinesterase at the surface of a lipid bilayer.

Addition of certain anionic detergents to one side of a membrane lowers the resistance from 10^8 ohm·cm^2 to a steady value of less than 10^6 ohm·cm^2 [84]. In the presence of a KCl gradient a potential develops across the membrane which is positive on the dilute side, indicating that the membrane has become cation selective. When cationic detergents are used, the resistance again drops by a factor of 10^2. In this case, however, no steady potentials are developed in the presence of an electrolyte concentration gradient. The authors proposed that the anionic detergents may lead to transitions from the lamellar to the micellar phases in the lipids and that the negative charges of the detergent head groups allow cations to diffuse through the micellar regions more quickly than anions. It seems likely that the cationic

TABLE 7-5 *Water Permeability of Black Membranes:* Permeability Coefficient in cm/sec $\times 10^4$

Membrane	THO Flux	Osmotic Flux	Temp.	Ref.
Phosphatidylcholine	4.3–4.4	17.3–104	36° C	44, 91
Phosphatidylcholine	—	18	20° C	32
Phosphatidylcholine	2.2	—	20° C	33
Phosphatidylcholine	5.1	—		103
Biological membrane	—	3–70	—	24

detergents produce similar transitions, but for some reason the micellar regions do not appear to be selective for anions.

The permeability coefficients obtained for tritiated water (THO) flux and for osmotic water flow are given in Table 7-5. The values are similar to those obtained for natural membranes. Furthermore, the osmotic permeability coefficients are several times larger than the isotopic permeability coefficients, a feature which is also observed in natural membranes. This difference has often been interpreted as indicating Poiseuille flow of water through aqueous pores under the influence of the osmotic pressure gradient. The ratio of the osmotic flow to the purely diffusive THO flow can be used to calculate the radius of the aqueous pores and the fractional membrane area occupied by the pores [51, 75]. (Dick [24] has summarized several other explanations for this discrepancy.) Huang and Thompson [44, 91] applied this treatment to their data and calculated effective pore radii of 5.3 to 16.6 Å with a fractional pore area of 10^{-5}. The pores were thought to arise as transient structures since the existence of a permanent fixed pore in a fluid membrane of finite surface tension was considered unlikely. Furthermore, the authors felt that such pores must have a mechanism for excluding ions from their interior because pores of the dimensions described above filled with electrolyte solution would yield a membrane resistance much lower than the observed value [44].

The latter considerations have led several authors [32, 33, 103] to reject the pore hypothesis and consider the osmotic flux in terms of diffusion of water within the hydrocarbon portion of the membrane. Using the measured value for the osmotic permeability coefficient and the value for the solubility of water in various hydrocarbons, Hanai and Haydon [32] calculated the diffusion coefficient of water in the membrane, assumed to be a layer of liquid hydrocarbon 48 Å thick. The values obtained agreed roughly with the diffusion coefficient of water in hydrocarbon as estimated by two different empirical relations. Similarly, Vreeman [103] found good agreement between the osmotic permeability coefficient and a value calculated from published values of the diffusion constant and solubility of water in *n*-hexadecane. The lower values for the isotopic permeability coefficients were thought to be due to the presence of stagnant water layers at the membrane surface [33, 103]. Hanai et al. [33] showed that the THO flux was increased by a factor of 3 when stirring of the water layers was achieved by osmotic flow or convention currents generated at the membrane surface owing to the density differences between transferred water and bulk solution.

CONCLUSION

The results presented above indicate that the rates of penetration of a variety of nonelectrolytes, including water, are quite similar in bilayer membranes and in natural membranes. Since the artificial membranes for which quantitative data are available appear to be without pores, these substances most likely penetrate by dissolution and diffusion within the hydrocarbon interior of the membrane.

A similar mechanism might be involved in the diffusion of ions across lipid membranes, a suggestion which is supported by the fact that low concentrations of ions and ion pairs can be found in hydrocarbon media of low dielectric constant [29]. An alternative proposal is that the ions diffuse through small pores in the membrane which are created either by thermal fluctuations in the spacing of the hydrocarbon chains [86] or by transitions between different micellar states in the lipids [56]. A third possibility is that ions—particularly cations—could interact with the membrane lipids to form a lipid-soluble complex which diffuses through the membrane. However, the low values observed for the conductance and ionic permeabilities of phospholipid bilayers suggest that none of these mechanisms can account for the relatively high rate of ionic diffusion through natural membranes.

As discussed above, the ionic permeability of phospholipid membranes can be greatly increased by the action of such substances as detergents, antibiotics, and Mueller's EIM, which appear to disturb the continuity of the lipid bilayer. Although the precise nature of the chemical interactions that bring about this disruption have not been defined, the primary interaction is probably a hydrophobic one in the case of the detergents and the macrocyclic or polyene antibiotics. Because the bilayer configuration is stabilized primarily by hydrophobic forces between the lipid hydrocarbon chains [19, 70], one might expect the same thing to be true for all substances which disrupt the bilayer. This suggestion finds some support in the fact that cytochrome *c*, which is known to act electrostatically with acidic phospholipids [20], has no effect on the rate of ion efflux from suspended phosphatidylserine spherules [74]. These considerations suggest that if the structure of natural membranes is at all similar to that proposed by Davson and Danielli [22] ion diffusion might occur primarily in regions where bilayer configuration has been disturbed as the result of hydrophobic interactions between the membrane proteins and lipids.

REFERENCES

1. Abramson, M. B., Katzman, R., and Gregor, H. P. Aqueous dispersions of phosphatidylserine—Ionic properties. *J. Biol. Chem.* 239:70, 1964.
2. Andreoli, T. Formation and properties of thin lipid membranes from sheep red cell lipids. *Science* 154:417, 1966.
3. Ansell, G. B., and Hawthorne, J. N. *Phospholipids: Chemistry, Metabolism and Function.* Amsterdam: Elsevier, 1964.
4. Babakov, A., Ermishkin, L. N., and Liberman, E. A. Influence of electric field on the capacity of phospholipid membranes. *Nature* (London) 210:953, 1966.
5. Baker, D., and Watson, S. Permeability of a model phospholipid membrane to sugars. *Fed. Proc.* 26:711, 1967.
6. Bangham, A. D., De Gier, S., and Greville, G. D. Osmotic properties and water permeability of phospholipid liquid crystals. *Chem. Phys. Lipids* 1:225, 1967.

6a. Bangham, A. D., and Horne, R. W. Negative staining of phospholipids and their structural modification by surface-active agents as observed in the electron microscope. *J. Molec. Biol.* 8:660, 1964.

7. Bangham, A. D., and Papahadjopoulos, D. Biophysical properties of phospholipids: I. Interaction of phosphatidylserine monolayers with metal ions. *Biochim. Biophys. Acta* 126:181, 1966.
8. Bangham, A. D., Standish, M. M., and Miller, N. Cation permeability of phospholipid model membranes: Effect of narcotics. *Nature* (London) 208:1295, 1965.
9. Bangham, A. D., Standish, M. M., and Watkins, J. C. Diffusion of univalent ions across the lamellae of swollen phospholipids. *J. Molec. Biol.* 13:238, 1965.
10. Bangham, A. D., Standish, M. M., and Weissmann, G. The actions of steroids and Streptolysin S on the permeability of phospholipid structures to cation. *J. Molec. Biol.* 13:253, 1965.
11. Bean, R. C., and Shepherd, W. C. Permeability of lipid, bimolecular leaflets to organic solutes. *Fed. Proc.* 26:862, 1967.
12. Bielawski, J., Thompson, T. E., and Lehninger, A. L. The effect of 2,4-dinitrophenol on the electrical resistance of phospholipid bilayer membranes. *Biochem. Biophys. Res. Commun.* 24:948, 1966.
13. Brinley, F. J., Jr., and Mullins, L. J. Ion fluxes and transference number in squid axons. *J. Neurophysiol.* 28:526, 1965.
14. Brown, A. D. Hydrogen ion titrations in intact and dissolved lipoprotein membranes. *J. Molec. Biol.* 12:491, 1965.
15. Chapman, D. Liquid Crystalline Nature of Phospholipids. In Gould, R. E. (Ed.), *Ordered Fluids and Liquid Crystals.* Washington: American Chemical Society, 1967. P. 157.
16. Chapman, D., and Fluck, D. J. Physical studies of phospholipids: III. Electron microscope studies of some pure fully saturated 2,3-diacyl-

DL-phosphatidyl ethanolamines and phosphatidylcholines. *J. Cell Biol.* 30:1, 1966.
17. Christensen, H. N., and Hastings, A. B. Phosphatides and inorganic salts. *J. Biol. Chem.* 136:387, 1940.
18. Cunningham, W. P., Prezbindowski, K., and Crane, F. L. The relation between structure and function in electron-transport systems: II. Effect of solvent treatment on membrane structure. *Biochim. Biophys. Acta* 135:614, 1967.
19. Danielli, J. F. On the thickness of lipid membranes. *J. Theor. Biol.* 12:439, 1966.
20. Das, M. L., Haak, E. D., and Crane, F. L. Proteolipids: IV. Formation of complexes between cytochrome *c* and purified phospholipids. *Biochemistry* (Washington) 4:859, 1965.
21. Davson, H. Growth of the concept of the paucimolecular membrane. *Circulation* 26:1022, 1962.
22. Davson, H., and Danielli, J. F. *The Permeability of Natural Membranes.* London: Cambridge University Press, 1952.
23. del Castillo, J., Rodriguez, A., Romero, C. A., and Sanchez, V. Lipid films as transducers for detection of antigen-antibody and enzyme-substrate reactions. *Science* 153:185, 1966.
24. Dick, D. A. T. *Cell Water.* Washington: Butterworth, 1966.
25. Feinstein, M. B. Reaction of local anesthetics with phospholipids. *J. Gen. Physiol.* 48:357, 1964.
26. Fleischer, S., Fleischer, B., and Stoeckenius, W. Fine structure of lipid depleted mitochondria. *J. Cell Biol.* 32:193, 1967.
27. Folch, J., Lees, M., and Sloane-Stanley, G. H. The Role of Acidic Lipids in the Electrolyte Balance of the Nervous System of Mammals. In Richter, D. (Ed.), *Metabolism of the Nervous System.* New York: Pergamon, 1957.
28. Gemant, A. *Liquid Dielectrics.* New York: Wiley, 1933.
29. Gemant, A. *Ions in Hydrocarbons.* New York: Interscience, 1962.
30. Goldman, D. E. Potential, impedance, and rectification in membranes. *J. Gen. Physiol.* 27:37, 1943.
31. Green, D. E., Allmann, D. W., Bachmann, E., Baum, H., Kopaczyk, K., Korman, E. F., Lipton, J., MacLennon, D. H., McConnell, D. G., Perdue, J. F., and Tzagoloff, A. Formation of membranes by repeating units. *Arch. Biochem.* 119:312, 1967.
32. Hanai, T., and Haydon, D. A. The permeability to water of bimolecular lipid membranes. *J. Theor. Biol.* 11:370, 1966.
33. Hanai, T., Haydon, D. A., and Redwood, W. R. The water permeability of artificial bimolecular leaflets: A comparison of radio-tracer and osmotic methods. *Ann. N.Y. Acad. Sci.* 137:731, 1966.
34. Hanai, T., Haydon, D. A., and Taylor, J. An investigation by electrical methods of lecithin-in-hydrocarbon films in aqueous solutions. *Proc. Roy. Soc.* A 281:377, 1964.
35. Hanai, T., Haydon, D. A., and Taylor, J. Polar group orientation and the electrical properties of lecithin bimolecular leaflets. *J. Theor. Biol.* 9:278, 1965.

36. Hanai, T., Haydon, D. A., and Taylor, J. Influence of lipid composition and of some adsorbed proteins on capacitance of black hydrocarbon membranes. *J. Theor. Biol.* 9:422, 1965.
37. Hanai, T., Haydon, D. A., and Taylor, J. The variation of capacitance and conductance of bimolecular lipid membranes with urea. *J. Theor. Biol.* 9:433, 1965.
38. Hanai, T., Haydon, D. A., and Taylor, J. Some further experiments on bimolecular lipid membranes. *J. Gen. Physiol.* 48 (pt. 2):59, 1965.
39. Haydon, D. A., and Taylor, J. Stability and properties of bimolecular lipid leaflets in aqueous solutions. *J. Theor. Biol.* 4:281, 1963.
40. Helfferich, F. *Ion Exchange.* New York: McGraw-Hill, 1962.
41. Henn, F. A., Decker, G. L., Greenawalt, J. W., and Thompson, T. E. Properties of lipid bilayer membranes separating two aqueous phases: Electron microscope studies. *J. Molec. Biol.* 24:51, 1967.
42. Hodgkin, A. L. The ionic basis of electrical activity in nerve and muscle. *Biol. Rev.* 26:339, 1951.
43. Howard, R. E., and Burton, R. M. Lipid bilayer membranes: Apparatus design and studies with adsorbed acetyl choline esterase. *Fed. Proc.* 26:403, 1967.
44. Huang, C., and Thompson, T. E. Properties of lipid bilayer membranes separating two aqueous phases: Water permeability. *J. Molec. Biol.* 15:539, 1966.
45. Huang, C., Wheeldon, L., and Thompson, T. E. The properties of lipid bilayer membranes separating two aqueous phases: Formation of a membrane of simple composition. *J. Molec. Biol.* 8:148, 1964.
46. Joos, R. W., and Carr, C. W. The binding of calcium in mixtures of phospholipids. *Proc. Soc. Exp. Biol. Med.* 124:1268, 1967.
47. Katzman, R., and Wilson, C. E. Extraction of lipid and lipid cation from frozen brain tissue. *J. Neurochem.* 7:113, 1961.
48. Kavanau, J. L. *Structure and Function in Biological Membranes.* San Francisco: Holden-Day, 1965. Vols. I and II.
49. Kinsky, S. C., Luse, S. A., and van Deenen, L. L. M. Interaction of polyene antibodies with natural and artificial membrane systems. *Fed. Proc.* 25:1503, 1966.
50. Kirschner, L. B. The cation content of phospholipids from swine erythrocytes. *J. Gen. Physiol.* 42:231, 1958.
51. Koefoed-Johnsen, V., and Ussing, H. H. The contributions of diffusion and flow to the passage of D_2O through living membranes. *Acta Physiol. Scand.* 28:60, 1953.
52. Läuger, P., Lesslauer, W., Marti, E., and Richter, J. Electrical properties of bimolecular phospholipid membranes. *Biochem. Biophys. Acta* 135:20, 1967.
53. Leitch, G. J., and Tobias, J. M. Phospholipid-cholesterol membrane model: Effects of Ca^{++}, K^+ or protamine on hydration, H_2O permeability and electrical resistance. *J. Cell. Comp. Physiol.* 63:225, 1964.
54. Leslie, R. B., and Chapman, D. Artificial phospholipid membranes and bioenergetics. *Chem. Phys. Lipids* 1:143, 1967.

55. Lesslauer, W., Richter, J., and Läuger, P. Some electrical properties of bimolecular phosphatidyl inositol membranes. *Nature* (London) 213:1224, 1967.
56. Lucy, J. A. Globular lipid micelles and cell membranes. *J. Theor. Biol.* 7:360, 1964.
57. Luzzati, V., and Husson, F. The structure of the liquid-crystalline phases of lipid-water systems. *J. Cell Biol.* 12:207, 1962.
58. MacDonald, R. Resistance and capacitance of lipid bilayer membranes. *Fed. Proc.* 26:863, 1967.
59. Maddy, A. H., Huang, C., and Thompson, T. E. Studies on lipid bilayer membranes: A model for the plasma membrane. *Fed. Proc.* 25:933, 1966.
60. Mikulecky, D. C., and Tobias, J. M. Phospholipid-cholesterol membrane model. *J. Cell. Comp. Physiol.* 64:151, 1964.
61. Morill, G. A. Competitive binding of calcium and various cations to phospholipids: Use of a 1-butanol:water complex as a model system for cation exchange. Abstract, 11th Annual Meeting, Biophysical Society, Houston, 1967. P. 66.
62. Mueller, P., and Rudin, D. O. Induced excitability in reconstituted cell membrane structure. *J. Theor. Biol.* 4:268, 1963.
63. Mueller, P., and Rudin, D. O. Action potential phenomena in experimental bimolecular lipid membranes. *Nature* (London) 213:603, 1967.
64. Mueller, P., and Rudin, D. O. Development of K^+-Na^+ discrimination in experimental bimolecular lipid membranes by macrocyclic antibiotics. *Biochem. Biophys. Res. Commun.* 26:398, 1967.
65. Mueller, P., Rudin, D. O., Tien, H. T., and Wescott, W. C. Reconstitution of cell membrane structure *in vitro* and its transformation into an excitable system. *Nature* (London) 194:979, 1962.
66. Mueller, P., Rudin, D. O., Tien, H. T., and Wescott, W. C. Reconstruction of excitable cell membrane structure *in vitro. Circulation* 26:1167, 1962.
67. Mueller, P., Rudin, D. O., Tien, H. T., and Wescott, W. C. Methods for the formation of single bimolecular lipid membranes in aqueous solution. *J. Phys. Chem.* 67:534, 1963.
68. Mueller, P., Rudin, D. O., Tien, H. T., and Wescott, W. C. Formation and properties of bimolecular lipid membranes. *Recent Progr. Surface Sci.* 1:379, 1964.
69. Nash, H. A., and Tobias, J. M. Phospholipid membrane model: Importance of phosphatidylserine and its cation exchanger nature. *Proc. Nat. Acad. Sci. U.S.A.* 51:476, 1964.
70. Ohki, S., and Fukuda, N. Interlayer-interaction for a lipid bilayer model. *J. Theor. Biol.* 15:362, 1967.
71. Overton, E. Über die allgemeinen osmotischen Eigenschaften der Zelle, ihre vermutlichen Ursachen und ihre Bedeutung für die Physiologie. *Vjschr. Naturforsch. Ges. Zürich* 44:88, 1899.
72. Palmer, K. J., and Schmitt, F. O. X-ray diffraction studies of lipide emulsions. *J. Cell. Comp. Physiol.* 17:385, 1941.

73. Papahadjopoulos, D., and Bangham, A. D. Biophysical properties of phospholipids: II. Permeability of phosphatidylserine liquid crystals to univalent ions. *Biochim. Biophys. Acta* 126:185, 1966.
74. Papahadjopoulos, D., and Watkins, J. C. Phospholipid model membranes: II. Permeability properties of hydrated liquid crystals. *Biochim. Biophys. Acta* 135:639, 1967.
75. Pappenheimer, J. R., Renkin, E. M., and Borrero, L. M. Filtration, diffusion and molecular sieving through peripheral capillary membranes. A contribution to the pore theory of capillary permeability. *Amer. J. Physiol.* 167:13, 1951.
76. Razin, S., Morowitz, H. J., and Terry, T. M. Membrane subunits of *Mycoplasma Laidlawii* and their assembly to membrane-like structures. *Proc. Nat. Acad. Sci. U.S.A.* 54:219, 1965.
77. Reiss-Husson, F. Structure des phases liquide-crystalline de différents phospholipides, monoglycerides, sphingolipides, anhydres ou en presence d'eau. *J. Molec. Biol.* 25:363, 1967.
78. Rendi, R. Water extrusion in isolated subcellular fractions: VI. Osmotic properties of swollen phospholipid suspensions. *Biochim. Biophys. Acta* 135:333, 1967.
79. Robertson, J. D. Unit Membranes: A Review with Recent New Studies of Experimental Alterations and a New Subunit Structure in Synaptic Membranes. In Locke, M. (Ed.), *Cellular Membranes in Development.* New York: Academic, 1964. P. 1.
80. Rojas, E., and Tobias, J. M. Membrane model: Association of inorganic cations with phospholipid monolayers. *Biochim. Biophys. Acta* 94:394, 1965.
81. Rosano, H. L., Schulman, J. H., and Weisbuch, J. B. Mechanism of the selective flux of salts and ions through nonaqueous liquid membranes. *Ann. N.Y. Acad. Sci.* 92:457, 1961.
82. Schulman, J. H., and Rosano, H. L. The Influence of Amphoteric Surface Active Agents on the Diffusion and Carrier Transport of Salts and Ions Through Liquid Non-Aqueous Membranes. In La Mer, V. K. (Ed.), *Retardation of Evaporation by Monolayers.* New York: Academic, 1962. P. 97.
83. Sessa, G., and Weissmann, G. Effect of polyene antibiotics on phospholipid spherules containing varying amounts of charged components. *Biochim. Biophys. Acta* 135:416, 1967.
84. Seufert, W. D. Induced permeability changes in reconstituted cell membrane structure. *Nature* (London) 207:174, 1965.
85. Shah, D. O., and Schulman, J. H. The ionic structure of lecithin monolayers. *J. Lipid Res.* 8:227, 1967.
86. Shanes, A. M. Electrochemical aspects of physiological and pharmacological action in excitable cells: I. The resting cell and its alteration by extrinsic factors. *Pharmacol. Rev.* 10:59, 1958.
87. Shimoso, T., and Ohnishi, T. Studies on membrane model: I. Surface pressure and surface potential of pure phospholipid monolayers. *J. Biochem.* 61:89, 1967.
88. Taylor, J., and Haydon, D. A. Stabilization of thin films of liquid

hydrocarbon by alkyl chain interaction. *Discuss. Faraday Soc.* 42:51, 1966.

89. Thompson, T. E. The Properties of Bimolecular Phospholipid Membranes. In Locke, M. (Ed.), *Cellular Membranes in Development.* New York: Academic, 1964. P. 83.
90. Thompson, T. E., and Huang, C. H. Thickness of bilayer membranes. *J. Molec. Biol.* 16:576, 1966.
91. Thompson, T. E., and Huang, C. The water permeability of lipid bilayer membranes. *Ann. N.Y. Acad. Sci.* 137:740, 1966.
92. Tien, H. T. Thickness and molecular organization of bimolecular lipid membranes in aqueous media. *J. Molec. Biol.* 16:577, 1966.
93. Tien, H. T. Black lipid membranes: Thickness determination and molecular organization by optical methods. *J. Theor. Biol.* 16:97, 1967.
94. Tien, H. T., Carbone, S., and Dawidowicz, E. A. Black lipid films: A new type of interfacial adsorption phenomenon. *Kolloid Z.* 212:165, 1966.
95. Tien, H. T., Carbone, S., and Dawidowicz, E. A. Formation of "black" lipid membranes by oxidation products of cholesterol. *Nature* (London) 212:718, 1966.
96. Tien, H. T., and Dawidowicz, E. A. Black lipid films in aqueous media: A new type of interfacial phenomenon. *J. Colloid Interface Sci.* 22:438, 1966.
97. Tobias, J. M. Further studies on the nature of the excitable system in nerve: I. Voltage-induced axoplasm movement in squid axons. II. Penetration of surviving, excitable axons by proteases. III. Effects of proteases and of phospholipases on lobster giant axon resistance and capacity. *J. Gen. Physiol.* 43 (Suppl. 1):57, 1960.
98. Tobias, J. M., Agin, D. P., and Pawlowski, R. Phospholipid-cholesterol membrane model control of resistance by ions or current flow. *J. Gen. Physiol.* 45:989, 1962.
99. Troshin, A. S. *Problems of Cell Permeability.* New York: Pergamon, 1966.
100. van Deenen, L. L. M. Phospholipids and Biomembranes. In Holman, R. T. (Ed.), *Progress in Chemistry of Fats and Other Lipids.* New York: Pergamon, 1965. Vol. 8.
101. van den Berg, H. V. A new technique of obtaining thin lipid films separating two aqueous media. *J. Molec. Biol.* 12:290, 1965.
102. Vandenheuvel, F. A. Lipid-protein interactions and cohesional forces in the lipoproteins systems of membranes. *J. Amer. Oil Chem. Soc.* 43:258, 1965.
103. Vreeman, H. J. Permeability of thin phospholipid films III. *Kon. Ned. Akad. Wetensch. Proc. Ser. B.* 69:564, 1966.
104. Wallach, D. F. H., and Zahler, P. H. Protein conformations in cellular membranes. *Proc. Nat. Acad. Sci. U.S.A.* 56:1552, 1966.
105. Weatherby, J. H. Permeability of the artificial phospholipid membrane. *J. Cell. Comp. Physiol.* 33:333, 1949.
106. Weissmann, G., and Sessa, G. The action of polyene antibiotics on phospholipid cholesterol structures. *J. Biol. Chem.* 242:616, 1967.
107. Weissmann, G., Sessa, G., and Bernheimer, A. W. Staphylococcal

alpha-toxin: Effects on artificial lipid spherules. *Science* 154:772, 1966.

108. Weissmann, G., Sessa, G., and Weissmann, S. Effect of steroids and "Triton X-100" on glucose filled phospholipid cholesterol structures. *Nature* (London) 208:649, 1965.
109. Weissmann, G., Sessa, G., and Weissmann, S. The action of steroids and Triton X-100 upon phospholipid/cholesterol structures. *Biochem. Pharmacol.* 15:1537, 1966.
110. Whittam, R. *Transport and Diffusion in Red Blood Cells.* Baltimore: Williams & Wilkins, 1964.
111. Wood, R. E., and Morgan, H. E. Permeability of lipid bilayers. Abstract, 11th Annual Meeting, Biophysical Society, Houston, 1967. P. 63.
112. Woolley, D. W., and Campbell, N. K. Tissue lipids as ion exchangers for cations and the relationship to physiological processes. *Biochim. Biophys. Acta* 57:384, 1962.

8
Hereditary Diseases with Membrane Defects

Leon E. Rosenberg

AN ANALYSIS OF MEMBRANE TRANSPORT and its genetic control in man is at once exciting and frustrating. The excitement is generated by the great importance of this subject to an understanding of human biology. The frustration stems from the inability to describe satisfactorily any of the vital biophysical, biochemical, or genetic mechanisms which regulate transport in man. As with all living cells, the human cell membrane acts as the barrier which permits the intracellular content to differ markedly from its extracellular surroundings. This universal function is complemented by several others which are unique to multicellular organisms and reach their highest level of development in mammalian species. Thus, not only must an understanding of membrane function in man include an analysis of the lipoprotein structure of membranes or the physical and chemical laws defining passive and active movement of solutes and solvents. In addition, we must explain the following: the effects of drugs, vitamins, and hormones whose actions may result, primarily or solely, from alterations in membrane function or structure; the processes by which tissues specialized for transport functions, such as the gut mucosa and kidney tubule, accomplish their unique functions of intestinal absorption and urine formation; and the mechanisms of tissue differentiation which make the processes of transport in the liver so different from those in muscle, erythrocyte, or brain.

GENERAL REMARKS

The current state of our knowledge of membrane transport in model systems, single-cell organisms, and mammalian species has been summarized in other portions of this book and in several recent reviews [24, 47, 61]. The present discussion will be restricted to an analysis of specific inborn errors of transport in man. These diseases are, in the main, rare, and one may ask why they should be discussed in a book of this kind. The answer is identical to that given as the reason for studying any genetic disturbance, namely, that the study of mutations has proved to be of great value in understanding the normal mechanisms of genetic control and the normal chemical and physiological expression of such control. Thus, studies of renal glycosuria, familial hypophosphatemic rickets, and cystinuria have contributed greatly to our understanding of normal renal transport mechanisms for glucose, phosphate, and dibasic amino acids, respectively.

One need only look at the insights provided by the study of transport mutants in bacteria (see Chap. 5) to appreciate this concept fully. The investigations of Rickenberg, Cohen, Monod, and Kepes [56, 92] on galactoside transport in *Escherichia coli* provided the impetus for studies of induction, repression, and feedback control of specific transport reactions. They also supplied the evidence for a specific galactoside transport locus which functioned coordinately with other genes controlling galactoside metabolism (the lac operon). We cannot point to any human mutations which indicate the presence of inducible or repressible transport systems in man, nor have mutations led to the discovery or even the hint of human operon systems. Nonetheless, crucial insights into such diverse pathophysiological processes as hemolytic anemia, renal lithiasis, polyuria, and rickets have come from the study of inherited disorders of membrane function. We can be confident that mechanisms of drug, vitamin, and hormone action will be clarified by similar future work.

Spectrum of Disorders

In this discussion the word *transport* will be used in a restrictive sense. That is, only those disorders involving specific, primary defects of either active or passive movement of small molecules across cell membranes will be discussed. Therefore, we shall not discuss inherited abnormalities of circulating macromolecules which bind substances and facilitate their transport through the blood stream.

Specific examples of such disorders include analbuminemia, abetalipoproteinemia, and deficiency of thyroxine-binding globulin, which result in defective plasma transport of free fatty acids, cholesterol, and thyroxine, respectively. Similarly, little mention will be made of inherited diseases which produce transport defects secondarily, e.g., generalized renal tubular dysfunction caused by the accumulation of galactose-1-phosphate in galactosemia, by copper accumulation in hepatolenticular degeneration, or by cystine deposition in cystinosis. Table 8-1 lists the disorders that will be discussed. Each of these conditions has been shown to be inherited and to result from a specific, primary transport defect. The "possible" primary transport errors compiled in Table 8-2 represent a second group of diseases whose etiology and pathogenesis will probably be shown to depend on specific, inherited transport abnormalities. The diseases in Table 8-1 will be considered in more detail than those noted in Table 8-2, not because they are more common or are of greater clinical importance, but because they point out with certainty particular facets or unique features of the genetic, biochemical, or physiological mechanisms which control transport mechanisms in man.

The diversity of substances and organs implicated in the diseases listed in Table 8-1 deserves comment. Most of the disorders involve transport abnormalities in the kidney and/or intestine, reflecting either a true propensity for inherited abnormalities in these tissues, which are highly specialized for transport functions, or the relative ease of detecting disorders of intestinal or renal transport compared with those of brain, muscle, or liver. Transport errors range from defects for single substances in one organ (renal glycosuria, vitamin B_{12} malabsorption), to defects for a group of chemically related compounds in more than one organ (cystinuria, Hartnup disease, imino-glycinuria), to a generalized transport abnormality for dissimilar substances in one organ (Fanconi syndrome). These inborn errors of transport involve many different substrates, including hydrogen ion, sodium, chloride, calcium, phosphorus, amino acids, sugars, vitamin B_{12}, and water. A specific defect in hormonal responsiveness characterizes one of the conditions (vasopressin-resistant diabetes insipidus), all the others apparently resulting from abnormalities in substrate-membrane site interaction not influenced by other cofactors or hormones.

The clinical consequences of these defects also vary greatly. Some have little or no clinical significance (imino-glycinuria, renal glycosuria). Others are responsible for anemia (hereditary spherocytosis,

TABLE 8-1. *Inherited Disorders of Membrane Transport*

Disorder	Substances Involved	Tissues Involved	Clinical Features	Mode of Inheritance	Ref.
Cystinuria	Cystine, lysine, arginine, and ornithine	Proximal tubule, small intestine	Cystine renal calculi	Autosomal recessive (3 types)	5, 59
Hartnup disease	Most monoamino-monocarboxylic acids	Proximal tubule, small intestine	Skin rash, ataxia	Autosomal recessive	53, 80
Imino-glycinuria	Glycine, proline, hydroxyproline	Proximal tubule, small intestine	Nonreproducible, mental retardation (?)	Autosomal recessive (2 types?)	112, 121
Renal glycosuria	Glucose	Proximal tubule	None	Autosomal dominant (?)	62, 78
Glucose-galactose malabsorption	Glucose and galactose	Small intestine, proximal tubule (?)	Diarrhea, dehydration	Autosomal recessive	68, 105
Familial hypophosphatemic rickets	Phosphate and calcium	Proximal tubule, small intestine	Rickets, osteomalacia	X-linked dominant	12, 128
Fanconi syndrome	Glucose, amino acids, phosphate, uric acid	Proximal tubule, distal tubule (?)	Osteomalacia	Autosomal recessive (?)	27, 66
Intestinal malabsorption of vitamin B_{12}	Vitamin B_{12}	Small intestine	Megaloblastic anemia	Autosomal recessive	81, 118
Renal tubular acidosis	Hydrogen ion	Distal tubule	Acidosis, nephrocalcinosis	Autosomal dominant	114

TABLE 8-1. *(Continued)*

Disorder	Substances Involved	Tissues Involved	Clinical Features	Mode of Inheritance	Ref.
Vasopressin-resistant diabetes insipidus	Water	Distal tubule	Polyuria, polydipsia, hyposthenuria	X-linked recessive	84
Hereditary spherocytosis	Sodium	Erythrocyte	Hemolytic anemia, splenomegaly, jaundice	Autosomal dominant	51, 52
Congenital chloridorrhea	Chloride	Colon, small intestine (?)	Diarrhea, dehydration	Autosomal recessive	32

TABLE 8-2. *Possible Inherited Disorders of Membrane Transport*

Disorder	Substances Involved	Tissues Involved	Clinical Features	Mode of Inheritance	Ref.
"Blue diaper" syndrome	Tryptophan, calcium	Small intestine	Failure to thrive, hypercalcemia	Not defined	30
Methionine malabsorption syndrome	Methionine	Small intestine	Mental retardation, convulsions, diarrhea	Not defined	48, 116
Lowe's syndrome	Amino acids, organic acids, glucose, phosphate	Proximal tubule, distal tubule (?)	Cataracts, mental retardation, glaucoma	X-linked recessive (?)	71, 73
Busby syndrome*	Amino acids	Proximal tubule	Growth retardation, muscle weakness	Not defined	101, 102
Gluco-glycinuria	Glucose, glycine	Proximal tubule	None	Not defined	55
Gluco-amino-aciduria	Glucose, amino acids	Proximal tubule	Growth retardation	Autosomal dominant (?)	72
Diabetes mellitus	Glucose, amino acids	Skeletal muscle, cardiac muscle, adipose tissue	Acidosis, infections, neuropathy, nephropathy	Not defined	67, 104
Pseudohypoparathyroidism	Phosphate, calcium	Proximal tubule, bone, gut (?)	Nephrocalcinosis, cataracts, mental retardation	Autosomal dominant	89
ATPase deficiency hemolytic anemia	Sodium, potassium	Erythrocyte	Anemia	Not defined	45
Cystic fibrosis	Sodium	Eccrine glands, pancreas, lungs	Intestinal malabsorption, pulmonary infections, growth failure	Autosomal recessive	69, 76

* Named after the family in which the disorder was described.

vitamin B_{12} malabsorption), bone disease (familial hypophosphatemic rickets, renal tubular acidosis), kidney stones (cystinuria, renal tubular acidosis), diarrhea and/or growth retardation (glucose-galactose malabsorption, congenital chloridorrhea), and neurological dysfunction (Hartnup disease). Some of the disorders listed in Table 8-2 are responsible for other significant aberrations in intermediary metabolism (diabetes mellitus), or diffuse eccrine gland dysfunction (cystic fibrosis).

Genetic Etiology

Thus far we have considered the physiological disturbances and their clinical consequences. Before turning to a detailed discussion of specific disorders, it may be worthwhile to ask what we know about the genetic features of this group of diseases. How are the diseases transmitted from one generation to the next? What kind of structural, enzymatic, or catalytic proteins are involved? What types of mutations have taken place? We have partially satisfactory answers to the first of these questions and can point out possible answers to the second and third. The fact that most of the disorders listed in Table 8-1 are inherited as autosomal recessive traits suggests that clinically affected patients have a "double dose" of the mutant gene (or are homozygous for the mutant gene). Carriers (or heterozygotes) for these mutations are clinically indistinguishable from normal individuals, but appropriate biochemical or physiological tests often allow their detection. Each of the other modes of Mendelian genetic transmission is represented in this group of disorders as well. Renal glycosuria, renal tubular acidosis, and hereditary spherocytosis appear to be inherited as autosomal dominant traits, the implication being that heterozygotes for these mutations are phenotypically indistinguishable from homozygotes. Vasopressin-resistant diabetes insipidus and familial hypophosphatemic rickets are transmitted as X-linked recessive and X-linked dominant traits, respectively.

Such information, of great value in diagnosis and in genetic counseling, is crucial to a definitive explanation of the genetic fine structure of the mutations and their chemical and physiological consequences. Let us assume that specific structural genes code for the synthesis of specific proteins or enzymes which make up the basic lipoprotein backbone of the membrane or which selectively catalyze the transport of one or more substrates. We can then focus on possible sites of the various genetic defects of transport. Transport reactions involve both passive and active movement across cell membranes. Passive processes depend on the concentration gradient of the substrate and the presence of aqueous channels or "pores" in the membrane. Active transport systems can function against an electrochemical gradient and are energy dependent, implying the existence of specific carrier molecules, enzymes, or binding sites in the cell membrane and a mechanism for coupling the transport system to a source of metabolic energy.

Thus, mutations which produce abnormalities in passive transport, such as renal tubular acidosis and hereditary spherocytosis, probably alter the basic lipoprotein structure of the renal tubular cell and the erythrocyte, respectively. In specific disorders resulting from defective active transport mechanisms, such as glucose-galactose malabsorption or cystinuria, the mutation very likely alters the rate of synthesis, the rate of turnover, or the structure of a carrier protein or enzyme in or near the cell membrane which preferentially binds structurally related hexoses or amino acids. The mutation could cause complete cessation of synthesis of the carrier protein or conformational changes markedly reducing the affinity or capacity of the carrier for its substrate. If, as has been proposed recently [15, 21], the carrier proteins for sugars and amino acids also have binding sites for sodium which influence substrate-carrier interaction by allosteric effects, defective binding of the substrate could be produced by structural changes in the "sodium site" as well as by defects in the "substrate site." Alternatively, the mutation could affect the energy-coupling mechanism. The latter possibility seems likely when several diverse transport systems are affected, as in Fanconi syndrome, but in the light of current evidence [15, 21], appears to be unlikely as a mechanism for specific transport defects because it demands the existence of many different mechanisms of energy-coupling for the different transport systems.

We have been discussing structural genes which direct the synthesis of specific structural or carrier proteins. Theoretically, regulator or operator mutations could also exist and would, in fact, aid in explaining certain transport mutations which are inherited in a dominant fashion. There is ample evidence in bacteria for the existence of such regulator and operator gene mutations which control the induction and repression of transport processes. In mammalian systems, however, induction and repression of transport mechanisms have not been demonstrated, and therefore, there is no compelling reason to postulate control gene mutations. Experiments designed to explore the presence or absence of inducible transport systems in man and other mammals would be of great interest.

Although it is generally agreed that specific substances exist in or near the cell membrane which bind certain substrates reversibly and are responsible for mediated transport, there is no unanimity about their chemical nature or about what they should be called. Some workers have used functional terms such as *carrier* or *pump*; others enzymatic terms: *permease, transferase, translocase,* and, more whimsically, *here-to-there-ase.* There seems to be little advantage in arguing over these words. All of them cloak our ignorance of the precise chemical events involved and are useful insofar as they help us describe phenomena observed in vivo or in vitro.

In discussing specific inherited diseases due to mutations of one or more transport systems, we shall emphasize the evidence for a transport abnormality, the relationship between the transport defect and the pathophysiology of the disorder, and the unique genetic characteristics of the disease. Clinical findings and therapeutic approaches will not be stressed since these aspects are well covered in the references noted in Table 8-1. We shall group the diseases according to the substrates involved, recognizing that they could also be classified by the tissue or tissues in which the transport defects exist, by the mode of inheritance, or by the presumed type of membrane defect. Certain diseases will be discussed in much greater detail than others either because more is known about their etiology and pathogenesis or because they present unique features of interest to this volume.

DISORDERS OF AMINO ACID TRANSPORT

The study of amino acid transport in mammalian systems has been pursued along several independent but related paths for the past twenty years. Some investigators have explored the relationship between amino acid structure and transport mechanisms and have shown that transmembrane movement of these vital building blocks of protein is markedly influenced by the length and composition of aliphatic side chains, by the cationic or anionic properties of the amino acids in solution, by the D (dextro) and L (levo) isomeric forms of these compounds, and by the α- or β-position of the amino groups which define this class of molecules [16, 17]. Other workers have stressed the features of amino acid transport which are similar to those of enzyme-catalyzed reactions: temperature dependence; saturability;

competitive and noncompetitive inhibition; energy dependence; and Michaelis-Menten kinetics [98, 113]. Both groups of investigators have relied on in vitro experiments with cells or tissues, while a third group of researchers has used in vivo techniques to study the physiological regulation of intestinal absorption or renal tubular reabsorption of amino acids [8, 125]. Finally, an increasing number of investigators have studied patients with specific transport mutations to test and extend the hypotheses formulated by the other approaches. It is the information obtained from this genetic vantage which we will now explore.

Cystinuria

This venerable inherited disorder, first described in 1810, holds a unique position in the study of human biochemical diseases. Cystinuria, pentosuria, albinism, and alkaptonuria comprised the disorders discussed by Archibald Garrod in his famous Croonian lectures in 1908, when human biochemical genetics may be said to have been born. Almost thirty years later Brand and Cahill [9] defined many of the basic chemical features of the sulfur amino acid pathway in studies with cystinuric subjects. In 1951 Dent and Rose [28] proposed that cystinuria is a specific inborn error of renal tubular transport, an etiological concept previously overlooked in the study of human inherited disease. Since this pioneering work, the study of cystinuria has continued to provide important insights into the biochemical and genetic mechanisms which regulate transport in man. Contributions include the demonstration that the transport defect in the kidney is also present in the gut, but not in leukocytes; the notion that dibasic amino acids are reabsorbed in the kidney by more than a single mechanism; the evidence for tubular secretion of amino acids by the mammalian kidney; and the demonstration that the abnormal phenotype in cystinuria may be produced by several allelic mutations.

Cystinuric patients are usually detected because they form cystine stones in the kidney, ureter, or bladder which lead to urinary tract obstruction or infection. The predisposition to cystine stone formation results from the large quantities of cystine in the urine (400–1500 mg per 24 hours compared to normal values of 20–60 mg) and the relative insolubility of this amino acid in aqueous solutions (maximum solubility is about 300 mg per liter). By 1951, however, it was known that cystinuric patients also excreted striking excesses of lysine, arginine, and ornithine [119, 131]. This specific aminoaciduria led Dent and Rose to their studies demonstrating markedly increased renal clearances of cystine, lysine, arginine, and ornithine, and they concluded that the disease was due to a specific inherited defect in renal tubular reabsorption of these four structurally related amino acids. The renal clearance of cystine in their patients approximated the glomerular filtration rate (GFR) while the clearances of the dibasic amino acids, lysine, arginine, and ornithine, were variable, but always less than the GFR. The latter results were taken as evidence that the dibasic amino acids have a greater affinity for the single tubular transport system which they share with cystine, and that the defect in cystinuria is not complete.

Several recent studies demand additional explanations for these findings. Fox et al. [36] performed in vitro experiments with human kidney slices from cystinuric subjects and demonstrated only a partial defect in the transport of lysine and arginine, no defect in cystine uptake, and failure of cystine to compete with the other dibasic amino acids for uptake in this system. Similar experiments by Rosenberg, Segal, and co-workers [94, 98] with normal human and rat kidney also failed to show competition between cystine and the dibasic amino acids and presented kinetic evidence for the presence of two lysine transport systems. Recent studies by Crawhall, Frimpter, and their colleagues [22, 37] have demonstrated cystine clearance values significantly greater than the GFR in at least seven patients. Finally, single families have been reported with renal tubular defects only for cystine transport [10] or with tubular defects for the reabsorption of arginine, lysine, and ornithine but *not* cystine [88].

These results effectively exclude the hypothesis that dibasic amino acids and cystine are reabsorbed by only a single, common renal tubular system. They suggest (Fig. 8-1) that cystine does, indeed, share a transport system with the dibasic amino acids but that at least two other transport mechanisms for these substances also exist in the kidney: one responsible for reabsorption and/or secretion of cystine; the second responsible for reabsorption of dibasic amino acids but not cystine. Such a schema has ample biological precedent [2] and can explain all the in vivo and in vitro observations made thus far. A genetic defect in the transport process shared by cystine and the dibasic amino acids, with sparing of the other systems, can explain the clearance data in cystinuria. The presence of two dibasic amino acid transport systems provides an explanation for the failure to demonstrate in vitro competition between cystine and the dibasic amino acids. The presence of a specific cystine transport process can

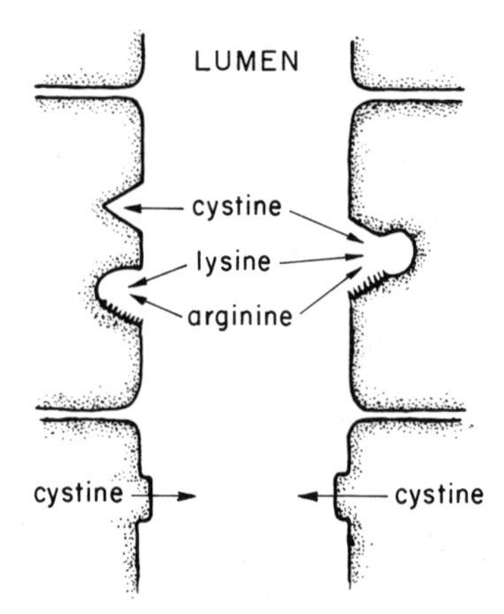

FIG. 8-1. Dibasic amino acid transport systems in the proximal renal tubule. Ornithine is omitted for artistic convenience, but it appears to be transported by the same mechanisms utilized by lysine and arginine. The transport system for lysine and arginine, not shared with cystine, is shown as a common site for purposes of simplicity. There is no evidence, however, which excludes transport sites specific for a single substrate. The arrows directed from the lumen to the tubular cell reflect reabsorption while the arrow from tubular cell to lumen depicts secretion.

explain both the apparent absence of an in vitro defect for cystine uptake in cystinuric kidney, since its directional component would not be distinguishable in such a system, and the presence of net tubular secretion of cystine in some cystinuric subjects. Finally, the families with defects in tubular reabsorption of cystine only or of dibasic amino acids only can be satisfactorily expained by postulating genetic defects in the specific transport mechanism for cystine and the specific mechanism for dibasic amino acids, respectively.

In 1961 Milne and co-workers [79] demonstrated a defect in intestinal absorption of dibasic amino acids in cystinuria. Their studies were extended by London and Foley [70] and by Rosenberg and co-workers [100], who demonstrated defective intestinal absorption of cystine as well. In 1964 Thier et al. [123, 124] and McCarthy et al. [74] showed, independently, that the gut transport defect for dibasic amino acids and cystine was easily shown in vitro using jejunal mucosa obtained by peroral biopsy.

These and subsequent in vitro experiments have proved to be of particular significance. First, they contrast sharply with results described previously using kidney tissue; cystine and the dibasic amino acids competed for transport in the gut [123], and uptake of cystine as well as dibasic amino acids was completely defective in most patients with cystinuria [124]. These findings indicate that a common transport system for cystine and dibasic amino acids, present in the kidney and defective in cystinuria, also exists in the intestine. They imply too that no other mediated transport process for these amino acids is present in the gut which can obscure the results of in vitro experiments or mitigate the defect in cystinuria. Second, Rosenberg and co-workers [97] demonstrated that the gut transport defect in most cystinuric patients was due to complete absence of the "carrier" system which mediates uptake of cystine, lysine, and arginine and not to a defect in coupling this transport process to sodium transport or oxidative metabolism. This study also showed that some cystinuric patients, otherwise phenotypically indistinguishable, had a less severe gut defect in vitro and in vivo than was noted in the majority of patients with the disease. Third, Rosenberg, Crawhall, and Segal [95] showed that cysteine is transported by a mechanism distinct from that utilized by cystine and the dibasic amino acids and is not implicated in the gut transport defect in cystinuria.

The results of the in vivo and in vitro studies in gut and kidney indicate that the transport of dibasic amino acids and cystine in these tissues is mediated by a protein, enzyme, or carrier under specific genetic control. Since further biochemical characterization of the carrier and its aberrations in cystinuria depends on the availability of large quantities of tissue, it seemed appropriate to determine whether the same mechanism also existed in a relatively accessible tissue such as peripheral leukocytes. In two separate studies [6, 96], no defect in dibasic amino acid transport was demonstrated in leukocytes from patients with cystinuria. Cystine was poorly transported, but lysine, arginine, and ornithine were transported actively by a common system in cells from controls and patients with cystinuria [96]. These results provide evidence for the variation of transport systems in tissues with different morphological and functional characteristics. They indicate that the expression of the common genetic information inherent in all human cells differs in the control of transport processes as it differs in the control of many other biochemical mechanisms.

The genetics of cystinuria have also proved to be challenging, reflecting the complexities shown in the study of the gut and kidney transport defect. Garrod noted that the disease was common in siblings of affected individuals but occurred very rarely in parents or children of cystinuric patients. Harris and co-workers [43, 44] studied 27 pedigrees extensively and concluded that cystinuria was inherited as an autosomal recessive trait. Their data indicated that subjects homozygous for the abnormal gene excreted large quantities of cystine, lysine, arginine, and ornithine in the urine and were likely to form cystine stones. Two distinct types of heterozygotes were noted. In most families studied, all parents and children of affected subjects (presumed heterozygotes) had urinary amino acid patterns indistinguishable from those of normal subjects. In the remaining families, all presumed heterozygotes excreted cystine and lysine in moderate excess. These findings led Harris to conclude that phenotypic cystinuria was caused by more than a single autosomal mutation. The findings have been extended by Rosenberg and co-workers, who have studied gut and kidney transport of dibasic amino acids and cystine in 20 patients with cystinuria and their close relatives. Their studies indicate that at least three different autosomal mutations may produce the same abnormal phenotype in cystinuria [97] and, furthermore, that these mutations are allelic [93].

The physiological, biochemical, and genetic studies in cystinuria have been presented in detail because they point out the complexities of what was once considered a simple transport defect and indicate the significant gaps in our information. The identification of the proposed dibasic amino acid permease, its rate of turnover, its expression in different mammalian tissues, and the nature of the mutations which alter its affinity or capacity for its substrates are unknown. These questions will be answered only by chemical and genetic studies of membrane structure similar to those currently in progress in microbial systems [35, 86]. Detailed in vivo and in vitro studies of other human transport mutants will almost surely reveal a degree of complexity similar to that noted in cystinuria.

Hartnup Disease

In 1956 Baron and co-workers [4] described four children in a single sibship with a bizarre illness characterized by a pellagra-like skin rash, temporary cerebellar ataxia, and aminoaciduria. At least 23 similar cases of this disorder, named for the first family in which it was identified, have been described subsequently. The high incidence of parental consanguinity and the frequent occurrence of the disease in siblings of affected patients indicate a genetic etiology. Chromatographic studies of urinary amino acids reveal marked increases in alanine, serine, threonine, asparagine, glutamine, valine, leucine, isoleucine, phenylalanine, tyrosine, trytophan, histidine, and citrulline. No abnormalities in urinary cystine, lysine, arginine, ornithine, glycine, proline, hydroxyproline, methionine, glutamic acid, or aspartic acid are observed. The findings point to a defect in renal tubular reabsorption of neutral amino acids with aliphatic or aromatic side chains, and renal clearance studies have corroborated this view. Less than 50 percent of the filtered histidine, serine, alanine, and glutamine are reabsorbed in patients with Hartnup disease [23], normal values exceeding 90 percent for each of these substrates. Nonetheless, plasma concentrations of these amino acids are normal or only slightly decreased, indicating that the clinical abnormalities are not due to a deficiency of one or more amino acids secondary to the exaggerated renal loss.

Milne and co-workers [80] demonstrated that tryptophan absorption from the intestine is defective in Hartnup disease. Scriver and Shaw [110] confirmed the observation and have indicated that intestinal transport of other neutral amino acids is also defective. In cystinuria the clinical abnormalities are a direct consequence of the renal tubular defect and the gut transport abnormality has little if any import. The converse is true in Hartnup disease. Current evidence suggests that the skin rash and cerebellar ataxia are due to the intestinal defect in tryptophan absorption, not by producing a specific tryptophan deficiency, but by decreasing the amount of tryptophan available for conversion to nicotinamide. This thesis is supported by the observation that a high protein diet supplemented with nicotinamide is efficacious in preventing or treating the clinical stigmata in the patients. Additional evidence for such a formulation has been provided by a child with a block in conversion of tryptophan to kynurenine (a precursor of nicotinamide), who exhibited a skin rash and cerebellar signs very much like those noted in Hartnup disease and responded to nicotinamide supplements [120]. In addition to producing a specific deficiency of nicotinamide, the gut transport defect in Hartnup disease is also responsible for the appearance of several indoles in the urine. Intestinal micro-organisms convert the unabsorbed tryptophan to indoles which are absorbed from the gut and excreted in the urine.

The pattern of the aminoaciduria and the sparing of the dibasic amino acids, the dicarboxylic amino acids, the imino acids, and glycine suggest that Hartnup disease is caused by a specific mutation in the transport system shared by most neutral and aromatic amino acids.

Available evidence indicates that this mechanism, like the dibasic amino acid one, is represented in both gut and kidney. Since in vitro studies have not been carried out, we do not know whether other tissues share this transport system, or whether there are other genetically controlled transport systems in gut and/or kidney which have preferential affinity for neutral amino acids. Such studies would be particularly valuable in the light of current information about the transport defect in cystinuria.

Consanguinity has been demonstrated in 4 of 14 families, suggesting that the gene for Hartnup disease is very rare. Eight male and 15 female patients have been described, without parent-to-child transmission, ruling out X-linked dominant, X-linked recessive, or autosomal dominant inheritance. The findings suggested that patients with Hartnup disease are homozygous for an autosomal mutation. Parents and children of these presumed homozygotes demonstrate no abnormality of urinary amino acid excretion or tryptophan absorption from the intestine, suggesting that a single dose of the mutant gene is not sufficient to impair gut or kidney transport in any easily detected way. Analysis of additional pedigrees will be of considerable value in determining whether genetic heterogeneity exists in this disease as it does in cystinuria, with which Hartnup disease shares so many other genetic and biochemical features.

Imino-Glycinuria

Patients with hyperprolinemia, due to a block in proline catabolism, were noted to excrete markedly increased quantities of hydroxyproline and glycine as well as proline. This particular aminoaciduria led Scriver, Schafer, and Efron [109] to postulate a common renal tubular reabsorptive mechanism for glycine and the imino acids, proline, and hydroxyproline. Renal clearance studies in normal adults [108] and in vitro experiments with rat kidney slices [111] gave additional evidence for the existence of such a system. Within the past three years several patients in Japan [121], Canada [112], and the United States [99] have been described who show unequivocally that the renal transport system utilized by the imino acids and glycine is under separate genetic control. All the patients have demonstrated a specific renal aminoaciduria characterized by the excretion of glycine, proline, and hydroxyproline. Since the imino acids are not normally excreted in detectable quantities beyond infancy, identification of these patients has been relatively easy. Biochemical features have been variable. Glycine clearance has ranged from 9 to 33 ml/min/1.73 m^2. Proline and hydroxyproline clearances have not been reported in all cases and seem to vary considerably with the filtered load of these imino acids. No reproducible clinical pattern has emerged. Four children were mentally retarded, but since their biochemical abnormalities were detected as part of screening programs in retarded children, it is far from certain that the mental retardation is related etiologically or pathophysiologically to the renal tubular defect. A Canadian adult male had no clinical abnormalities while a 6-year-old American boy had complete, congenital nerve deafness, which was responsible for his detection.

Scriver and Wilson [112] have described in detail the renal tubular lesion in their Canadian patient. The fact that infusions of large quantities of proline failed to alter the already elevated glycine clearance in their patient indicates that glycine is reabsorbed from the renal tubule by at least two different mechanisms: one shared with the imino acids and essentially completely defective in their patient and a second not shared with the imino acids. Proline and hydroxyproline were also shown to be reabsorbed by a separate mechanism not shared with glycine. Similar results were noted by Rosenberg and Durant [99] in their study of a 6-year-old boy with biochemical findings very similar to those noted by Scriver and Wilson.

The latter workers have also investigated gut transport extensively in their patient. Gut biopsy, oral loading, and fecal amino acid excretion studies have shown no defect in intestinal transport of either proline or glycine. These findings, confirmed by Scriver, differ from those reported by Morikawa [83], Goodman [41], and their coworkers, who demonstrated a gut defect for proline and glycine in some of their cases.

A genetic etiology for this disorder seems clear from the Canadian and American studies. In both families, parents and children of the index case excreted distinctly increased amounts of glycine, but not proline or hydroxyproline. The data suggest that imino-glycinuria is inherited as an autosomal recessive trait. Individuals homozygous for the mutation excrete glycine, proline, and hydroxyproline in excess; heterozygotes excrete only excess glycine. Despite the paucity of cases reported thus far, there is already considerable evidence for genetic heterogeneity: the widely different clinical patterns; the presence of a gut defect in some but certainly not all patients; the normal urine pattern of presumed heterozygotes in two Japanese families [121] contrasted with the hyperglycinuria observed in presumed heterozygotes from Canada and the United States.

One other facet of the genetics of the condition is noteworthy. In 1957 DeVries and associates [29] described an Ashkenazi Jewish family with "dominantly inherited" hyperglycinuria. Although possibly they were studying a completely different condition, it seems more likely that they were describing patients heterozygous for the mutation which produces imino-glycinuria. The "dominant" pattern of inheritance is consistent with this hypothesis, as is the ethnic background, since the Canadian and American families with imino-glycinuria both came, at least in part, from Ashkenazi Jewish backgrounds.

DEFECT OF HEXOSE TRANSPORT

It is not surprising that the mechanisms by which glucose and related hexose molecules enter cells have intrigued investigators for more than 100 years. The key role played by glucose in adenosine triphosphate (ATP) generation, glycogen synthesis, fatty acid formation, and other vital intracellular processes makes knowledge about its transport into and out of cells of great importance. In addition, much current thinking about diabetes mellitus and the mechanism of insulin action has centered about the control of glucose transport. Despite these powerful stimuli to investigation, our knowledge is fragmentary. Experiments in vitro using mammalian intestine [19, 20], kidney [58, 63], erythrocytes [126], and muscle [57, 82] indicate that glucose and structurally related hexoses are transported by specific, energy-dependent, sodium-sensitive, saturable systems which conform to Michaelis-Menten kinetics and are thus assumed to be catalyzed by carrier molecules or permeases in the region of the cell membrane. Studies with competitive inhibitors of hexose transport, such as phlorizin, and with nonutilizable hexose analogs indicate that glucose and other actively transported sugars are not altered chemically during transport. Crane and his colleagues [19] have defined the structural requirements for mediated hexose transport in the gut: the presence of a D-pyranose ring, a methyl or hydroxy-methyl group at carbon 5, and a hydroxy group in the glucose configuration at carbon 2. Crane [21] has suggested that the proposed hexose carrier also contains a specific sodium binding site, but there is, at present, no information on the structure, location, or control of the postulated carrier.

The renal physiologist has established conclusively that glucose is freely filtered at the glomerulus and is actively reabsorbed in the proximal convoluted tubule, where its transport is characterized by a threshold and a tubular reabsorptive maximum (T_{m_G}). Much emphasis has been placed on the shape of the curve relating filtered glucose to reabsorbed glucose (Fig. 8-2). The glucose titration curve in dogs, Shannon [115] pointed out, conformed to the theory that glucose was reabsorbed identically by all nephrons until the concentration in proximal tubular fluid exceeded the capacity of the proposed carrier sites and caused tubular rejection and excretion of all additional glucose (Fig. 8-2, "theoretical" plot). However, careful experiments in normal human subjects [117] failed to yield such glucose titration curves, showing instead the "splay" indicated in Figure 8-2 (normal human). The chemical or anatomic basis for this splay is the object of much controversy. Some investigators believe that the splay is produced by the known structural heterogeneity of the nephron population in human kidney, which leads to saturation of the glucose reabsorptive mechanisms in some tubules at lower blood glucose concentrations than that observed in other tubules. But, as other researchers point out, there is no reason to assume that glucose reabsorption is an "all or none" phenomenon as saturation conditions are approached; the shape of the glucose titration curve observed in normal subjects, they contend, is compatible with classic Michaelis-Menten kinetics. These workers point out that T_{m_G} can be equated with the V_{max} of a specific enzymatic reaction and have

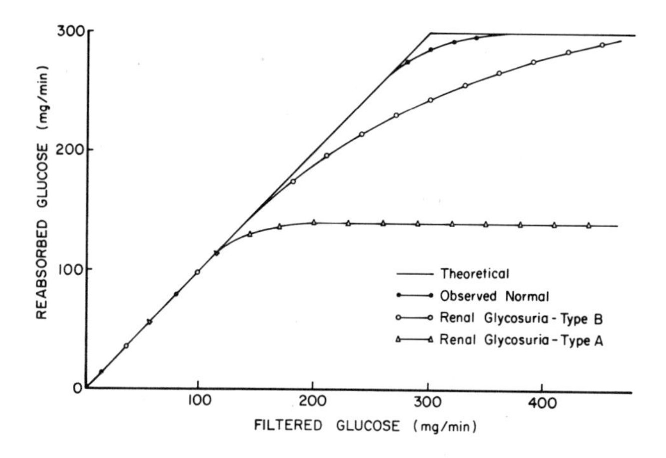

FIG. 8-2. Idealized glucose titration curves in normal man and in patients with renal glycosuria. The qualitative and quantitative differences shown are discussed in the text.

derived mathematical expressions from glucose titration curves with which to estimate the affinity (K_m) of glucose for its proposed saturable carrier system. These opposing views are prominently displayed in the discussion of renal glycosuria to follow.

Renal Glycosuria

This benign disorder of renal tubular transport has been known as long as diabetes mellitus, with which it has often been confused [62]. Estimates of its incidence vary widely, depending on the criteria used to establish the diagnosis. If the rigid criteria of Marble [77] are employed (glycosuria in virtually all urine samples tested, normal oral glucose tolerance curve, quantitative excretion of less than 10 to more than 100 gm of glucose per 24 hours), the disorder is very rare (94 out of 50,000 patients with melituria). If a less stringent criterion is employed (appearance of glucose in the urine during an otherwise normal oral glucose tolerance test), the incidence of the disorder may be as high as 1 out of 200 in the population. Regardless of which criteria are used for diagnosis, there is no doubt about the existence of such an entity nor about the validity of classifying renal glycosuria as a specific, inherited disturbance of renal transport. Patients with this isolated disorder have no other functional or morphological evidence of renal disease nor, with rare exceptions, does the condition lead to clinical symptoms. Renal glycosuria and diabetes mellitus have been described in the same family and even in the same person. The very high incidence of diabetes in the population makes these associations difficult to interpret, and for the present discussion we will assume that the two conditions are totally unrelated.

Glucose titration curves in patients with renal glycosuria have yielded two very different patterns. In one group, both the threshold for glucose excretion and the T_{m_G} are markedly reduced (Fig. 8-2, curve marked renal glycosuria, Type A). Other patients (curve marked renal glycosuria, Type B) have been demonstrated to have a low threshold, a markedly exaggerated "splay," and a normal or near normal T_{m_G}. Patients with Type A renal glycosuria suffer from either a qualitative or a quantitative abnormality in the carrier mechanism for glucose reabsorption. Type B patients have occasioned the controversy mentioned above. The proponents of the "nephron heterogeneity" theory of renal glycosuria contend that the low threshold and increased splay exhibited by these patients reflect simply an exaggeration of the nephron heterogeneity observed in normal individuals.

The proponents of the "enzyme saturation" theory of splay argue that the curves observed in Type B can be explained by postulating the existence of an altered carrier for glucose whose affinity is distinctly less than normal but whose capacity (T_{m_G}) is unaltered.

The surprising thing about this argument is that so little has been done to resolve it. Three approaches seem warranted. First, studies of intestinal glucose transport in such patients would be very interesting, for if a defect were also noted in the gut, the nephron heterogeneity argument would be shattered. Second, detailed studies of phosphate and amino acid reabsorption in such patients should also demonstrate prominent splay if heterogeneity of the nephron population is at fault. Third, careful genetic studies might well provide a direct answer to the question. Thus, it is well known that renal glycosuria commonly occurs in family members and is passed on from parent to child in a phenotypically dominant fashion [46, 49]. With but one exception [38], however, glucose titration curves in several affected members of a single sibship have not been carried out to firmly establish the mode of inheritance or the constancy of the defect within a family. Several important findings could emerge from family studies. If both Type A and Type B renal glycosuria occurred in the same sibship, the nephron heterogeneity argument would appear very weak indeed; the indication would be that Type A and Type B patients were, respectively, homozygous and heterozygous for a mutant gene leading to an altered glucose carrier. Similarly, if two parents with Type B had a child with Type A, the evidence for a single mutation with different expressions in the homozygous and heterozygous form would be extremely strong. If, on the other hand, Type A and Type B patients were never found in the same family, it would suggest, as Woolf has postulated [129], that these designations depend on different mutations which alter the glucose carrier system in specific ways: by reducing the capacity but not affinity in Type A renal glycosuria and by decreasing the affinity without affecting the capacity in Type B patients. The genetic approach has been fruitful in cystinuria and imino-glycinuria. It will almost surely be revealing in this disorder as well.

Glucose-Galactose Malabsorption

Eight patients with this previously unrecognized inborn error of hexose transport have been described since 1962 [3, 64, 68, 105]. The clinical findings in each patient were similar, consisting of profuse, watery diarrhea following the ingestion of glucose, galactose, or disaccharides containing one or both of these hexoses. Seven of the patients presented in the newborn period with diarrhea severe enough to be life threatening. The eighth was a 35-year-old woman who controlled her diarrhea by avoiding glucose or galactose in any form. The nature of the transport defect in these patients has been defined by oral loading tests and by in vitro studies with jejunal mucosa. Oral loading with glucose resulted in little or no increase in blood glucose and in the prompt excretion of large quantities of free glucose in the diarrheal feces. Similar results have been obtained following ingestion of galactose and 3-0-methyl glucose. In contrast, ingestion of fructose was followed by a prompt rise in blood fructose and blood glucose and no diarrhea.

These findings suggested a specific absorptive defect for glucose and galactose, a formulation strengthened by studies of gut mucosa from the same patients. Disaccharidase deficiency was ruled out by normal lactase and maltase activity in the biopsy specimens. Schneider et al. [105], using autoradiographic techniques, demonstrated a total lack of uptake of ^{14}C galactose by gut mucosa from one patient. This study also showed that phlorizin markedly reduced the uptake of labeled galactose by normal intestinal mucosa but had no effect on the mucosal labeling in the patient. Eggermont and Loeb [31] reported that gut mucosa from one patient failed to accumulate glucose in excess of the medium concentration and that the presence or absence of sodium in the incubation medium failed to influence glucose uptake. These studies provide direct evidence for a profound defect in galactose and glucose transport by the gut mucosa and suggest that, in the glucose-galactose malabsorption syndrome, the proposed specific carrier for these hexoses is either completely absent or so altered as to have no affinity for its substrates. The data also imply that the intestinal transport mechanism for glucose and galactose is under specific genetic control not shared by fructose or other monosaccharide molecules.

Two other facets of this disorder are worth mentioning. Glycosuria without hyperglycemia has been demonstrated in three of these patients. Here again, we may be dealing with a combined renal and intestinal transport error. Glucose titration curves of the kind used in the study of renal glycosuria would be particularly interesting. No clinical or chemical abnormalities have been noted in the parents of affected children, but the occurrence of the disease in siblings of both sexes and the presence of consanguinity in one family with three affected subjects suggest that the disorder is due to a rare autosomal mutation inherited in a recessive fashion. Additional biochemical and genetic studies are needed to confirm this view.

DEFECTIVE PHOSPHATE TRANSPORT

Inorganic phosphate, like amino acids and glucose, appears to be completely filtered by the glomerulus and reabsorbed in the proximal renal tubule. It is not a "threshold" substance and is excreted in significant concentrations even when the filtered load is normal or decreased. Tubular reabsorption of phosphate is characterized by saturation kinetics with a tubular maximum T_{m_P} of about 150 μmoles per minute. Proximal tubular reabsorption of this anion differs significantly from that of glucose and amino acids in that it is regulated by parathyroid hormone, a distinction which will become of obvious importance in our discussion. Few studies of renal phosphate transport have been carried out in vitro, nor have the mechanisms of intestinal absorption or transport of this vital anion been defined. Although abnormalities of phosphate reabsorption are well documented as secondary events owing to renal insufficiency or to disturbances of parathyroid hormone secretion, there is only one inherited disorder which may be due to a *primary* defect in phosphate transport.

Familial Hypophosphatemic Rickets

This well-studied hereditary disorder is characterized by X-linked inheritance, hypophosphatemia, reduced renal tubular reabsorption of phosphate, diminished gastrointestinal absorption of calcium and phosphate, and a high incidence of rickets or osteomalacia [12, 128]. The disease has been called vitamin D-resistant rickets to differentiate it from simple vitamin D deficiency, but this term should be discarded since there is good evidence that the relative resistance to vitamin D is not related to the primary defect and bone disease is not a constant feature. Hypophosphatemia is the most useful biochemical and genetic marker for the disease. Thus, until it was pointed out by Burnett and co-workers [12] that hypophosphatemia without bone disease could be the only manifestation of the condition, neither the mode of inheritance nor the currently held theories of pathogenesis was clearly defined. These workers showed that the familial hypophosphatemia followed the appropriate Mendelian laws for X-linked dominant inheritance and suggested, therefore, that the hypophosphatemia was the primary manifestation of the genetic disturbance.

Many investigators have shown that the hypophosphatemia in these patients is accompanied by hyperphosphaturia and reduced tubular maximum for phosphate reabsorption (T_{m_P}). The controversy over the pathogenesis of familial hypophosphatemic rickets has centered about the primary or secondary nature of this impairment in renal tubular transport for phosphate and has been complicated by the demonstration that many patients also have distinct impairment of calcium absorption from the intestine. Several hypotheses have been proposed [128]. Two will be discussed in some detail.

One group of investigators [12] has favored the following sequence: A genetically induced disturbance in the carrier or enzyme which mediates renal tubular reabsorption of phosphate leads to hypophosphatemia and hyperphosphaturia; the hypophosphatemia results in a reduced calcium-phosphorus ion product at the site of bone crystal formation, predisposing to rickets or osteomalacia, depending on the age of the patient. This thesis is attractive because it proposes a kind of defect frequently observed in other inborn errors of transport, because it describes the biochemical hallmark of the disease (hypophosphatemia), and because it is simple. It has one major flaw: its inability to explain the impaired intestinal absorption of calcium. Theoretically, however, there is one simple explanation which has not received much attention. If the defect in phosphate transport involved the gut as well as the kidney, the impaired calcium absorption could be admirably explained by postulating that calcium is complexed in the gut in the form of calcium phosphate and hence is not available for absorption. Several groups [40] have demonstrated reduced intestinal phosphate absorption in these patients but have pointed out that the reduced phosphate absorption could be secondary to impaired calcium absorption rather than a primary defect. In vitro studies of phosphate transport by gut mucosa might help settle this important question.

A second group of workers [1] has proposed a very different schema. They suggest that the defect in calcium absorption from the intestine is primary and that this disturbance stimulates secretion of parathyroid hormone, which in turn reduces renal tubular reabsorption of phosphate and is responsible for the hypophosphatemia and reduced T_{m_P}. They point out that the parathyroid glands of a few of the patients have been hyperplastic at autopsy. The bone disease is presumed to reflect this secondary hyperparathyroidism. Although the theory adequately explains the absorptive defect for calcium and the hypophosphatemic rickets, it does not account for the reproducibly normal serum calcium levels in these patients and the presence of many individuals with hypophosphatemia and hyperphosphaturia without demonstrable bone disease or hypercalcemia.

It does not seem likely that this controversy will be resolved by serum assays for parathyroid hormone since elevated values could be found with either formulation. The requirement for large doses of vitamin D to treat the rickets or osteomalacia may, likewise, reflect only the impaired calcium absorption—whether primary or secondary. Direct investigation of calcium transport in gut mucosa from these patients would be of considerable interest and should be possible in view of previous animal experiments reported by several groups of investigators [103].

GENERALIZED RENAL TUBULAR DEFECTS

Each of the disorders discussed thus far has been characterized by very specific, isolated transport abnormalities which can logically be attributed to mutations involving a single carrier protein or enzyme. However, there are a number of other syndromes in which generalized abnormalities of renal tubular reabsorption are prominent. The clinical and biochemical spectra noted with this group of conditions are broad. Some have abnormalities in virtually every renal tubular function (aminoaciduria, phosphaturia, glycosuria, uricosuria, hypercalciuria, and renal tubular acidosis). This constellation has been called the Fanconi syndrome [27, 66]. Other disorders involve defects in reabsorption of amino acids and sugars [55, 72], of amino acids and phosphate [54], of glucose and phosphate [26], or of amino acids alone [71, 102]. These biochemical abnormalities may be of no clinical significance in some cases and of great physiological consequence in others, e.g., in the severe osteomalacia frequently seen in patients with the Fanconi syndrome. Generalized proximal tubular abnormalities may be acquired or inherited. Intoxication with lead, uranium, cadmium, maleic acid, mercury, Lysol, and outdated tetracycline can produce the Fanconi syndrome. Renal tubular deposition of copper, galactose-1-phosphate, glycogen, and cystine in Wilson's disease, galactosemia, von Gierke's disease, and cystinosis, respectively, may also result in the Fanconi syndrome, which is often reversible with appropriate therapy of the primary metabolic disease. In many other patients the absence of a specific etiology and the appearance of a similar disorder in siblings have been interpreted as evidence for a primary inherited defect in tubular function [7, 27], it being realized that the defect could be secondary to some unidentified toxic substance, inherited or acquired.

There is only one feature shared by all of these syndromes. None of them can be explained adequately by postulating specific defects in membrane carrier molecules. The very existence of these conditions points out the need to characterize active transport mechanisms not only by the chemical or physiological nature of specific carrier molecules but also by the mechanism(s) coupling these transport systems to a source of metabolic energy. It seems likely that many, if not all, transport systems are coupled to the same source of energy and that diffuse abnormalities of the kind described in the Fanconi syndrome are due to metabolic disturbances which uncouple the transport systems from their source of energy. What is this source (or sources) of energy? How is the energy supplied to specific transport reactions? Why do we encounter the very broad spectrum of abnormalities noted above? Are all the transport abnormalities noted in the Fanconi syndrome caused by the same toxic or metabolic disturbance or does a defect in the tubular reabsorption of one group of compounds produce a "cascade" effect resulting in the observed diffuse abberation? These are only a few of the disturbing and challenging questions raised by these disorders.

DEFECTIVE VITAMIN TRANSPORT

The vitamins are a unique group of chemical cofactors which the human organism cannot synthesize. Hence, they must be supplied in the diet and absorbed from the gut. Vitamins A, D, E, and K are fat-soluble substances whose intenstinal absorption depends on the same mechanisms that facilitate lipid digestion and absorption. Abnormalities in lipid transport or digestion may lead to defective absorption of the fat-soluble vitamins, but we know of no primary, inherited defects in the absorption of these substances. However, recent studies have identified two inherited disorders of intestinal absorption of one of the water-soluble vitamins, vitamin B_{12}. One of these disorders may be classified as a primary transport defect.

Hereditary Intestinal Malabsorption of Vitamin B_{12}

Vitamin B_{12} has intrigued students of human physiology for much of this century because of its vital role in normal erythropoiesis. Its absorption from the gut depends on two processes: the elaboration of an "intrinsic factor" by the gastric mucosa and the binding of vitamin B_{12} to specific receptors in the ileum, this binding being catalyzed in some way by the intrinsic factor. Classic pernicious anemia in the adult is caused by deficiency of intrinsic factor leading to malabsorption of vitamin B_{12}. The lack of intrinsic factor is usually secondary to some other gastric lesion such as atrophic gastritis or gastric carcinoma, but primary, congenital, isolated intrinsic factor deficiency has been well documented in at least 20 cases [75]. It led to typical megaloblastic anemia corrected by oral administration of intrinsic factor or parenteral administration of vitamin B_{12}.

Since 1960 [42, 50, 81, 118] evidence has accumulated for a very different kind of inherited defect of vitamin B_{12} absorption. About 20 children have been described with megaloblastic anemia, appearing during the first two years of life, who responded completely to parenteral therapy with vitamin B_{12} but who have *no* response to intrinsic factor. The fact that gastric juice from these children corrects vitamin B_{12} malabsorption in patients with typical adult pernicious anemia indicates that intrinsic factor is present and is normal. No other functional abnormalities in intestinal absorption have been noted in the children, and morphological studies of their gut mucosa have yielded normal results. The data suggest that the basic defect in these children involves the specific receptor mechanism or carrier by which vitamin B_{12} is absorbed from the ileum and imply that this mechanism is the product of a single gene. Suboptimal absorption of vitamin B_{12} in both parents of one of the children [81], sibling involvement in 6 families, and a high incidence of parental consanguinity suggest that the disorder is inherited as an autosomal recessive trait. Persistent proteinuria has been reported in most of the patients, but the relationship between this finding and the specific intestinal malabsorption of vitamin B_{12} is obscure.

DEFECTS OF WATER AND ION TRANSPORT

Each disease discussed so far has involved the small intestine and/or the proximal renal tubule. Well-defined, inherited transport defects exist in three other tissues, the distal renal tubule, the colon, and the erythrocyte. These disorders affect the transport of water, hydrogen ion, chloride, and sodium. The physiological derangements produced are often very serious, affecting such vital processes as acid-base balance, osmotic regulation, and red cell life-span.

Renal Tubular Acidosis

The kidneys and lungs maintain the plasma pH near 7.4 by controlling the excretion of the acid end products of metabolism. The lungs regulate plasma carbon dioxide content. The kidneys maintain the sodium bicarbonate concentration of plasma within narrow limits by reabsorbing filtered bicarbonate and by reabsorbing sodium ions in exchange for hydrogen ions secreted by the renal tubule, the latter process depending on the formation of titratable acid and the secretion of ammonia. Renal tubular acidosis is the name given to the entity characterized by low concentrations of bicarbonate and elevated concentrations of chloride in plasma and by a urinary pH which never falls below about 6 (normal urine pH falls to 4.5 under acid load conditions) [114]. In addition to acidosis, the disorder leads to hypokalemia, hyperkaluria, hypercalciuria, nephrocalcinosis, impairment of urinary concentrating ability, and polyuria. This constellation may be acquired secondary to heavy metal poisoning, hypercalcemia, disorders of gamma globulin metabolism, and cystinosis. It may also occur primarily, either as a sporadic, transient finding in the newborn or as a persistent disease with a striking familial pattern. Extensive family studies have demonstrated an autosomal dominant mode of inheritance with variable degrees of expression [114], in contrast to the recessive mode of inheritance exhibited by most other specific transport defects.

The numerous physiological derangements in familial renal tubular acidosis probably result from a single defect in urinary acidification. Bicarbonate reabsorption and ammonia excretion are normal, thus militating against an abnormality of hydrogen ion secretion. Abnormal diffusion of bicarbonate out of the proximal tubule has not been demonstrated. Reynolds and others [91, 130], therefore, suggested that the primary defect appears to be an inability of the distal tubule to generate steep hydrogen ion gradients between blood and urine. Seldin and co-workers [114] postulated that the distal tubular cells in these patients have an increased permeability to hydrogen ions, leading to excessive back-diffusion of hydrogen from tubular urine to blood and an abnormal hydrogen ion gradient. Although this hypothesis is consistent with the known facts, it remains one of exclusion rather than direct demonstration. The nature of the proposed hydrogen ion diffusion system is completely unknown and will probably be elucidated only by identification and study of appropriate animal models or by isolation and characterization of the cell membrane of the distal tubule. There is no information at present which suggests the presence of specific mediated mechanisms for hydrogen ion transfer in the distal tubule, nor is there evidence to the contrary.

Vasopressin-Resistant Diabetes Insipidus

This rare inherited disorder of distal tubular function is unique among the primary disorders of transport because it involves hormonal responsiveness. The polyuria, polydipsia, and hypotonic urine characteristic of the condition are caused not by failure to secrete vasopressin, the antidiuretic hormone of the posterior pituitary gland, but by failure of the distal tubule and collecting ducts of the kidney to respond to the hormone. Vasopressin, a cyclic octapeptide released in response to an elevation of the osmotic pressure of plasma, increases water reabsorption in the distal tubule and leads to the formation of hypertonic urine. The hormone is one of the fundamental regulators of body water and tonicity, and a good deal is known about its mode of action in biochemical and physiological terms. Studies utilizing the isolated frog skin [60] or toad bladder [65] demonstrated that vasopressin produces a marked increase in net water movement, presumably by increasing the size or number of aqueous channels or pores in the cell membrane. Similar phenomena are presumed to take place in the distal tubule and collecting duct of the mammalian kidney. Fong, Rasmussen, Schwartz, and co-workers [33, 107] reported that vasopressin binds to renal and toad bladder epithelial membrane, probably by forming a covalent bond between its cystine residues and free sulfhydryl groups in the membrane. They suggested initially that this disulfide-sulfhydryl exchange is directly responsible for the opening of aqueous channels, but experiments with physiologically active analogs of vasopressin containing no disulfide bonds failed to support this notion [106]. Orloff and Handler [85] demonstrated that adenosine-3′,5′-monophosphate (cyclic AMP) induces an increase in osmotic flow of water in the toad bladder similar in magnitude to that observed with vasopressin. They and others [11] have also shown that vasopressin accelerates the conversion of ATP to cyclic AMP, presumably by activating the enzyme adenyl cyclase.

Thus, they propose that vasopressin stimulates the formation of (or activates) adenyl cyclase, which accelerates the conversion of ATP to cyclic AMP, the latter compound being responsible for the increased permeability observed. A similar mechanism of action has been proposed for epinephrine, glucagon, and ACTH, which activate adenyl cyclase in other tissues.

Males with vasopressin-resistant diabetes insipidus have absolutely no response to the hormone and markedly outnumber affected females. These observations, plus the lack of male-to-male transmission, led Forssman [34] and Williams and Henry [127] to conclude that the disease is transmitted as an X-linked recessive trait. Other workers have shown that presumed female carriers may manifest partial vasopressin resistance [14], thus supporting this view. However, Cannon [13] reported a large pedigree in which four affected males had affected sons, a finding which precludes X-linked inheritance in this pedigree. It is possible, therefore, that more than one mutation leads to the characteristic abnormal phenotype in this disorder—one X-linked and one autosomal.

There is no information about the nature of the tubular lesion. An abnormal vasopressin is virtually excluded by the complete lack of response to large amounts of exogenous hormone. A defect in vasopressin binding or in one of the steps leading to the formation or utilization of cyclic AMP seems likely but will be very difficult to test in man. As with many human diseases, an appropriate animal model would be of tremendous assistance in the definition of biochemical abnormality.

Hereditary Spherocytosis

This disease, due to an intrinsic defect in the membrane of circulating erythrocytes, is one of the best studied errors of human membrane function. The ready availability of circulating red blood cells has enabled investigators to describe the physiological abnormalities of these cells, to define their mechanism, and to relate the results to the study of the genetics of the disorder. Family studies have shown direct parent-to-child transmission and an occurrence in siblings of affected patients of about 50 percent. Approximately equal numbers of males and females are affected. These findings indicate that the disorder is transmitted as an autosomal dominant trait and imply that affected subjects are heterozygous for an abnormal gene. Current evidence, recently reviewed by Jandl [52] and Jacob [51], suggests that the product of this gene is intimately associated with the intrinsic structure of the erythrocyte membrane. Such evidence will be summarized briefly.

As the name implies, the circulating red blood cells in hereditary spherocytosis tend to assume a more spherical shape than normal. It is this morphological aberration which appears to predispose the cells to splenic sequestration and hemolysis. These spherocytes also have an excessive osmotic fragility and are prone to premature autohemolysis when incubated in vitro. Jandl, Jacob, and others [51, 52] have performed a series of elegant studies indicating that the primary defect in the disease is related to a distinctly increased permeability of the erythrocyte membrane to sodium. Passive influx of sodium and water into these "leaky" cells is increased, necessitating a compensatory increase in active, "pump"-mediated outflux of sodium. The increased "pump" activity is associated with stimulation of Na,K–activated ATPase and with acceleration of glycolysis. These increased metabolic demands can be met in vivo, but under in vitro conditions or in the absence of glucose the cells are prone to accumulate sodium and water, to swell, and ultimately to hemolyze.

Several investigators [51] have shown that the turnover of phospholipids in the spherocytes is increased and that they have a tendency to lose phospholipids and cholesterol. These changes appear to be secondary to the enhanced fluxes described above, but the point remains controversial [90]. Current information satisfactorily relates the morphology of these cells to their altered metabolic and transport characteristics. The exact nature of the chemical defect which makes the cells more permeable to sodium and water is not understood and remains a fertile field for study.

Congenital Chloridorrhea

This very rare disorder, first recognized in 1945 [25, 39], is characterized by diarrhea from birth on, fecal chloride concentrations in excess of the combined concentrations of sodium and potassium (usually more than 150 mEq per liter), metabolic alkalosis, and almost complete absence of urinary chloride. Twelve cases have been reported in males and females. Two siblings were definitely affected in one family [87] and probably in several others. Parents of affected children have shown no clinical or chemical abnormalities, nor has consanguinity been documented in any of the families. The findings suggest autosomal recessive inheritance, but conclusive evidence is lacking.

The disorder presents two important facets when viewed as a transport defect. First, the osmotic diarrhea and very high fecal chloride concentrations suggest that the primary defect is an abnormality in reabsorption or secretion of chloride in the colon and terminal ileum. This is significant since chloride movement is usually considered a passive phenomenon which merely follows transport of cations to maintain osmotic balance and electrochemical neutrality. The absence of any defect in cation transport in congenital chloridorrhea and the marked aberration in intestinal chloride flux offer the only evidence in man for a specific, genetically controlled chloride transport process. This process may be a facilitated diffusion or a specific "pump" [18, 122], but, regardless of its chemical nature, its primary function appears to be that of chloride transport.

The second interesting finding concerns tissue specificity. We have commented repeatedly on shared transport systems in the small intestine and kidney. In this disorder chloride reabsorption in the renal tubule is very efficient, in contrast to the defective chloride transport in the colon. This dissociation indicates that the common genetic control of certain transport systems in the proximal tubule and upper small intestine does not extend to processes in the ileum or colon, again pointing out the great complexities of tissue differentiation, specific chemical or physiological processes, and genetic control.

POSSIBLE PRIMARY INHERITED MEMBRANE DEFECTS

The disorders just discussed have been studied extensively enough to warrant the conclusion that they are inherited and are primary disorders of transport. Those listed in Table 8-2 may be additional examples of primary inherited transport defects, but at present convincing evidence that they are *both* inherited and primary is lacking. Thus, the defects in intestinal absorption of methionine [48, 116] and tryptophan [30] have been documented in too few patients to justify the conclusion that the disorders have a genetic etiology. Similarly, the several generalized proximal tubular abnormalities listed in Table 8-2 (Lowe's syndrome [71, 73], Busby syndrome [101, 102], glucoglycinuria [55]) unequivocally demonstrate transport defects which are probably primary, but further chemical and genetic data are required.

While there is no doubt that the transport of glucose and amino acids into muscle is defective in diabetes [67, 104], convincing evidence is lacking that these are the primary abnormalities in this common metabolic disease, which almost surely represents a large number of genetically and biochemically distinct diseases. Pseudohypoparathyroidism [89] is a specific disorder characterized by resistance to the metabolic effects of parathyroid hormone and by autosomal dominant inheritance. This end-organ failure appears to affect both kidney and bone and hence may well be due to a primary defect in membrane receptor sites for parathyroid hormone in these tissues. Since it is also possible that the primary defect is in the structure of the hormone rather than in the end-organs, the disease has not been classified as a primary transport error. Similarly, the report by Harvald et al. [45] showing reduced activities of cation-stimulated ATPase activity in erythrocytes of patients with hemolytic anemia from two families does not demonstrate convincingly that the observed enzyme defect is directly related to the hemolysis, nor do the findings indicate a consistent mode of inheritance.

Finally, we must point out the recent report by Mangos and McSherry [76] which showed that the sweat of patients with cystic fibrosis inhibited sodium transport in rat salivary glands. This is the first direct evidence suggesting that the high concentrations of sodium and chloride in the sweat of patients with this very common and serious inherited disease results from a specific transport defect. Future results in this system will be awaited with interest.

CONCLUDING REMARKS

This chapter has attempted to point out the wide variety of genetic defects known to affect specific transport processes in man. In no instance do we have sufficient information to define the molecular processes responsible for specific transport defects and relate them to the abnormal clinical findings. However, it seems very likely that, as our knowledge of membrane structure and function increases, investigation of these human transport mutants will provide fundamental insights into the biophysical and biochemical mechanisms by which cells control transmembrane movement and regulate their intracellular contents. This is the goal toward which the many different approaches summarized in the present volume are directed.

REFERENCES

1. Albright, F., and Sulkowitch, H. W. The effect of vitamin D on calcium and phosphorus metabolism; studies on four patients. *J. Clin. Invest.* 17:305, 1938.
2. Ames, G. F. Uptake of amino acids by *Salmonella typhimurium. Arch. Biochem.* 104:1, 1964.
3. Anderson, C. M., Kerry, K. R., and Townley, R. R. An inborn defect of intestinal absorption of certain monosaccharides. *Arch. Dis. Child.* 40:1, 1965.
4. Baron, D. N., Dent, C. E., Harris, H., Hart, E. W., and Jepson, J. B. Hereditary pellagra-like skin rash with temporary cerebellar ataxia, constant renal aminoaciduria, and other bizarre biochemical features. *Lancet* 1:421, 1956.
5. Bartter, F. C., Lotz, M., Thier, S., Rosenberg, L. E., and Potts, J. T. Cystinuria. *Ann. Intern. Med.* 62:796, 1965.
6. Becker, F. F., and Green, H. Incorporation of cystine and lysine by normal and "cystinuric" leukocytes. *Proc. Soc. Exp. Biol. Med.* 99:694, 1958.
7. Ben-Ishay, D., Dreyfuss, F., and Ullman, T. D. Fanconi syndrome with hypouricemia in an adult. *Amer. J. Med.* 31:793, 1961.
8. Beyer, K. H., Wright, L. D., Skeggs, H. R., Russo, H. F., and Shaner, G. A. Renal clearance of essential amino acids: Their competition for reabsorption by the renal tubules. *Amer. J. Physiol.* 15:202, 1947.
9. Brand, E., Cahill, G. F., and Harris, M. M. Cystinuria: II. The metabolism of cystine, cysteine, methionine and glutathione. *J. Biol. Chem.* 109:69, 1935.
10. Brodehl, J., Gellison, K., and Kowalewski, S. Isolated cystinuria (without lysine-ornithine-argininuria) in a family with hypocalcemic tetany. *Proceedings of the Third International Congress on Nephrology* (in press).
11. Brown, E., Clarke, D. L., Roux, V., and Sherman, G. H. The stimulation of adenosine 3′,5′-monophosphate production by antidiuretic factors. *J. Biol. Chem.* 238:852, 1963.
12. Burnett, C. H., Dent, C. E., Harper, C., and Warland, B. J. Vitamin D-resistant rickets: Analysis of twenty-four pedigrees with hereditary and sporadic cases. *Amer. J. Med.* 36:222, 1964.
13. Cannon, J. F. Diabetes insipidus: Clinical and experimental studies with consideration of genetic relationships. *Arch. Intern. Med.* (Chicago) 96:215, 1955.
14. Carter, C., and Simpkiss, M. The "carrier" state in nephrogenic diabetes insipidus. *Lancet* 2:1069, 1956.
15. Chez, R. A., Palmer, R. P., Schultz, S. G., and Curran, P. F. Effect of inhibitors on alanine transport in isolated rabbit ileum. *J. Gen. Physiol.* 50:2357, 1967.

16. Christensen, H. N. *Biological Transport.* New York: Benjamin, 1962.
17. Christensen, H. N. Methods for distinguishing amino acid transport systems of a given cell or tissue. *Fed. Proc.* 25:850, 1966.
18. Cooperstein, I. L., and Hogben, C. A. M. Ionic transfer across the isolated frog large intestine. *J. Gen. Physiol.* 42:461, 1959.
19. Crane, R. K. Intestinal absorption of sugars. *Physiol. Rev.* 40:789, 1960.
20. Crane, R. K. Hypothesis for mechanism of intestinal active transport of sugars. *Fed. Proc.* 21:891, 1962.
21. Crane, R. K. Gradient coupling and the transport of water soluble compounds: A general biological mechanism? In *The Fifteenth Annual Symposium on Protides of the Biological Fluids* (in press).
22. Crawhall, J. C., Scowen, E. F., Thompson, C. J., and Watts, R. W. E. The renal clearance of amino acids in cystinuria. *J. Clin. Invest.* 46:1162, 1967.
23. Cusworth, D. C., and Dent, C. E. Renal clearances of amino acids in normal adults and in patients with aminoaciduria. *Biochem. J.* 74:550, 1960.
24. Czaky, T. Z. Transport through biologic membranes. *Ann. Rev. Physiol.* 27:415, 1965.
25. Darrow, D. C. Congenital alkalosis with diarrhea. *J. Pediat.* 26:519, 1945.
26. Dent, C. E. Rickets and osteomalacia from renal tubule defects. *J. Bone Joint Surg.* [Brit.] 34:266, 1952.
27. Dent, C. E., and Harris, H. The genetics of "cystinuria." *Ann. Eugenics* 16:60, 1951.
28. Dent, C. E., and Rose, G. A. Aminoacid metabolism in cystinuria. *Quart. J. Med.* 20:205, 1951.
29. DeVries, A., Kochwa, S., Lazebnik, J., Frank, M., and Djaldetti, M. Glycinuria, a hereditary disorder associated with nephrolithiasis. *Amer. J. Med.* 23:408, 1957.
30. Drummond, K., Michael, A., Ulstrom, A., and Good, R. The blue diaper syndrome: Familial hypercalcemia with nephrocalcinosis and indicanuria. *Amer. J. Med.* 37:928, 1964.
31. Eggermont, E., and Loeb, H. Letter to editor. *Lancet* 2:343, 1966.
32. Evanson, J. M., and Stanbury, S. W. Congenital chloridorrhea or so-called congenital alkalosis with diarrhea. *Gut* 6:29, 1965.
33. Fong, C. T., Silver, L., Christman, D. R., and Schwartz, I. L. On the mechanism of action of the antidiuretic hormone (vasopressin). *Proc. Nat. Acad. Sci. U.S.A.* 46:1273, 1960.
34. Forssman, H. On hereditary diabetes insipidus: With special reference to a sex-linked form. *Acta Med. Scand.* 121 (Suppl. 159):1, 1956.
35. Fox, C. F., Carter, J. R., and Kennedy, E. P. Genetic control of the membrane protein component of the lactose transport system of *Escherichia coli. Proc. Nat. Acad. Sci. U.S.A.* 57:698, 1967.
36. Fox, M. S., Thier, S., Rosenberg, L. E., Kiser, W., and Segal, S. Evidence against a single renal transport defect in cystinuria. *New Eng. J. Med.* 270:556, 1964.

37. Frimpter, G. W., Horwith, M., Furth, E., Fellows, R. E., and Thompson, D. O. Inulin and endogenous amino acid renal clearances in cystinuria: Evidence for tubular secretion. *J. Clin. Invest.* 41:281, 1962.
38. Froesch, E. R., Winegrad, A. I., and Renold, A. E. Die tubulare nierenfunktion bei verscheidenen formen des renalen Diabetes mellitus. *Helv. Med. Acta* 24:548, 1957.
39. Gamble, J. L., Fahey, K. R., Appleton, J., and MacLachlan, E. Congenital alkalosis with diarrhea. *J. Pediat.* 26:509, 1945.
40. Gerbeaux-Balsan, S. L'Absorption intestinale du phosphore dans le rachitisme vitamins-résistant hypophosphatémique héréditaire. *Rev. Franc. Etud. Clin. Biol.* 10:65, 1965.
41. Goodman, S. I., McIntyre, C. A., and O'Brien, D. Impaired intestinal transport of proline in a patient with familial iminoaciduria. *J. Pediat.* 71:246, 1967.
42. Grasbeck, R., Gordin, R., Kantero, I., and Kuhlback, B. Selective vitamin B_{12} malabsorption and proteinuria by young people. *Acta Med. Scand.* 167:289, 1960.
43. Harris, H., Mittwoch, U., Robson, E. B., and Warren, F. L. The pattern of amino acid excretion in cystinuria. *Ann. Hum. Genet.* 19:196, 1955.
44. Harris, H., Mittwoch, U., Robson, E. B., and Warren, F. L. Phenotypes and genotypes in cystinuria. *Ann. Hum. Genet.* 20:57, 1955.
45. Harvald, B., Hanel, K. H., Squires, R., and Trap-Jensen, T. Adenosine triphosphatase deficiency in patients with non-spherocytic hemolytic anemia. *Lancet* 2:18, 1964.
46. Hjarne, V. A. A study of orthoglycemic glycosuria with particular reference to its hereditability. *Acta Med. Scand.* 67:422, 1927.
47. Hokin, L. E., and Hokin, M. R. Biological transport. *Ann. Rev. Biochem.* 32:553, 1963.
48. Hooft, C., Timmermans, J., Snoeck, J., Antener, I., Oyaert, W., and van den Hende, C. Methionine malabsorption syndrome. *Ann. Paediat.* (Basel) 205:73, 1965.
49. Houston, J. C., and Merrivale, W. H. H. Renal glycosuria in a family. *Guy Hosp. Rep.* 98:233, 1949.
50. Immerslund, O. Idiopathic chronic megaloblastic anemia in children. *Acta Paediat. Scand.* 49 (Suppl. 119):1, 1960.
51. Jacob, H. S. Abnormalities in the physiology of the erythrocyte membrane in hereditary spherocytosis. *Amer. J. Med.* 41:734, 1966.
52. Jandl, J. H. Hereditary Spherocytosis. In Stanbury, J. B., Wyngaarden, J. B., and Fredrickson, D. S. (Eds.), *The Metabolic Basis of Inherited Disease* (2nd ed.). New York: McGraw-Hill, 1966.
53. Jepson, J. B. Hartnup Disease. In Stanbury, J. B., Wyngaarden, J. B., and Fredrickson, D. S. (Eds.), *The Metabolic Basis of Inherited Disease* (2nd ed.). New York: McGraw-Hill, 1966.
54. Jonxis, J. H. P. Aminoaciduria and rickets. *Helv. Paediat. Acta* 10:245, 1955.
55. Kaser, H., Cottier, P., and Antener, I. Glucoglycinuria, a new familial syndrome. *J. Pediat.* 61:386, 1962.
56. Kepes, A., and Cohen, G. N. Permeation. In Gunsalus, I. C., and

Stanier, R. Y. (Eds.), *The Bacteria*. New York: Academic, 1962. Vol. IV.
57. Kipnis, D. M., and Cori, C. F. Studies of tissue permeability: III. The effect of insulin on pentose uptake by the diaphragm. *J. Biol. Chem.* 224:681, 1957.
58. Kleinzeller, A., Kolinska, K., and Benes, I. Transport of glucose and galactose in kidney cortex cells. *Biochem. J.* 104:843, 1967.
59. Knox, W. E. Cystinuria. In Stanbury, J. B., Wyngaarden, J. B., and Fredrickson, D. S. (Eds.), *The Metabolic Basis of Inherited Disease* (2nd ed.). New York: McGraw-Hill, 1966.
60. Koefoed-Johnsen, V., and Ussing, H. H. The contributions of diffusion and flow to the passage of water through living membranes: Effect of neurohypophyseal hormone on isolated Anuran skin. *Acta Physiol. Scand.* 28:60, 1953.
61. Korn, E. D. Structure of biological membranes. *Science* 153:1491, 1966.
62. Krane, S. M. Renal Glycosuria. In Stanbury, J. B., Wyngaarden, J. B., and Fredrickson, D. S. (Eds.), *The Metabolic Basis of Inherited Disease* (2nd ed.). New York: McGraw-Hill, 1966.
63. Krane, S. M., and Crane, R. K. The accumulation of D-galactose against a concentration gradient by slices of rabbit kidney cortex. *J. Biol. Chem.* 234:211, 1959.
64. Laplane, R., Polonovski, C., Etienne, M., Debray, P., Lods, J.-C., and Pissaro, B. L'Intolérance aux sucres à transfert intestinal actif. *Arch. Franc. Pediat.* 19:895, 1962.
65. Leaf, A. Action of neurohypophyseal hormones on the toad bladder. *Gen. Comp. Endocr.* 2:148, 1962.
66. Leaf, A. The Syndrome of Osteomalacia, Renal Glycosuria, Aminoaciduria and Increased Phosphorus Clearance (the Fanconi Syndrome). In Stanbury, J. B., Wyngaarden, J. B., and Fredrickson, D. S. (Eds.), *The Metabolic Basis of Inherited Disease* (2nd ed.). New York: McGraw-Hill, 1966.
67. Levine, R. Concerning the mechanisms of insulin action. *Diabetes* 10:421, 1961.
68. Lindquist, B., Meeuwisse, G. W., and Melin, K. Glucose-galactose malabsorption. *Lancet* 2:666, 1962.
69. Lobeck, C. C. Cystic Fibrosis of the Pancreas. In Stanbury, J. B., Wyngaarden, J. B., and Fredrickson, D. S. (Eds.), *The Metabolic Basis of Inherited Disease* (2nd ed.). New York: McGraw-Hill, 1966.
70. London, D. R., and Foley, T. H. Cystine metabolism in cystinuria. *Clin. Sci.* 29:129, 1965.
71. Lowe, C. U., Terrey, M., and MacLachlan, E. A. Organic-aciduria, decreased renal ammonia production, hydrophthalmos, and mental retardation; a clinical entity. *Amer. J. Dis. Child.* 83:164, 1952.
72. Luder, J., and Sheldon, W. A familial tubular absorption defect of glucose and amino acids. *Arch. Dis. Child.* 30:160, 1955.
73. McCance, R. A., Matheson, W. J., Gresham, G. A., and Elkinton, J. R. The cerebro-ocular-renal dystrophies: A new variant. *Arch. Dis. Child.* 35:240, 1960.
74. McCarthy, C. F., Borland, J. L., Lynch, H. J., Owen, E. E., and Tyor,

M. P. Defective uptake of basic amino acids and L-cystine by intestinal mucosa of patients with cystinuria. *J. Clin. Invest.* 43:1518, 1964.

75. McIntyre, O. R., Sullivan, L W., Jeffries, G. H., and Silver, R. H. Pernicious anemia in childhood. *New Eng. J. Med.* 272:981, 1965.
76. Mangos, J. A., and McSherry, N. R. Sodium transport: Inhibitory factor in sweat of patients with cystic fibrosis. *Science* 158:135, 1967.
77. Marble, A. Non-diabetic Mellituria. In Joslin, E. P., Root, H. F., White, P., and Marble, A., *The Treatment of Diabetes Mellitus.* Philadelphia: Lea & Febiger, 1959.
78. Milne, M. D. Renal Tubular Dysfunction. In Strauss, M. B., and Welt, L. G. (Eds.), *Diseases of the Kidney.* Boston: Little, Brown, 1963.
79. Milne, M. D., Asatoor, A. M., Edwards, K. D. G., and Loughridge, L. W. The intestinal absorption defect in cystinuria. *Gut* 2:323, 1961.
80. Milne, M. D., Crawford, M. A., Girdo, C. B., and Loughridge, L. W. The metabolic disorder in Hartnup disease. *Quart. J. Med.* 29:407, 1960.
81. Mohamed, S. D., McKay, E., and Galloway, W. H. Juvenile familial megaloblastic anemia due to selective malabsorption of vitamin B_{12}. *Quart. J. Med.* 35:433, 1966.
82. Morgan, H. E., Post, R. L., and Park, C. R. Glucose Transport and Phosphorylation in the Perfused Heart of Normal and Diabetic Rats. In Kleinzeller, A., and Kotyk, A. (Eds.), *Membrane Transport and Metabolism.* New York: Academic, 1961.
83. Morikawa, T., Tada, K., Ando, T., Yoshida, T., Yokoyama, Y., and Arakawa, T. Prolinuria: Defect in intestinal absorption of imino acids and glycine. *Tohoku J. Exp. Med.* 90:105, 1966.
84. Orloff, J., and Burg, M. B. Vasopressin-Resistant Diabetes Insipidus. In Stanbury, J. B., Wyngaarden, J. B., and Fredrickson, D. S. (Eds.), *The Metabolic Basis of Inherited Disease* (2nd ed.). New York: McGraw-Hill, 1966.
85. Orloff, J., and Handler, J. S. The similarity of effects of vasopressin, adenosine-3′,5′-phosphate (cyclic AMP) and theophylline on the toad bladder. *J. Clin. Invest.* 41:702, 1962.
86. Pardee, A. B. Crystallization of a sulfate binding protein (permease) from *Salmonella typhimurium. Science* 156:1627, 1967.
87. Perheentupa, J., Eklund, J., and Kojo, N. Familial chloride diarrhea. *Acta Paediat. Scand.* (Suppl. 159):119, 1965.
88. Perheentupa, J., and Visakorpi, J. K. Protein intolerance with deficient transport of basic amino acids. *Lancet* 2:813, 1965.
89. Rasmussen, H., and Reifenstein, E. C. The Parathyroid Glands. In Williams, R. H. (Ed.), *Textbook of Endocrinology* (3rd ed.). Philadelphia: Saunders, 1962.
90. Reed, C. F., and Swisher, S. N. Abnormalities of *in vitro* behavior of structural lipids of red blood cells from patients with hereditary spherocytosis. *J. Clin. Invest.* 39:1019, 1960.
91. Reynolds, T. B. Observations on the pathogenesis of renal tubular acidosis. *Amer. J. Med.* 25:503, 1958.
92. Rickenberg, H. W., Cohen, G. N., Buttin, G., and Monod, J. La galactoside-permease d'*Escherichia coli. Ann. Inst. Pasteur* (Paris) 91:829, 1956.

93. Rosenberg, L. E. Genetic heterogeneity in cystinuria: Evidence for allelism. *Science* 154:1341, 1966.
94. Rosenberg, L. E., Albrecht, I., and Segal, S. Lysine transport in human kidney: Evidence for two systems. *Science* 155:1426, 1967.
95. Rosenberg, L. E., Crawhall, J. C., and Segal, S. Intestinal transport of cystine and cysteine in man: Evidence for separate mechanisms. *J. Clin. Invest.* 46:30, 1967.
96. Rosenberg, L. E., and Downing, S. J. Transport of neutral and dibasic amino acids by human leukocytes: Absence of defect in cystinuria. *J. Clin. Invest.* 44:1382, 1965.
97. Rosenberg, L. E., Downing, S. J., Durant, J. L., and Segal, S. Cystinuria: Biochemical evidence for three genetically distinct diseases. *J. Clin. Invest.* 45:365, 1966.
98. Rosenberg, L. E., Downing, S. J., and Segal, S. Competitive inhibition of dibasic amino acid transport in rat kidney. *J. Biol. Chem.* 237:2265, 1962.
99. Rosenberg, L. E., and Durant, J. L. Iminoglycinuria: An inborn error of renal transport. *J. Clin. Invest.* 46:1110, 1967.
100. Rosenberg, L. E., Durant, J. L., and Holland, J. M. Intestinal absorption and renal extraction of cystine and cysteine in cystinuria. *New Eng. J. Med.* 273:1239, 1965.
101. Rosenberg, L. E., Mueller, P. S., and Watkin, D. M. A new syndrome: Familial growth retardation, renal aminoaciduria and cor pulmonale: II. Investigation of renal function, amino acid metabolism and genetic transmission. *Amer. J. Med.* 31:205, 1961.
102. Rowley, P. T., Mueller, P. S., Watkin, D. M., and Rosenberg, L. E. Familial growth retardation, renal aminoaciduria and cor pulmonale: I. Description of a new syndrome with case reports. *Amer. J. Med.* 31:187, 1961.
103. Schachter, D. Vitamin D and the Active Transport of Calcium by the Small Intestine. In Wasserman, R. H. (Ed.), *The Transfer of Calcium and Strontium Across Biological Membranes.* New York: Academic, 1963.
104. Scharff, R., and Wool, I. G. Effect of diabetes on the concentration of amino acids in plasma and heart muscle of rats. *Biochem. J.* 99:173, 1966.
105. Schneider, A. J., Kinter, W. B., and Stirling, C. E. Glucose-galactose malabsorption. *New Eng. J. Med.* 274:305, 1966.
106. Schwartz, I. L., Rasmussen, H., and Rudinger, J. Activity of neurohypophyseal hormone analogues lacking a disulfide bridge. *Proc. Nat. Acad. Sci. U.S.A.* 52:1044, 1964.
107. Schwartz, I. L., Rasmussen, H., Schoessler, M. A., Silver, L., and Fong, C. T. Relation of chemical attachment to physiologic action of vasopressin. *Proc. Nat. Acad. Sci. U.S.A.* 46:1288, 1960.
108. Scriver, C. R., Efron, M. L., and Schafer, I. A. Renal tubular transport of proline, hydroxyproline and glycine in health and in familial hyperprolinemia. *J. Clin. Invest.* 43:374, 1964.
109. Scriver, C. R., Schafer, I. A., and Efron, M. L. New renal tubular amino acid transport system and a new hereditary disorder of amino acid metabolism. *Nature* (London) 192:672, 1961.

110. Scriver, C. R., and Shaw, K. N. F. Hartnup disease: An example of genetically determined defective cellular amino acid transport. *Canad. Med. Ass. J.* 86:232, 1962.
111. Scriver, C. R., and Wilson, O. H. Possible locations for a common gene product in membrane transport of imino acids and glycine. *Nature* (London) 202:92, 1964.
112. Scriver, C. R., and Wilson, O. H. Amino acid transport: Evidence for genetic control of two types in human kidney. *Science* 155:1428, 1967.
113. Scriver, C. R., and Wilson, O. H. Specificity of transport of neutral and basic amino acids in rat kidney. *Amer. J. Physiol.* 213:185, 1967.
114. Seldin, D. W., and Wilson, J. D. Renal Tubular Acidosis. In Stanbury, J. B., Wyngaarden, J. B., and Fredrickson, D. S. (Eds.), *The Metabolic Basis of Inherited Disease* (2nd ed.). New York: McGraw-Hill, 1966.
115. Shannon, J. A., and Fisher, S. The renal tubular reabsorption of glucose in the normal dog. *Amer. J. Physiol.* 122:765, 1938.
116. Smith, A. J., and Strang, L. B. An inborn error of metabolism with the urinary excretion of α-hydroxy-butyric acid and phenylpyruvic acid. *Arch. Dis. Child.* 33:109, 1958.
117. Smith, H. W. *The Kidney: Structure and Function in Health and Disease.* New York: Oxford University Press, 1951.
118. Spurling, C. L., Sacks, M. J., and Jiji, R. M. Juvenile pernicious anemia. *New Eng. J. Med.* 271:995, 1964.
119. Stein, W. H. Excretion of amino acids in cystinuria. *Proc. Soc. Exp. Biol. Med.* 78:705, 1951.
120. Tada, K., Ito, H., Wada, Y., and Arakawa, T. Congenital tryptophanuria with dwarfism. *Tohoku J. Exp. Med.* 80:118, 1963.
121. Tada, K., Morikawa, T., Ando, T., Yoshido, T., and Minagawa, A. Prolinuria: A new renal tubular defect in transport of proline and glycine. *Tohoku J. Exp. Med.* 87:133, 1965.
122. Taylor, A. E., Wright, E. M., Schultz, S. G., and Curran, P. F. Effect of sugars on ion fluxes in intestine. *Amer. J. Physiol.* (in press).
123. Thier, S., Fox, M. S., Segal, S., and Rosenberg, L. E. Cystinuria: *In vitro* demonstration of an intestinal transport defect. *Science* 143:482, 1964.
124. Thier, S., Segal, S., Fox, M., Blair, A., and Rosenberg, L. E. Cystinuria: Defective intestinal transport of dibasic amino acids and cystine. *J. Clin. Invest.* 44:442, 1965.
125. Webber, W. A., Brown, J. L., and Pitts, R. F. Interactions of amino acids in renal tubular transport. *Amer. J. Physiol.* 200:380, 1961.
126. Wilbrandt, W. The Sugar Transport Across the Red Cell Membrane. In Kleinzeller, A., and Kotyk, A. (Eds.), *Membrane Transport and Metabolism.* New York: Academic, 1961.
127. Williams, R. H., and Henry, C. Nephrogenic diabetes insipidus: Transmitted by females and appearing during infancy in males. *Ann. Intern. Med.* 27:84, 1957.
128. Williams, T. F., Winters, R. W., and Burnett, C. H. Familial Vitamin D-Resistant Rickets with Hypophosphatemia. In Stanbury, J. B., Wyngaarden, J. B., and Fredrickson, D. S. (Eds.), *The Metabolic Basis of Inherited Disease* (2nd ed.). New York: McGraw-Hill, 1966.

129. Woolf, L. I., Goodwin, B. L., and Phelps, C. E. T_m-limited renal tubular reabsorption and the genetics of renal glycosuria. *J. Theor. Biol.* 11:10, 1966.
130. Wrong, O., and Davies, H. E. The excretion of acid in renal disease. *Quart. J. Med.* 28:259, 1959.
131. Yeh, H. L., Frankl, W., Dunn, M. S., Parker, P., Hughes, B., and Gyorgy, P. The urinary excretion of amino acids by a cystinuric subject. *Amer. J. Med. Sci.* 214:507, 1947.

Index